VOLUME SEVENTY SEVEN

Current Topics in
MEMBRANES

Dynamic Plasma Membranes:
Portals Between Cells and
Physiology

CURRENT TOPICS IN MEMBRANES, VOLUME 77

Series Editors

ROBERT BALABAN
National Heart, Lung and Blood Institute
National Institutes of Health
Bethesda, Maryland, USA

SIDNEY A. SIMON
Department of Neurobiology
Duke University Medical Centre
Durham, North Carolina, USA

VOLUME SEVENTY SEVEN

Current Topics in
MEMBRANES

Dynamic Plasma Membranes: Portals Between Cells and Physiology

Edited by

VANN BENNETT

HHMI, and Department of Biochemistry,
Duke University Medical Center,
Durham, NC, USA

Amsterdam • Boston • Heidelberg • London
New York • Oxford • Paris • San Diego
San Francisco • Singapore • Sydney • Tokyo

Academic Press is an imprint of Elsevier

Academic Press is an imprint of Elsevier
50 Hampshire Street, 5th Floor, Cambridge, MA 02139, USA
525 B Street, Suite 1800, San Diego, CA 92101-4495, USA
125 London Wall, London EC2Y 5AS, UK
The Boulevard, Langford Lane, Kidlington, Oxford OX5 1GB, UK

First edition 2016

Notices

Knowledge and best practice in this field are constantly changing. As new research and experience broaden our understanding, changes in research methods, professional practices, or medical treatment may become necessary.

Practitioners and researchers must always rely on their own experience and knowledge in evaluatingand using any information, methods, compounds, or experiments described herein. In using such information or methods they should be mindful of their own safety and the safety of others, including parties for whom they have a professional responsibility.

To the fullest extent of the law, neither the Publisher nor the authors, contributors, or editors, assume any liability for any injury and/or damage to persons or property as a matter of products liability, negligence or otherwise, or from any use or operation of any methods, products, instructions, or ideas contained in the material herein.

ISBN: 978-0-12-805404-8
ISSN: 1063-5823

For information on all Academic Press publications
visit our website at http://store.elsevier.com/

CONTENTS

CONTRIBUTORS

Vann Bennett
HHMI and Department of Biochemistry, Duke University Medical Center, Durham, NC, USA

Max Berkowitz
Department of Chemistry, University of North Carolina at Chapel Hill, Chapel Hill, NC, USA

Alexis R. Demonbreun
Center for Genetic Medicine, Northwestern University, Chicago, IL, USA

Nicolas Destainville
Laboratoire de Physique Theorique (IRSAMC), Universite Toulouse 3-Paul Sabatier, UPS/CNRS, Toulouse, France

Masaki Fukata
Division of Membrane Physiology, Department of Cell Physiology, National Institute for Physiological Sciences, National Institutes of Natural Sciences; Department of Physiological Sciences, School of Life Science, SOKENDAI (The Graduate University for Advanced Studies), Okazaki, Japan

Yuko Fukata
Division of Membrane Physiology, Department of Cell Physiology, National Institute for Physiological Sciences, National Institutes of Natural Sciences; Department of Physiological Sciences, School of Life Science, SOKENDAI (The Graduate University for Advanced Studies), Okazaki, Japan

Thorsten Lang
Department of Membrane Biochemistry, Life & Medical Sciences (LIMES) Institute, University of Bonn, Bonn, Germany

Christophe Leterrier
Aix Marseille Université, CNRS, CRN2M UMR 7286, Marseille, France

Damaris N. Lorenzo
HHMI and Department of Biochemistry, Duke University Medical Center, Durham, NC, USA

Elizabeth M. McNally
Center for Genetic Medicine, Northwestern University, Chicago, IL, USA

Tatsuro Murakami
Division of Membrane Physiology, Department of Cell Physiology, National Institute for Physiological Sciences, National Institutes of Natural Sciences; Department of Physiological Sciences, School of Life Science, SOKENDAI (The Graduate University for Advanced Studies), Okazaki, Japan

Thomas H. Schmidt
Department of Membrane Biochemistry, Life & Medical Sciences (LIMES) Institute,
University of Bonn, Bonn, Germany

Norihiko Yokoi
Division of Membrane Physiology, Department of Cell Physiology, National Institute
for Physiological Sciences, National Institutes of Natural Sciences; Department
of Physiological Sciences, School of Life Science, SOKENDAI (The Graduate
University for Advanced Studies), Okazaki, Japan

PREFACE

It is just a little ironic that I find myself compiling and editing a book focused on basement membranes, a topic about which I was never officially taught anything in graduate school or in college. In my experience, I find that students are easily underwhelmed by contemplating the function of molecules found outside of cells—in the domain of the extracellular matrix (ECM)—when so much more seems to be going inside of cells. The realm of biological inquiry can appear boundless to a young scientist searching for a thesis or a postdoctoral training laboratory. What could be appealing about basement membranes as a topic for 4–6 years, if not a lifetime, of study?

I once told a nonscientist acquaintance at a party that I worked on basement membranes. He replied "Oh, so you work in construction?" As silly as the question seemed to me at the time and for years thereafter, in hindsight I now realize that for 23 years I have indeed been studying a type of biological "construction." Basement membranes are required for the construction of all multicellular organisms, exclusive of plants and biofilms (as I was once taught by a sharp student); they are required for the architecture of tissues, organs, and for the construction of "highways" upon which cells migrate; and they sometimes serve as well-constructed barriers, whether to block the passage of macromolecules in the kidney filter or to prevent the movement of cells from one tissue compartment into another.

Sometimes basement membranes can be defective, either due to genetic, mechanical, or environmental insults that impact the structure and/or function of the basement membrane or its components, in the same way that defective construction materials or a tornado can impact the integrity of a building. For basement membranes, the consequences of such defectiveness can either be lethal for the organism at an early developmental stage or cause an array of ongoing functional problems that we commonly refer to as diseases. And it is the diverse group of basement membrane-based diseases that has gotten (and kept) me, and I presume many others, so excited about the field. I hope this book will help to excite others and expand the cadre of investigators focused on basement membranes.

The opening chapter by Peter Yurchenco discusses the importance of laminins for the initiation of basement membrane assembly. A milestone in the basement membrane field was the seminal discovery that in knockout mice lacking laminin γ1, a component of most laminin heterotrimers, there were no basement membranes. No other protein has been shown to be as important for constructing a basement membrane.

Chapter 2 by Madeleine Durbeej focuses on the importance of one particular laminin trimer, laminin-211 ($\alpha2\beta1\gamma1$), in the function and health of skeletal muscle. Although the Duchenne's form of muscular dystrophy with defects in dystrophin may be the best known and most common (due to X-linkage), studies of congenital muscular dystrophy due to mutation of the gene encoding laminin $\alpha2$ (termed MDC1A) have provided important insights into how the linkage between the myofiber basement membrane and the actin cytoskeleton (via dystrophin and dystroglycan) is so critical. I use this story when teaching the importance of cell/matrix interactions to first year PhD students.

In Chapter 3, Mao Mao, Marcel Alavi, Cassandre Labelle-Dumais, and Doug Gould focus on another major component of all basement membranes, type IV collagen. There are six type IV collagen chains, two of which ($\alpha1$ and $\alpha2$) are considered ubiquitous and are required for life. However, heterozygous mutations in these two chains have been shown to cause defects in multiple organs in mice and humans, especially in the vasculature, brain, and skeletal muscle. In contrast, mutations that impact the $\alpha3$, $\alpha4$, and $\alpha5$ chains cause kidney disease and eye and hearing defects.

The focus of Chapter 4, coauthored by Cristina Has and Alex Nyström, is the epidermal basement membrane. As with the skeletal muscle fiber basement membrane, laminin plays a critical role in cell/matrix interactions, but in the skin it is laminin-332 ($\alpha3\beta3\gamma2$) that is so important for establishing and maintaining tight adhesion between the epidermis and its basement membrane through the establishment of hemidesmosomes. In addition, a special form of collagen, type VII, is critical for linking the basement membrane to the underlying dermis. Defects in either of these connections can cause severe skin blistering and in some cases, death.

In Chapter 5, coauthors Michael Randles and Rachel Lennon present the details of a state-of-the-art methodology for investigating the composition of ECMs. They focus on methods for extracting and enriching ECM components from tissue and the use of proteomics (mass spectrometry) to identify novel components of the ECM in multiple tissues. They also discuss how these methods have been used to investigate ECM composition in both health and disease.

The focus of Chapter 6 by Masashi Yamada and Kiyotoshi Sekiguchi turns to receptors for basement membrane laminins, with a particular focus on integrins. Defining exactly how cells bind to and respond to basement membranes is an important goal for a complete understanding of cell/matrix interactions. This chapter summarizes the three known laminin-binding integrins and discusses the sites on the laminin α, β, and γ chains that are critical for mediating the adhesion of integrins to the basement membrane. Also discussed are the tetraspanins, transmembrane proteins that can modulate the affinity of integrins for their laminin ligands.

In Chapter 7, Corina Borza, Xiwu Chen, Roy Zent, and Ambra Pozzi address the basement membrane and integrin proteins that are known to be important in the kidney. They address the heparan sulfate proteoglycans and nidogens found in the kidney. They also summarize data about how the collagen-binding integrins $\alpha 1 \beta 1$ and $\alpha 2 \beta 1$ influence the production of collagen in normal and diseased kidney glomeruli and how important integrin $\beta 1$ is for development and function of the kidney.

Kevin McCarthy focuses on the heparan sulfate proteoglycans perlecan and agrin in Chapter 8. These widely expressed macromolecules have been shown to be important both in basement membranes and in nonbasement membrane contexts. The different types of glycosaminoglycan side chains that can be appended to the core proteins and extent of sulfation that occurs makes these molecules and their functions especially complex and interesting.

Chapters 9 and 10 cover basement membranes in two invertebrates. Adam Isabella and Sally Horne-Badovinac tackle the role of basement membranes in *Drosophila* development in Chapter 9. They discuss the powerful genetic tools that allow the study of the important roles of basement membranes in live, maturing *Drosophila* oocytes. They also discuss how basement membranes influence axonal pathfinding, BMP signaling, and stem cells.

In Chapter 10, Matthew Clay and David Sherwood summarize what is known about the importance of basement membranes in the nematode *Caenorhabditis elegans*. Worms carrying a variety of mutations in genes encoding basement membrane proteins have shown their importance in multiple processes, including those that involve the nervous system, muscle cells, and the germ stem cell niche. In addition, the worm has been especially useful for investigating how cells penetrate a basement membrane.

As might be obvious from the preceding chapter summaries, when planning this volume I chose to invite authors who share my own appreciation for the importance of specific basement membrane proteins in human disease, while at the same time recognize that it is basic studies of basement membranes, in both higher and lower organisms, that will bring us a complete understanding of how they assemble and the mechanisms by which cells interact and respond to them.

I wish to express my sincere thanks to all the authors who generously gave of their precious time to write their outstanding contributions. I also appreciate the invitation from Sid Simon and Bob Balaban to compile and edit this volume and the dedication of the staff at Elsevier—Mary Ann Zimmerman, Helene Kabes, and Roshmi Joy—who ensured the time-liness and the quality of the product you hold in your hands (or read on your screen). Finally, I wish to thank the many funding organizations around the world who have supported and will continue to support studies of basement membranes.

Jeffrey H. Miner
Washington University School of Medicine
St. Louis, Missouri, USA

PREVIOUS VOLUMES IN SERIES

Current Topics in Membranes and Transport

★Part of the series from the Yale Department of Cellular and Molecular Physiology

CHAPTER ONE

A Molecular Look at Membranes

Max Berkowitz
Department of Chemistry, University of North Carolina at Chapel Hill, Chapel Hill, NC, USA
E-mail: maxb@unc.edu

Contents

Abstract

Due to the recent advances in computer hardware and software, we can now use molecular dynamics and Monte Carlo computer simulation techniques to study systems with large conformational spaces. It is demonstrated here that computer simulations allow us to get a glimpse at the structural and dynamical properties of membranes and also at the interaction of membranes with other molecules. Specifically two examples are considered: (1) structural properties of lipid rafts in model membranes and (2) interaction of model membranes with an antimicrobial peptide, melittin.

1. COMPUTER SIMULATIONS AS A MICROSCOPE

Cells are universal units of life, and biological membranes separate the interior of the cells from their exterior. In case when cells are more complex, like eukaryotic cells, that have organelles, membranes separate the interior of organelles from the rest of the cells. But cell (organelle) membranes are not just passive fences; they are active participants in the process of communication of cells (organelles) with the surrounding. Cell membranes contain mostly two molecular components: lipids and proteins. Because different cells (organelles) perform different missions, their membranes also differ, and this may explain why a large number of different lipids and also membrane-associated proteins can be found in nature.

To study the structural and dynamical properties of membranes, a large variety of experimental techniques are used. These include traditional

Current Topics in Membranes, Volume 77
ISSN 1063-5823
http://dx.doi.org/10.1016/bs.ctm.2015.10.002

1

scattering techniques such as X-ray and neutron scattering (Nagle & Tristram-Nagle, 2000), atomic force microscopy (Roduit, Longo, Dietler, & Kasas, 2015), NMR (Tyler, Clarke, Seddon, & Law, 2015), and fluorescent spectroscopy of different flavors (Demchenko, Duportail, Oncul, Klymchenko, & Mely, 2015). Polarity-sensitive fluorescent dyes are used to study the heterogeneity of the molecular arrangement in the membrane plane (Golfetto, Hinde, & Gratton, 2015). To study the dynamics of molecular motion in the membrane plane, one can use fluorescent recovery after photobleaching (Carnell, Macmillan, & Whan, 2015) or scanning fluorescent correlation spectroscopy (Hermann, Ries, & Garcia-Saez, 2015).

It is not straightforward to get a molecular look at a cell membrane using experimental data, since the accuracy limitations of the measurements and also interpretation of the involved models complicate the picture. Approximately 60 years ago a new tool appeared that provided a direct way to have a molecular look at systems under study. This tool is based on usage of computer simulation techniques such as molecular dynamics (MD) or Monte Carlo (MC) (Wood & Erpenbeck, 1976), and this chapter will concentrate on the results obtained by performing computer simulations. To use any of the two simulation methods, one needs to know the form and parameters of the potential energy function that describes the interaction between molecules. In MC the program creates Boltzmann distribution in the configuration space of the molecules and uses this distribution to perform statistical analysis of the data to extract information about structural and thermodynamic properties of the system. In MD the program moves the molecules according to dynamical equations (mostly classical equations of motion, when biosystems are considered) creating trajectories; averages over trajectories provide information about system properties. While both MC and MD can be effectively used to study thermodynamics of the system, when utilizing the MD program one also produces the time history of the system. This information can be used to study kinetics, an important subject when biological systems are investigated. It is clear from the outset that the form of the potential function together with the assigned parameters appearing in this function, what we call a force field (FF) and supply as an input into the MC or MD program, directly determines the output. Therefore, a large amount of effort was devoted to the development of the FFs. Comparing the results from simulations with the results obtained from experiments usually does the validation of the developed FF.

The level of molecular resolution distinguishes different FFs. In case one describes the interaction of molecules as a combination of interactions among every atom belonging to these molecules, the FF is called all-atom (AA). Often, to save on computational time, light atoms (such as hydrogen) are combined with heavier atoms (like carbon) and presented as one particle called a united atom (UA) and the FF is called, correspondingly, a UA FF. Examples of AA and UA FFs that are most often used for simulations of membrane bilayers are well-known fields such as CHARMM (Brooks et al., 1983), AMBER (Case et al., 2005), and GROMOS (Scott et al., 1999). These FFs were initially constructed to perform simulations on systems containing biomolecules such as proteins, but were later extended to include molecules found in membranes (for recent examples, see Lee et al., 2014; Siu, Vacha, Jungwirth, & Boeckmann, 2008; Wu et al., 2014). Since biological systems possess a rich configuration space, to explore it one needs to perform many MC moves or perform a long MD run. The majority of the simulations that were performed to study membrane systems were MD simulations, perhaps because the history of molecular motion can be directly obtained from the simulations. Given that one uses standard computational equipment, which today is most probably a cluster of Linux nodes, one can simulate a membrane patch of a size of few tens of nanometer over a period of time of up to a microsecond, if the UA or even AA FF are used in MD. To understand the majority of structural and dynamical properties of membranes, one needs to expand both these time and length scales. One of the ways to do this is to combine several heavier atoms, three or four of them, into a particle. This is equivalent to coarse-graining the molecular system and therefore the corresponding FF is called a coarse-grained (CG) FF. MARTINI FF is a popular CG FF used to perform membrane simulations (Marrink & Tieleman, 2013; Marrink, Risselada, Yefimov, Tieleman, & de Vries, 2007). Simulations performed with MARTINI substantially speed up the MD run and therefore provide a possibility to explore more regions in the configuration space during the MD run. The price one pays for using MARTINI is a loss in fine resolution due to combining three to four atoms together into one particle. Also, the potentials in CG FF are softer in their nature, and therefore the dynamics occurs faster; this means that to compare time duration of the process simulated using CG FF and using detailed FF one needs to scale up the value of the time run of the CG simulation. Thus in case of MARTINI, it is recommended to multiply the duration time of the simulation by a factor of 4 (Marrink, de Vries, & Mark, 2004) when such a comparison is desired. The soft nature of the potential may also change the barriers of the activated processes and therefore produce distortions in their kinetics.

Since the data output from MC and MD programs is provided in the form of coordinates of every atom or particle, one can use this output as an input to an image-rendering program and produce molecular-resolved images of the system, thus making MC and MD programs act as a microscope, a computational microscope. Recent development of a special hardware to perform MD on special computers produces long trajectories reaching milliseconds timescale (Shaw et al., 2008). Thus, one can think of these new powerful machines as powerful computational microscopes; appropriately the new computer is called Anton, in honor of Anton van Leeuwenhoek who is considered to be the father of optical microscopy. Below we discuss briefly how the use of computational microscopy provided us with a molecular look at two important issues connected to cell membranes: structural properties of rafts in mammalian membranes and action of antimicrobial peptides (AMPs) defending organism from bacteria.

2. A MOLECULAR LOOK AT A LIPID RAFT

Although biological membranes were studied for quite some time, modern ideas about molecular arrangement of membranes constituents were developed relatively late. It was only in early 1970s that a picture of a membrane, a so-called "fluid mosaic model," was proposed by Singer and Nicolson (1972). In this model the membrane was considered to be a liquid mixture of proteins and lipids displaying heterogeneous ordering on a short-range scale. The concept of membrane heterogeneity is emphasized today; thus in case of mammalian membranes we use a concept of "functional rafts" (Simons & Ikonen, 1997). According to this concept, mammalian membranes can be described as a heterogeneous dynamical substance containing nanoscale domains. These domains consist of a mixture of cholesterol and saturated lipids that have a high temperature, Tm, of main transition from liquid state to solid (gel) state. The saturated lipids in the domains, although translationally mobile, have their orientational conformations in a more ordered state and, therefore, are considered to be in a liquid-ordered state, Lo. The Lo domains might "float" in a surrounding liquid-disordered phase, Ld, which is dominated by unsaturated lipids that have a low Tm. Lipid rafts play an important role in selecting out proteins interacting with membranes and also in the signaling processes. As of today there are still many questions that remain to be answered about the nature of lipid rafts (Jacobson, Mouritsen, & Anderson, 2007; Lingwood & Simons, 2010). For example, are rafts in their equilibrium states or are they nonequilibrium phenomenon? What is their size distribution? What is their lifetime? What is the role of proteins in their creation? Lipids that are found in mammalian membranes belong to three major classes: glycerophospholipids, sphingolipids, and cholesterol, and the bilayer leaflets are asymmetric in their lipid composition. For example, in the cell plasma mammalian membrane the outer leaflet contains domains of cholesterol and high Tm lipid, sphingomyelin, belonging to a class of sphingolipids. A low Tm lipid found in the outer leaflet is an unsaturated phosphatidylcholine molecule belonging to the class of glycerophospholipids. Inner leaflets contain different mixtures of mostly glycerophospholipids and cholesterol, and raft structures can be also formed in this leaflet. How does the distribution of rafts on the outer leaflet influence the distribution on the inner leaflet?

The fact that biomembranes are so rich in their composition substantially complicates their study, and therefore model lipid bilayers are often used in experiments. Initially, a large amount of experimental work was done to study binary mixtures of lipids and their phase diagrams (Luna & McConnell, 1977). Of special interest was to study binary mixtures containing cholesterol as one of the components, since cholesterol plays such an important role in mammalian membranes (McMullen & McElhaney, 1996). It was suggested in the literature that cholesterol can make special complexes with lipids that have high Tm and the propensity to make such complexes can play an important role in explaining the existence of lipid rafts (McConnell & Radhakrishnan, 2003). Checking the special affinity of cholesterol to complex with high Tm lipids is a task that seemed to be tailored for study that uses detailed AA simulations. Indeed, such simulations performed using AA or UA FFs (Pandit, Bostick, & Berkowitz, 2004) were among the first ones done to study properties of lipid mixtures containing cholesterol. The results did not indicate any existence of special complexes. Nevertheless, as it was shown using special free-energy calculation techniques, cholesterol does prefer the neighborhood of sphingomyelin lipids to unsaturated lipids (Berkowitz, 2009).

To mimic the process of raft creation occurring in biological membranes, in model membranes one needs to work with a bilayer that contains at least three lipid components: high Tm lipid, low Tm lipid, and cholesterol. Experiments with model membranes containing this kind of three-component mixtures were performed, and phase diagrams of these systems were studied (Feigenson, 2009; Feigenson & Buboltz, 2001; Veatch & Keller, 2005). For certain mixtures, like the one containing a saturated high Tm lipid DPPC (16:0, 16:0), an unsaturated DOPC lipid (18:1, 18:1), and cholesterol, a mixture considered to be "canonical" to study raft forming ternary mixtures, the initial studies indeed showed the existence of "lipid rafts," but the sizes of the domains in the artificial membranes were much larger than the sizes of the domains observed in the biological membranes (Veatch & Keller, 2003). While domains in biological membranes were nanoscopic in size, the three-component lipid membranes contained microscopic-sized domains. Later studies of domains obtained in three-component lipid mixtures showed that it is possible to regulate the size of the domains in these mixtures. This can be achieved by changing the type of the unsaturated, low Tm lipid. When the degree of unsaturation in the unsaturated lipid is not that high, the Lo domains in mixtures are nanoscopic in their size. An example of a mixture containing nanoscopic domains is a ternary mixture of DSPC (18:0, 18:0), POPC (16:0, 18:1), and cholesterol.

To use computer simulations to study ternary mixtures of lipids and domains in these mixtures, one needs to simulate patches of relatively large sizes over relatively long periods of time. Therefore, in most cases, one resorts to use of the CG FF to describe the lipids, and the MARTINI FF is often used to simulate lipid mixtures. We start by briefly describing the results from some of the simulations reported in the literature (Baoukina, Mendez-Villuendas, Bennett, & Tieleman, 2013; Davis, Kumar, Sperotto, & Laradji, 2013) of a "canonical" mixture that was used to study "raft"-forming mixtures: a three-component mixture DPPC/DOPC/cholesterol (Veatch & Keller, 2003). The ratio of DPPC:DOPC:cholesterol in the simulation of Baoukina et al. was 5:3:4 and the simulated bilayer unit cell contained 4608 lipids (2304 lipids per leaflet) solvated by $\sim$ 100,000 water particles. The size of the unit cell was $\sim$ 40 nm, and the time range for the simulations was $\sim$ 20 μs. To observe and understand the mechanism of phase separation, the simulations were performed at different temperatures. At T = 340 K the bilayer did not display any phase separation, but as the temperature was lowered to T = 323 K strong compositional fluctuations appeared. Interestingly, the compositional fluctuations on the two opposing each other leaflets were disconnected at T = 340, while they appeared to be connected at T = 323 K. When the temperature was lowered to T = 290 K the mixture phase separated into Lo and Ld phases. Simulation snapshots of the mixture shown in Figure 1 depict the time progress of the mixture separation into Lo and Ld phases. The Lo phase contained mostly cholesterol and DPPC lipids, while unsaturated DOPC lipids were mostly found in Ld phase. During the growth of Lo domains did not maintain a continuous network; they were somewhat isolated at intermediate times, but

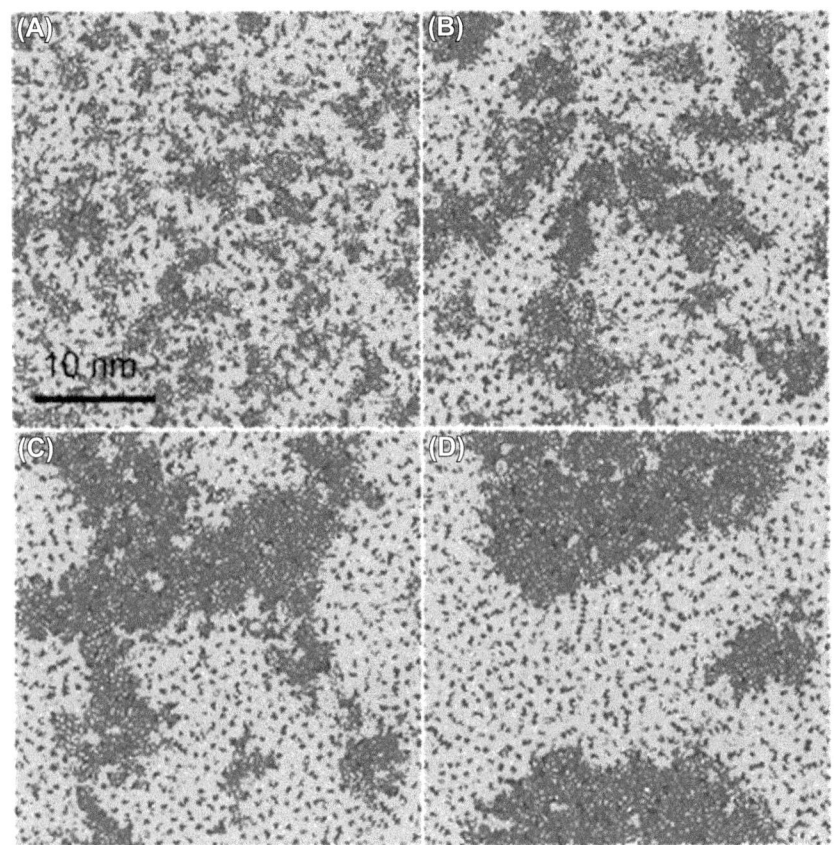

Figure 1 Transformation into Ld and Lo phases in a DPPC:DOPC:cholesterol 5:3:4 bilayer at 290 K. Snapshots of top view were made at (A) 10 ns, (B) 500 ns, (C) 2 μs, and (D) 11 μs. DPPC is shown in green, cholesterol in purple, DOPC in orange. *Reproduced with permission from Baoukina et al. (2013).*

reconnected as the simulation proceeded. The boundaries between the Lo domains and the Ld phase had a rough shape on the level of the molecular resolution. It was also observed that the Lo domains on the opposing leaflets were strongly coupled.

Davis et al. (2013) also used CG FF MARTINI to perform simulations on ternary systems containing DPPC, unsaturated PC, and cholesterol. Four different systems were simulated, each with a different unsaturated PC component. The ratio of molar amounts of DPPC:unsaturated PC: cholesterol was 35%:35%:30% (notice this was a different composition than the one studied in simulations of Baoukina et al.). The unsaturated PC molecules in the simulations were POPC, DOPC, DUPC, and

DAPC possessing one, two, four, and eight unsaturated carbon—carbon bonds, respectively. The simulations were performed for "small- and big-sized" systems, and the "big" systems contained 1040 lipid molecules and 8320 solvent particles. Temperature in all simulations was kept at $T = 295$ K. No phase separation was observed in the DPPC/POPC/cholesterol mixture and this is consistent with the experimental observation. Surprisingly, no phase separation was observed in case of the "canonical" for rafts studies ternary DPPC/DOPC/cholesterol mixture. As the degree of unsaturation increased, phase separation was observed for cases containing mixtures DPPC/DUPC/cholesterol and DPPC/DAPC/cholesterol. By tweaking the values of the interaction between the hydrophobic groups of PC molecules, it was determined that when MARTINI FF is used, the phase separation occurs due to interaction between saturated and unsaturated particles of the FF, i.e., the transition is enthalpically driven. Simulations of Davis et al. also illustrated that one needs to be careful when studies of phase separations of mixtures are done using CG FF, especially in cases when delicate balance of forces determines the degree of phase separation.

Since use of CG FF MARTINI may produce some issues when a ternary system DPPC/DOPC/cholesterol is studied in simulations, it seems like a good idea to employ detailed AA FF to study this mixture. Indeed, more recent MD simulation that used AA FF provided a more detailed picture of the Lo domain created in this ternary mixture (Sodt, Sandar, Gawrisch, Pastor, & Lyman, 2014). As mentioned, to perform a simulation of a larger-sized membrane patch over microseconds time period describing lipids with AA FF, one needs to use a special computer, such as Anton. The simulations by Sodt et al. (2014) used Anton and were performed on a ternary system containing 114 DPPC, 174 DOPC, and 60 cholesterol molecules, and the bilayer was solvated in a water box containing 16,261 water molecules. The Lo domain was prepared initially, and its time development was monitored over a time period of 10 µs. The data from the simulations showed that a substructure was present within the Lo phase. The substructure consisted of local regions of hexagonally ordered saturated chains of DPPC molecules separated by interstitial regions enriched in cholesterol. The presence of this substructure provided an explanation for many previously unexplained experimental data. A careful look at the substructure of the Lo domains showed that instead of complexes of DPPC with cholesterol, DPPC created complexes with itself. It was also observed that cholesterol preferred to interact with other cholesterol in the next-nearest-neighbor solvation shell. The pictorial representation that shows the substructure of Lo domain observed in the simulation is shown in Figure 2.

Further experiments performed on three-component lipid system mixtures, containing high Tm lipid, low Tm lipid, and cholesterol, showed that two general types of phase diagrams are needed to describe the behavior of these mixtures (Feigenson, 2009). One type of phase diagrams, called Type I, describes mixtures for which nanoscopic-sized Lo domains can be observed. An example of such a mixture is a mixture of a saturated lipid, DSPC (18:0, 18:0), a lipid with one unsaturated chain, POPC (16:0, 18:1), and cholesterol. If POPC lipid with one unsaturated chain is replaced by a lipid with two unsaturated chains, DOPC (18:1, 18:1) lipid, the phase diagram for the ternary system DPPC/DOPC/cholesterol looks substantially different from the phase diagram of the mixture DPPC/POPC/cholesterol. The phase diagram for the mixture DPPC/DOPC/cholesterol, called Type II diagram, shows at least three two-phase coexistence regions and one three-phase coexistence region. At low cholesterol concentration the two-phase coexistence region is between liquid and gel phases. When cholesterol is added, the new phase, Lo phase, is formed and a three-phase coexisting region appears. This happens because addition of cholesterol promotes the formation of Lo phase in expense of the gel phase. When even more cholesterol is added, a three-phase region disappears; instead the system finds itself in a two-phase coexistence region—a region where

Figure 2 The molecular details of lipid chains in Lo/Ld simulation at T = 298 K. DPPC chains are shown in red, DOPC in blue, cholesterol in yellow, and methyl groups on the β face of cholesterols are in gray. The area of locally hexagonal order is highlighted in yellow. *Reproduced with permission from Sodt et al. (2014). (See color plate)*

Ld and Lo phases coexist. The important factor is that in Type II mixture the domains of Lo are macroscopic in their size. To see how nanodomains were transformed to macrodomains, very revealing experiments were recently performed on four-component lipid mixtures containing DSPC, POPC, DPPC, and cholesterol (Goh, Amazon, & Feigenson, 2013). While the fraction of DSPC and cholesterol was fixed, the fraction of POPC and DOPC was changed, so that the composition of the change could be characterized by the parameter ρ defined as the ratio [DOPC]/ ([DOPC] + [POPC]). It was observed that as ρ changed from 0 to 1 the size of the domains created in giant unilamellar vesicles (GUVs) changed from nanoscopic to macroscopic. Interestingly, at intermediate values of ρ morphologies termed "modulated phases" were observed in experiments with GUVs. These "modulated phases" that had morphologies of stripes, honeycomb and broken-up domains were not observed in experiments with three-component mixtures.

Very recently, to get a molecular understanding of how the degree of lipid unsaturation changes the Lo domain size in four-component lipid mixtures, MD simulations were performed on mixtures containing DPPC, PUPC, DUPC, and cholesterol (Ackerman & Feigenson, 2015). The unsaturated lipids were PUPC (16:0, 18:2) and DUPC (18:2, 18:2). The choice of unsaturated lipids as more unsaturated than POPC and DOPC was dictated by the fact that in CG simulations using MARTINI to observe the phase separations it is more reliable to use lipids with higher degree of unsaturation (Davis et al., 2013). For the simulated mixture with $\rho = $ [DUPC]/([DUPC] + [PUPC]), 11 different simulations with different ρ values (from 0 to 1 in an interval of 0.1) were performed using CG FF initially. In all simulations the ratio DPPC:[DUPC + PUPC]:cholesterol was $\sim 0.4:0.4:02$. Each simulation contained a membrane patch consisting of 4608 lipids. To get a more detailed molecular picture of lipids in the mixture the CG simulation were stopped, and the description of all the molecules was transferred from the CG description to the UA. Once the procedure of such transfer, which is in a way a procedure of improving the computational microscope resolution, was finished, the simulations were continued using UA FF. The approach when part of the simulations was performed using CG description and part using more detailed AA or UA description is called multiscale. Every simulation from a set of 11 simulations was a multiscale one.

From the results of CG simulations it was observed that the composition of Ld and Lo phases depended on the value of ρ. Specifically, the amount of low Tm lipid in Lo phase decreased as ρ went from 0 to 1, and so did the amount of DPPC in the Ld phase. Therefore, as DUPC was replacing PUPC the degree of demixing increased: DPPC and cholesterol enriched the Lo phase, while low Tm lipid enriched the Ld phase. Figure 3 reproduced from reference Ackerman and Feigenson (2015) depicts the patches of Lo and Ld phases obtained at different values of ρ. The degree of ordering in membranes was determined by the calculation of intraleaflet and interleaflet correlation functions. These functions showed the probability of finding two Ld lipids at a certain distance. For small values of ρ the intraleaflet correlation decayed quickly, indicating that correlation was short ranged. At $\rho \sim 1$ the correlations decayed slowly, indicating the presence of large-scale patches limited by the size of the simulation box. The interleaflet correlation function provided information on domain alignment in apposed leaflets. The results showed that domains were not aligned at $\rho \sim 0$ and the alignment increased as ρ increased toward the value of 1. By looking at the interface length, it was determined that a sharp transition in phase morphology occurred in CG simulations when ρ changed in the interval between 0.5 and 0.8. Interestingly, the alignment between domains in the apposing leaflets also underwent a sharp increase when ρ changed in the same interval. This points out that the intraleaflet and interleaflet transitions are interconnected. This is happening because there is a large energetic penalty for the phase mismatch in different leaflets, and this penalty is increasing with the size of the domains. To avoid it, a significant change

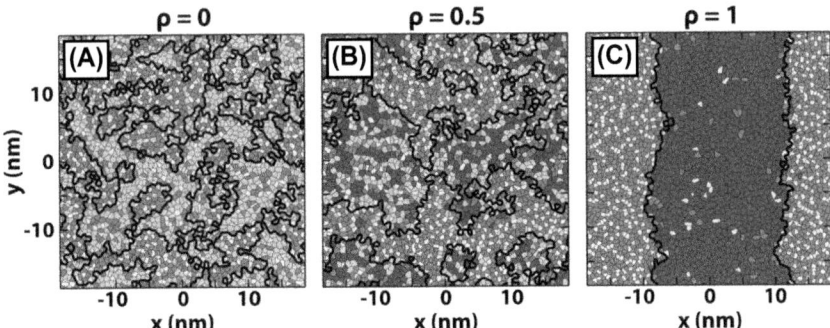

Figure 3 Patches of Lo and Ld phases at (A) $\rho = 0$, (B) $\rho = 0.5$, and (C) $\rho = 1$. DPPC is shown in blue, PUPC in green, DUPC in red, and cholesterol in yellow. Black lines demarcate coexisting patches of different phases. *Reproduced with permission from Ackerman and Feigenson (2015).* (See color plate)

in alignment is happening at the same value of $\rho \sim 0.6$ which is the value characterizing a significant change in the domain size.

Why do we observe nanoscale domains at $\rho \sim 0$ and macroscale domains at $\rho \sim 1$ and "modulated phases" at intermediate ρ? According to the explanation provided by Ackerman and Feigenson (2015) the creation of domains in mixtures is a result of a competition between bending energy and line tension. Due to the existence of curvature in GUVs, the creation of domains of more rigid Lo phase increases the bending energy, while creation of many small patches of Lo phase increases the size of interface between domains and therefore the energy due to line tension. When $\rho \sim 0$ lipids with low Tm reduce the line tension between domains, thus favoring the creation of many small domains. At the opposite side of the spectrum, when $\rho \sim 1$, the value of line tension is high, and therefore the penalty for the creation of many small domains will be too big. As a result macroscopic-sized domains appear in GUVs.

The CG simulations showed many features also observed in experiments. These are the following:

1. Both experiments and simulations showed an increase in the domain size as ρ increased.
2. Both in experiments and simulations the transition in domain size occurred rather sharply in a small interval of ρ.
3. Both experiments and simulations showed that compositional difference between Lo and Ld domains increased with the increase of ρ.
4. Both experiments and simulations displayed an increased domains thickness mismatch with the increase of ρ. There were also discrepancies between the simulation and experimental results:
   a. Simulations did not reproduce the modulated phase pattern observed in experiments at intermediate values of ρ.
   b. Simulations displayed a much stronger demixing pattern.
   c. At high values of ρ the Lo phase underwent gelation. The discrepancies can be attributed to the defects of CG potentials. Also the small size (compared to the size of GUV) and absence of curvature played a role in not being able to reproduce the modulated phase pattern (Ackerman & Feigenson, 2015).

UA simulations were performed to study the details of the molecular arrangements in the phases and especially at the interface between the Lo and Ld phase. It was observed that Lo/Ld interface extended by ~ 2 nm in terms of order and extend of a lipid tilt and the largest changes occurred within 1 nm. This size of the interface was independent of the value of ρ.

It was also observed that for a circular domain of radius 7 nm around 50% of lipids was part of the interface; when the radius of the domain went up to 400 nm only 1% belonged to interface. So when the domains are small the interface represents a large part of the membrane and this should be taken into account when the experimental data are considered. The UA simulations also showed that lipids displayed a preferential tilt toward the Ld phase. Thus, at $\rho = 1$ lipids in Lo phase that were as far as 6 nm from the interface significantly tilted toward the Ld phase. This indicates that lipids in a phase far from the interface may be affected by a distant coexisting phase.

The results of the simulations briefly described above illustrate the power and limitations of the computational techniques used to get a molecular look at the membrane. Recent substantial improvement in hardware and software of computers allows much better understanding of the nature of mixtures containing phospholipids and cholesterol, mixtures that are used as a model to study a "lipid raft" phenomenon. Further advancement in the computational power will allow the study of larger membrane patches that will also contain proteins and this will provide us with the possibility to gain further detailed molecular understanding of rafts.

3. A MOLECULAR LOOK AT MELITTIN ACTION ON MEMBRANES

As a second example of how use of computer simulations can help us in gaining a molecular view on processes occurring in biological membranes, we will consider the interactions of lipid bilayers with peptides. Specifically, we will consider interactions of membranes with AMPs that often represent the first line of defense against invading bacteria. The AMPs adsorb on bacterial cell membranes and create pores in them or rupture them, causing death of the bacterial cell. Consider, for example, the action of a well-known AMP: melittin. The experiments indicate that melittin creates toroidal pores in membranes (Allende, Simon, & McIntosh, 2005; Gordon-Grossman, Zimmermann, Wolf, Shai, & Goldfarb, 2012; Lee, Sun, Hung, & Huang, 2013; Matsuzaki, Yoneyama, & Miyajima, 1997; Raghuraman & Chattopadhyay, 2007). Because of the large body of the simulation work was done on melittin, we will consider below this AMP.

AMPs are usually small in size and positively charged (cationic). Melittin contains 26 residues (the amino acid sequence of melittin is GIGA-VLKVLTTGLPALISWIKRKRQQ) and its charge is either $+5$ or $+6$

depending on the charge of the N-terminus that can be +1 when protonated (Raghuraman & Chattopadhyay, 2007). The rest of the positive charge comes from three lysine and two arginine residues. Since most of melittin's charge is concentrated in the vicinity of the C-terminus, this terminus can serve as an anchor to the bilayer headgroup region by creating salt bridges with the negatively charged phosphates on phospholipids. In aqueous solutions, melittin is disordered, but it adsorbs on a membrane surface as an α-helix with its main axis parallel to the membrane surface. The one side of the adsorbed helix is hydrophobic, while the other side is hydrophilic, and therefore helix is located on the interface between hydrophobic and hydrophilic regions of the membrane. The crystal structure of melittin shows that it actually consists of two helices separated by a hinge at the PRO14 residue. Melittin adsorbs to lipid bilayer in a form of a monomer, but its activity involves a cooperative effect and its activity depends on the amount adsorbed. When the peptide to lipid ratio (P/L ratio) reaches some critical value (P/L)*, pores in membranes appear (Lee et al., 2013). Initially, it was suggested that melittin creates pores using "wedge and edge" mechanism (Terwilliger, Weissman, & Eisenberg, 1982). When melittin is adsorbed on the membrane surface, it acts as a wedge and increases the area of the leaflet on which it is adsorbed. To reduce the energetic penalty for presence of the wedge, a pore opens up in the membrane and peptides move into the pore edge. The motion of a peptide to the pore edge can be described is its reorientation from a surface, S state to a transmembrane (TM), T state. Based in the recent experimental results, it was proposed that pore creation by melittins actually involves two stages (Lee et al., 2013). In the first stage the adsorbed peptides create transient pores by using the "wedge and edge" mechanism, and melittins also move across the bilayer, eventually appearing on the opposite leaflet of the bilayer. After peptides reach equal concentrations on both leaflets, they produce tension in the membrane and to release this tension permanent larger-sized pores appear.

MD simulations that were performed on melittin interacting with the lipid bilayer were among the first simulations performed to study the action of AMPs on membranes. Although initial simulations (Bachar & Becker, 1999; Berneche, Nina, & Roux, 1998; Lin & Baumgaertner, 2000) provided useful information on the structure of melittin and its interactions with lipid bilayer, computational abilities at that time were too limited and it was not possible to study how this peptide spontaneously creates pores in membranes. Poration of membranes by melittin was observed in a series of more recent works. In one of these, a series of simulations on systems containing melittin peptides adjacent to the surface of a DPPC bilayer were performed (Sengupta, Leontiadou, Mark, & Marrink, 2008). In these simulations the P/L ratio was in a range 1/128–6/128, and pores were observed when the P/L ratio was equal or above 3/128, which is above the critical concentration. The results showed that pores had a disordered structure; moreover the melittin peptides in the pores were not helical. Based on these results, another model of AMP action through pore formation was proposed which was called a disordered toroidal pore model. Because the simulations were performed using detailed (AA) description, they were performed for time periods of only few hundred nanoseconds. Since initially all peptides were placed on one side of the bilayer and not on both sides, the simulations could not observe creation of permanent pores, but could, possibly, depict the creation of transient pores. The fact that relatively short time was needed to create pores may indicate that initial arrangement of peptides was biased toward pore creation.

Very recently a set of five simulations using AA FF was performed, each of the simulations lasting between 2.6 and 4.2 μs (Upadhyay, Wang, Zhao, & Ulmschneider, 2015). The membrane in these simulations was a one-component membrane containing DMPC (14:0, 14:0) lipids. In these simulations, peptides were placed either asymmetrically (all on one of the leaflets) or symmetrically (the same amount on the each leaflet). In all of the simulations no spontaneous pore formation was observed, even when a defect in a membrane was prepared initially for the purpose to assist in pore creation. Based on this observation the authors concluded that it takes a long time, at least larger than ∼ 10 μs for melittins to spontaneously create a pore in a membrane.

In spite that it may take a very long time to observe pore creation in simulations that use AA FF, one can still study possible architecture of melittin-created pores by preassembling porelike structures and letting them relax (Irudayam & Berkowitz, 2011; Mihajlovic & Lazaridis, 2010). Starting with the prearranged initial conditions the simulation let the peptides to adjust to neighboring lipids and reach a conformation that provided information about the structure of a possible pore. Since charges in melittin are distributed in an inhomogeneous way, with most of the charge located in the vicinity of the C-terminus, it was proposed that to have a symmetric charge distribution in the pore, some of the peptides may anchor their C-termini at one of the leaflets, while other will have C-termini anchoring at the opposing leaflet (Irudayam & Berkowitz, 2011). It was argued that symmetric distribution of charges results in a more stable pore. Also, in the same work the issue of melittin helicity in the membrane and in the pore, and its influence on pore stability was investigated (Irudayam & Berkowitz, 2011). Therefore, to study the effect of mutual arrangement of peptides and their helicity on the pore stability, four systems were simulated, each containing four melittins. The number of melittins was chosen as four because in one of the earlier influential works on melittin pores, it was suggested that pores contain four peptides (Vogel & Jahnig, 1986). Thus, melittin tetramer was placed in a pore by either placing four peptide molecules in a symmetric or in an asymmetric arrangement (Irudayam & Berkowitz, 2011). In the symmetric arrangement, two melittin molecules had their C-termini anchored to the top leaflet and the other two had their C-termini anchored to the lower leaflet. In the asymmetric arrangement, all four peptides had their C-termini anchored to the top leaflet. The four systems studied were system A, where four copies of the less helical melittin were symmetrically arranged; system B, where four copies of the less helical melittin were in the asymmetric arrangement; system C, where four copies of the helical melittin were symmetrically arranged; and system D, with four copies of the helical melittin arranged asymmetrically. The bilayer in these simulations contained only one component, POPC lipid. Of the four cases considered, formation of a stable toroidal pore was observed only in system A, where four less helical peptides were placed in a symmetric arrangement. Figure 4 shows snapshots from simulations of systems A–D. From this figure, we observe that although initially pore existed in all four systems it remained stable only in system A. Based on the result of the simulations, it was proposed that the formation of a symmetric toroidal pore is possible and also that maintaining the helical secondary structure by melittin is not required for the toroidal pore formation. At the same time, it was emphasized that the results of the simulations do not exclude the possibility that toroidal pores with asymmetric arrangements of peptides exist in nature, since four considered simulations were performed for only 250 ns each, and during this time it is not possible to explore configuration space of pores in full.

To allow the observation of spontaneous transient pores creation, their opening and closing, longer simulation times are required. Recently, results from a set of 8 CG simulations using MARTINI FF performed on systems containing melittin peptides interacting with a phospholipid bilayer were reported (Santo, Irudayam, & Berkowitz, 2013). In four of the simulations, the peptide to lipid ratio was relatively small (P/L $\sim$ 1/100), so these simulations were done below the critical P/L (the value of the critical P/L is not

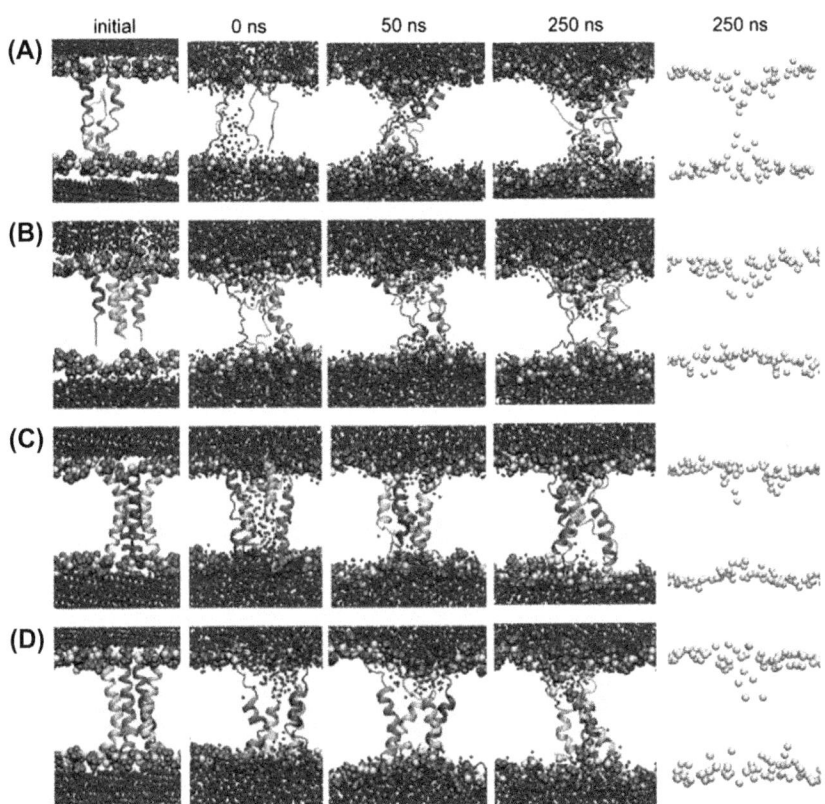

Figure 4 Snapshots of systems A—D at various time points during the simulations. The 0 ns snapshot is the starting conformation obtained after equilibration. Peptide 1 is shown in violet, peptide 2 in magenta, peptide 3 in green, peptide 4 in cyan, phosphorus atoms in yellow, choline groups in gray, and water oxygen atoms in blue. The last column shows positions of phosphorus atoms at 250 ns, to show clearly the toroidal nature of the pore in simulation A. *Reproduced with permission from Irudayam and Berkowitz (2011).* (See color plate)

known accurately and it depends on many factors such as the composition of lipid bilayer, but it is usually higher than 1/100). In other four simulations, P/L was relatively large (P/L ~ 1/21), higher than usual (P/L)*. The initial structure of peptides was obtained from their crystal structure (Terwilliger & Eisenberg, 1982a, 1982b), and since the simulations used CG MARTINI FF (Marrink et al., 2004, 2007; Monticelli et al., 2008) that keeps the helicity of the peptide helix fixed, the peptide helicity remained fixed at 88%. The bilayer in the simulations was a mixture of zwitterionic (DPPC) and anionic (POPG) phospholipids taken in a 7:3 ratio.

The four systems containing peptides in the low P/L ratio were simulated for a time of 5.4 μs each. This run time corresponds to ~22 μs of run time, if simulations would be performed using atomically detailed FF. No peptide translocation across the bilayer or pore formation was observed in any of these four simulations. It was observed that hydrophobic residues of some peptides were located deeper in the bilayer, at distances nearly 1 nm from the bilayer center, while their hydrophilic residues were mostly located in the headgroup region. This separation of residues was achieved because these melittin peptides assumed a bent shape, anchoring residues located next to termini in the bilayer headgroup region. Moreover, the presence of PRO14 residue in melittin facilitated creation of such bent conformations. These conformations were observed in other CG simulations of melittin interacting with a pure DPPC bilayer and were called "U-shaped" (Santo & Berkowitz, 2012). The existence of bent conformations of melittin called "banana-shaped" was also inferred from the experimental study using pulse EPR method (Gordon-Grossman et al., 2009). Very recent results from long simulations that used AA FF also showed a large presence of U-shaped melittin conformers (Upadhyay et al., 2015).

In two of the four simulations with P/L ~ 1/21 it was observed that peptides slowly penetrated into the bilayer interior. In one of the runs this event occurred after ~255 ns (corresponding to ~1 μs of AA run), while in another run penetration into the bilayer interior was observed after ~1.6 μs (corresponding to ~6 μs of AA run). During the penetration melittin initially inserted its N-terminus into the bilayer and following this step assumed a TM orientation. After that, assisted by the reorientation of the neighboring lipid headgroups, peptides started to aggregate in the bilayer interior, creating small aggregates. Not all of the peptides in the aggregate were in a TM conformation. In some cases, melittin penetrated deeply into the membrane still remaining in the U-shaped conformation aggregating with other melittins, some of them also in a U-shaped conformation. For example, an aggregate containing three peptides was observed and two of its peptides were in the U-shaped conformation. By checking the presence of water particles in the aggregate region it was decided if the aggregates were pores. In both simulations where pores in the membranes appeared, it happened after a time period of ~3 μs (~12 μs in AA run). The pores had a transient character and were present in the membrane for a relatively short period of time. Figure 5 presents the snapshots made from one of the simulations and they show the appearance of a transient toroidal pore, its dynamics, and decay. The number of peptides in the

(A)   **(B)**

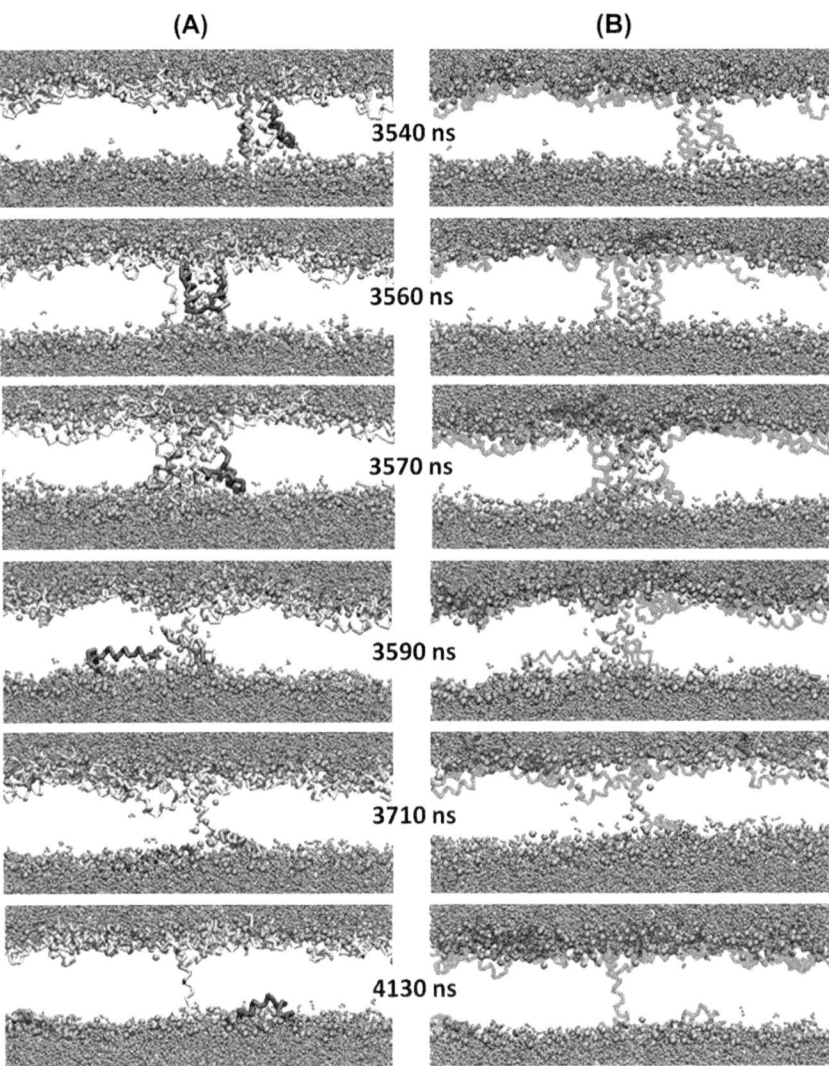

Figure 5 Transient pore formation and translocation by melittins in a run with P/L = 1/21. In panels (A) peptides are shown in yellow, except the two peptides that translocate, which are shown in purple and cyan. In panels (B) peptides are in transparent red, so that water flux can be visible. Water particles are shown in green; pink spheres are PO_4 groups. *Reproduced with permission from Santo et al. (2013).* (See color plate)

pore region did not remain stable; it was changing from three to four in one simulation and changed from three to five in another. Interestingly, many peptides in the pore and also in peptide aggregates in the bilayer were observed to be in the U-shaped conformation. These U-shaped peptides helped in forming the wedge containing few peptides that initiated pore creation.

As we can see simulations using AA FF such as CHARMM that were performed for time period ∼4 μs did not show any spontaneous pore creation, while CG simulations that used MARTINI showed creation of pores. Why is there such a difference in the results observed from these simulations? We do not know the answer to this question and more work is needed to understand this issue. With all that, it should be noticed that pore creation occurred not in every CG simulation, while in those when it occurred it took a time of ∼3.6 μs (corresponding to ∼14 μs in AA simulations).

To quantify the probability of melittin reorientation from the S state to T state, it is necessary to determine the free energy (potential of mean force, PMF) as a function of a coordinate (or set of coordinates) that characterize(s) this process. Such PMFs describing melittin reorientation in POPC were calculated recently by performing a set of simulations using AA FF (Irudayam, Pobandt, & Berkowitz, 2013). The results of these calculations are shown in Figure 6. The PMFs from Figure 6 are functions of a coordinate that measures the distance between the center of mass of three residues at the N-terminus and the plane parallel to the bilayer surface and passing through the bilayer center. Three PMFs were calculated: (A) in case when P/L was 1/128, (B) when P/L was 1/32, and (C) when P/L was also equal to 1/32, but for the reorientation of a melittin molecule when another melittin was already reoriented. The last simulation was done to study the cooperative effect. It was observed that when P/L was low, the PMF had a barrier of ∼22 kT. The value of the barrier got reduced to a value of ∼16 kT when P/L increased to a value of 1/32. If we assume that the transition rate can be obtained by using usual Arrhenius kinetics formula $k = A \exp(-W/kT)$, where W is the value of the barrier in the PMF plot, the reduction of the barrier by ∼6 kT increases the rate by a factor ∼400. Unfortunately, we cannot determine the absolute value for the rates or the times for melittin reorientation, since we do not know the value of the preexponential factor, A. Nevertheless, the calculated value of the barrier of 22 kT indicates that the reorientation of a single melittin at low P/L has a low probability. This probability increases by two to three orders of magnitude at P/L above critical. The third PMF shows a barrier of

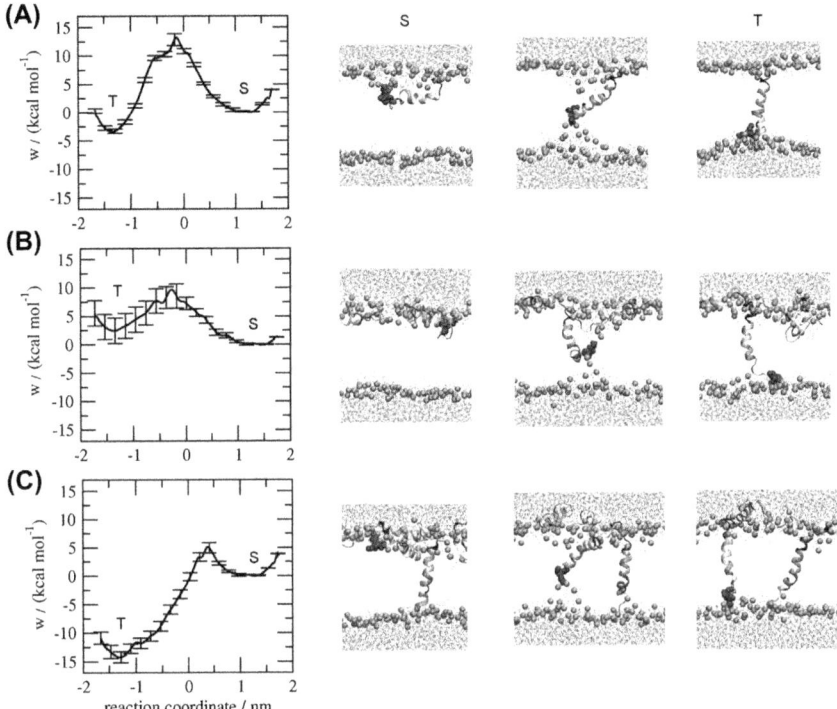

Figure 6 Potential of mean force (PMFs) for the reorientation of melittin from S state to T state for simulations A (A), B (B), and C (C) along with snapshots in the PMFs wells and the barrier peak. The first three residues at the N-terminus are shown in magenta (black in print versions). Phosphorus atoms are shown as orange spheres, water molecules as blue (lighter gray in print versions) dots. Peptide residues are colored by their type: hydrophobic in yellow (white in print versions), hydrophilic in green (gray in print versions), charged in blue (lighter gray in print versions). *Reproduced with permission from Irudayam et al. (2013).*

∼8.5 kT, pointing out that when one melittin molecule has already reoriented, it is much easier for the second melittin in its neighborhood to reorient, indicating the existence of a strong cooperative effect. The structures of the transition states (see Figure 6 for these), when the values of the barrier in the PMF are at their maxima were discussed, and the importance of water defects/pores in stabilizing the transition states was also studied (Irudayam et al., 2013).

The simulations of AMP interacting with membranes, as the simulations of lipid mixtures to imitate appearance of lipid rafts, also illustrate the power and limitations of simulation techniques. Because we need to simulate the membrane patch over relatively large space and over large time period, we either simplify the FF or use special purpose computers. Special purpose computers allow us to perform simulations with more accurate FF and get few simulations over microseconds or even milliseconds time period. Nevertheless, having few trajectories still does not allow performing quantitative analysis of thermodynamics and kinetics of the processes we consider and, therefore, fully understand the mechanism of AMP action. Refining special techniques to calculate PMFs that require accurate sampling of rare events (Kumar, Bouzida, Swendsen, Kollman, & Rosenberg, 1992; Laio & Gervasio, 2008), understanding of how to choose correctly the set of coordinates describing PMFs in full (Yu & Schatz, 2013), and also developing the methodology to study kinetics of the membrane permeation processes (Cardenas & Elber, 2013; Makarov, 2014) are necessary theoretical developments to advance further in our study of AMPs interacting with membranes. In general, combining theoretical developments with advances in computer software and hardware by building computers similar to Anton will be required to make computational tools effective in solving complicated problems in biophysics, like understanding the character of protein—membrane interactions or structure and dynamics of lipid rafts.

ACKNOWLEDGMENTS

I am very grateful to all my collaborators who contributed immensely to the work performed in the laboratory and that is referenced here. The work was funded by the Molecular and Cell Biology Division of the NSF. I am thankful to Professors G. Feigenson and S. Simon for reading the manuscript and providing useful comments.

REFERENCES

Ackerman, D. G., & Feigenson, G. W. (2015). Multiscale modeling of four-component lipid mixtures: domain composition, size, alignment, and properties of the phase interface. *Journal of Physical Chemistry B, 119*(11), 4240—4250.

Allende, D., Simon, S. A., & McIntosh, T. J. (2005). Melittin-induced bilayer leakage depends on lipid material properties: evidence for toroidal pores. *Biophysical Journal, 88*(3), 1828—1837.

Bachar, M., & Becker, O. M. (1999). Melittin at a membrane/water interface: effects on water orientation and water penetration. *Journal of Chemical Physics, 111*(18), 8672—8685.

Baoukina, S., Mendez-Villuendas, E., Bennett, W. F. D., & Tieleman, D. P. (2013). Computer simulations of the phase separation in model membranes. *Faraday Discussions, 161*, 63—75.

Berkowitz, M. L. (2009). Detailed molecular dynamics simulations of model biological membranes containing cholesterol. *Biochimica et Biophysica Acta-Biomembranes, 1788*(1), 86—96.

Berneche, S., Nina, M., & Roux, B. (1998). Molecular dynamics simulation of melittin in a dimyristoylphosphatidylcholine bilayer membrane. *Biophysical Journal, 75*(4), 1603—1618.

Brooks, B. R., Bruccoleri, R. E., Olafson, B. D., States, D. J., Swaminathan, S., & Karplus, M. (1983). Charmm — a program for macromolecular energy, minimization, and dynamics calculations. *Journal of Computational Chemistry, 4*(2), 187–217.

Cardenas, A. E., & Elber, R. (2013). Computational study of peptide permeation through membrane: searching for hidden slow variables. *Molecular Physics, 111*(22–23), 3565–3578.

Carnell, M., Macmillan, A., & Whan, R. (2015). Fluorescence recovery after photobleaching (FRAP): acquisition, analysis, and applications. In D. M. Owen (Ed.), *Methods in membrane lipids* (2nd ed., Vol. 1232, pp. 255–271).

Case, D. A., Cheatham, T. E., Darden, T., Gohlke, H., Luo, R., Merz, K. M., et al. (2005). The Amber biomolecular simulation programs. *Journal of Computational Chemistry, 26*(16), 1668–1688.

Davis, R. S., Kumar, P. B. S., Sperotto, M. M., & Laradji, M. (2013). Predictions of phase separation in three-component lipid membranes by the MARTINI force field. *Journal of Physical Chemistry B, 117*(15), 4072–4080.

Demchenko, A. P., Duportail, G., Oncul, S., Klymchenko, A. S., & Mely, Y. (2015). Introduction to fluorescence probing of biological membranes. In D. M. Owen (Ed.), *Methods in membrane lipids* (2nd ed., Vol. 1232, pp. 19–43).

Feigenson, G. W. (2009). Phase diagrams and lipid domains in multicomponent lipid bilayer mixtures. *Biochimica et Biophysica Acta-Biomembranes, 1788*(1), 47–52.

Feigenson, G. W., & Buboltz, J. T. (2001). Ternary phase diagram of dipalmitoyl-PC/dilauroyl-PC/cholesterol: nanoscopic domain formation driven by cholesterol. *Biophysical Journal, 80*(6), 2775–2788.

Goh, S. L., Amazon, J. J., & Feigenson, G. W. (2013). Toward a better raft model: modulated phases in the four-component bilayer, DSPC/DOPC/POPC/CHOL. *Biophysical Journal, 104*(4), 853–862.

Golfetto, O., Hinde, E., & Gratton, E. (2015). The Laurdan spectral phasor method to explore membrane micro-heterogeneity and lipid domains in live cells. In D. M. Owen (Ed.), *Methods in membrane lipids* (2nd ed., Vol. 1232, pp. 273–290). Totowa: Humana Press Inc.

Gordon-Grossman, M., Gofman, Y., Zimmermann, H., Frydman, V., Shai, Y., Ben-Tal, N., et al. (2009). A combined pulse EPR and Monte Carlo simulation study provides molecular insight on peptide–membrane interactions. *Journal of Physical Chemistry B, 113*(38), 12687–12695.

Gordon-Grossman, M., Zimmermann, H., Wolf, S. G., Shai, Y., & Goldfarb, D. (2012). Investigation of model membrane disruption mechanism by melittin using pulse electron paramagnetic resonance spectroscopy and cryogenic transmission electron microscopy. *Journal of Physical Chemistry B, 116*(1), 179–188.

Hermann, E., Ries, J., & Garcia-Saez, A. J. (2015). Scanning fluorescence correlation spectroscopy on biomembranes. In D. M. Owen (Ed.), *Methods in membrane lipids* (2nd ed., Vol. 1232, pp. 181–197).

Irudayam, S. J., & Berkowitz, M. L. (2011). Influence of the arrangement and secondary structure of melittin peptides on the formation and stability of toroidal pores. *Biochimica et Biophysica Acta-Biomembranes, 1808*(9), 2258–2266.

Irudayam, S. J., Pobandt, T., & Berkowitz, M. L. (2013). Free energy barrier for melittin reorientation from a membrane-bound state to a transmembrane state. *Journal of Physical Chemistry B, 117*(43), 13457–13463.

Jacobson, K., Mouritsen, O. G., & Anderson, R. G. W. (2007). Lipid rafts: at a crossroad between cell biology and physics. *Nature Cell Biology, 9*(1), 7–14.

Kumar, S., Bouzida, D., Swendsen, R. H., Kollman, P. A., & Rosenberg, J. M. (1992). The weighted histogram analysis method for free-energy calculations on biomolecules. 1. The method. *Journal of Computational Chemistry, 13*(8), 1011–1021.

Laio, A., & Gervasio, F. L. (2008). Metadynamics: a method to simulate rare events and reconstruct the free energy in biophysics, chemistry and material science. *Reports on Progress in Physics, 71*(12).

Lee, M. T., Sun, T. L., Hung, W. C., & Huang, H. W. (2013). Process of inducing pores in membranes by melittin. *Proceedings of the National Academy of Sciences of the United States of America, 110*(35), 14243−14248.

Lee, S., Tran, A., Allsopp, M., Lim, J. B., Henin, J., & Klauda, J. B. (2014). CHARMM36 united atom chain model for lipids and surfactants. *Journal of Physical Chemistry B, 118*(2), 547−556.

Lin, J. H., & Baumgaertner, A. (2000). Stability of a melittin pore in a lipid bilayer: a molecular dynamics study. *Biophysical Journal, 78*(4), 1714−1724.

Lingwood, D., & Simons, K. (2010). Lipid rafts as a membrane-organizing principle. *Science, 327*(5961), 46−50.

Luna, E. J., & McConnell, H. M. (1977). Lateral phase separations in binary-mixtures of phospholipids having different charges and different crystalline-structures. *Biochimica et Biophysica Acta, 470*(2), 303−316.

Makarov, D. E. (2014). Computational and theoretical insights into protein and peptide translocation. *Protein and Peptide Letters, 21*(3), 217−226.

Marrink, S. J., de Vries, A. H., & Mark, A. E. (2004). Coarse grained model for semiquantitative lipid simulations. *Journal of Physical Chemistry B, 108*(2), 750−760.

Marrink, S. J., Risselada, H. J., Yefimov, S., Tieleman, D. P., & de Vries, A. H. (2007). The MARTINI force field: coarse grained model for biomolecular simulations. *Journal of Physical Chemistry B, 111*(27), 7812−7824.

Marrink, S. J., & Tieleman, D. P. (2013). Perspective on the Martini model. *Chemical Society Reviews, 42*(16), 6801−6822.

Matsuzaki, K., Yoneyama, S., & Miyajima, K. (1997). Pore formation and translocation of melittin. *Biophysical Journal, 73*(2), 831−838.

McConnell, H. M., & Radhakrishnan, A. (2003). Condensed complexes of cholesterol and phospholipids. *Biochimica et Biophysica Acta, 1610*, 159−173.

McMullen, T. P. W., & McElhaney, R. N. (1996). Physical studies of cholesterol−phospholipid interactions. *Current Opinion in Colloid and Interface Science, 1*, 83−90.

Mihajlovic, M., & Lazaridis, T. (2010). Antimicrobial peptides in toroidal and cylindrical pores. *Biochimica et Biophysica Acta-Biomembranes, 1798*(8), 1485−1493.

Monticelli, L., Kandasamy, S. K., Periole, X., Larson, R. G., Tieleman, D. P., & Marrink, S. J. (2008). The MARTINI coarse-grained force field: extension to proteins. *Journal of Chemical Theory and Computation, 4*(5), 819−834.

Nagle, J. F., & Tristram-Nagle, S. (2000). Structure of lipid bilayers. *Biochimica et Biophysica Acta-Reviews on Biomembranes, 1469*(3), 159−195.

Pandit, S. A., Bostick, D., & Berkowitz, M. L. (2004). Complexation of phosphatidylcholine lipids with cholesterol. *Biophysical Journal, 86*(3), 1345−1356.

Raghuraman, H., & Chattopadhyay, A. (2007). Melittin: a membrane-active peptide with diverse functions. *Bioscience Reports, 27*(4−5), 189−223.

Roduit, C., Longo, G., Dietler, G., & Kasas, S. (2015). Measuring cytoskeleton and cellular membrane mechanical properties by atomic force microscopy. In D. M. Owen (Ed.), *Methods in membrane lipids* (2nd ed., Vol. 1232, pp. 153−159).

Santo, K. P., & Berkowitz, M. L. (2012). Difference between magainin-2 and melittin assemblies in phosphatidylcholine bilayers: results from coarse-grained simulations. *Journal of Physical Chemistry B, 116*(9), 3021−3030.

Santo, K. P., Irudayam, S. J., & Berkowitz, M. L. (2013). Melittin creates transient pores in a lipid bilayer: results from computer simulations. *Journal of Physical Chemistry B, 117*(17), 5031−5042.

Scott, W. R. P., Hunenberger, P. H., Tironi, I. G., Mark, A. E., Billeter, S. R., Fennen, J., et al. (1999). The GROMOS biomolecular simulation program package. *Journal of Physical Chemistry A, 103*(19), 3596–3607.

Sengupta, D., Leontiadou, H., Mark, A. E., & Marrink, S. J. (2008). Toroidal pores formed by antimicrobial peptides show significant disorder. *Biochimica et Biophysica Acta (BBA)-Biomembranes, 1778*(10), 2308–2317.

Shaw, D. E., Deneroff, M. M., Dror, R. O., Kuskin, J. S., Larson, R. H., Salmon, J. K., et al. (2008). Anton, a special-purpose machine for molecular dynamics simulation. *Communications of the ACM, 51*(7), 91–97.

Simons, K., & Ikonen, E. (1997). Functional rafts in cell membranes. *Nature, 387*(6633), 569–572.

Singer, S. J., & Nicolson, G. L. (1972). The fluid mosaic model of structure of cell-membranes. *Science, 175*(4023), 720–731.

Siu, S. W. I., Vacha, R., Jungwirth, P., & Boeckmann, R. A. (2008). Biomolecular simulations of membranes: physical properties from different force fields. *Journal of Chemical Physics, 128*(12).

Sodt, A. J., Sandar, M. L., Gawrisch, K., Pastor, R. W., & Lyman, E. (2014). The molecular structure of the liquid-ordered phase of lipid bilayers. *Journal of the American Chemical Society, 136*(2), 725–732.

Terwilliger, T. C., & Eisenberg, D. (1982a). The structure of melittin. 1. Structure determination and partial refinement. *Journal of Biological Chemistry, 257*(11), 6010–6015.

Terwilliger, T. C., & Eisenberg, D. (1982b). The structure of melittin. 2. Interpretation of the structure. *Journal of Biological Chemistry, 257*(11), 6016–6022.

Terwilliger, T. C., Weissman, L., & Eisenberg, D. (1982). The structure of melittin in the form-I crystals and its implication for melittins lytic and surface-activities. *Biophysical Journal, 37*(1), 353–361.

Tyler, A. I. I., Clarke, J. A., Seddon, J. M., & Law, R. V. (2015). Solid state NMR of lipid model membranes. In D. M. Owen (Ed.), *Methods in membrane lipids* (2nd ed., Vol. 1232, pp. 227–253).

Upadhyay, S. K., Wang, Y. K., Zhao, T. Z., & Ulmschneider, J. P. (2015). Insights from micro-second atomistic simulations of melittin in thin lipid bilayers. *Journal of Membrane Biology, 248*(3), 497–503.

Veatch, S. L., & Keller, S. L. (2003). Separation of liquid phases in giant vesicles of ternary mixtures of phospholipids and cholesterol. *Biophysical Journal, 85*(5), 3074–3083.

Veatch, S. L., & Keller, S. L. (2005). Miscibility phase diagrams of giant vesicles containing sphingomyelin. *Physical Review Letters, 94*(14).

Vogel, H., & Jahnig, F. (1986). The structure of melittin in membranes. *Biophysical Journal, 50*(4), 573–582.

Wood, W. W., & Erpenbeck, J. J. (1976). Molecular-dynamics and Monte-Carlo calculations in statistical-mechanics. *Annual Review of Physical Chemistry, 27*, 319–348.

Wu, E. L., Cheng, X., Jo, S., Rui, H., Song, K. C., Davila-Contreras, E. M., et al. (2014). CHARMM-GUI membrane builder toward realistic biological membrane simulations. *Journal of Computational Chemistry, 35*(27), 1997–2004.

Yu, T., & Schatz, G. C. (2013). Free energy profile and mechanism of self-assembly of peptide amphiphiles based on a collective assembly coordinate. *Journal of Physical Chemistry B, 117*(30), 9004–9013.

CHAPTER TWO

Where Biology Meets Physics—A Converging View on Membrane Microdomain Dynamics

Nicolas Destainville[1],*, Thomas H. Schmidt[2] and Thorsten Lang[2],*

[1]Laboratoire de Physique Theorique (IRSAMC), Universite Toulouse 3-Paul Sabatier, UPS/CNRS, Toulouse, France
[2]Department of Membrane Biochemistry, Life & Medical Sciences (LIMES) Institute, University of Bonn, Bonn, Germany
*Corresponding authors: E-mail: destain@irsamc.ups-tlse.fr and thorsten.lang@uni-bonn.de

Contents

Abstract

For several decades, the phenomenon of membrane component segregation into microdomains has been a well-known and highly debated subject, and varying concepts including the raft hypothesis, the fence-and-picket model, hydrophobic-mismatch, and specific protein–protein interactions have been offered as explanations. Here, we review the level of insight into the molecular architecture of membrane domains one is capable of obtaining through biological experimentation. Using SNARE proteins as a paradigm, comprehensive data suggest that several dozens of molecules crowd together into almost circular spots smaller than 100 nm. Such clusters are highly dynamical as they constantly capture and lose molecules. The

Current Topics in Membranes, Volume 77
ISSN 1063-5823
http://dx.doi.org/10.1016/bs.ctm.2015.10.004

organization has a strong influence on the functional availability of proteins and likely provides a molecular scaffold for more complex protein networks. Despite this high level of insight, fundamental open questions remain, applying not only to SNARE protein domains but more generally to all types of membrane domains. In this context, we explain the view of physical models and how they are beneficial in advancing our concept of micropatterning. While biological models generally remain qualitative and descriptive, physics aims towards making them quantitative and providing reproducible numbers, in order to discriminate between different mechanisms which have been proposed to account for experimental observations. Despite the fundamental differences in biological and physical approaches as far as cell membrane microdomains are concerned, we are able to show that convergence on common points of views is in reach.

1. HISTORY AND CONCEPTUAL FRAMEWORK OF THE MICROCOMPARTMENTAL PHENOMENON

The first popular model for a biological membrane was proposed in 1972 by Singer and Nicolson (1972). At that time, only crystal structures from water-soluble proteins were available, showing extended hydrophobic patches exposed on their surfaces. Relying upon basic thermodynamic principles and the assumption that membrane proteins possess a hydrophobic belt, the fluid mosaic model was proposed. This model assumes that the hydrophobic tails of lipids would embed the proteins into the membrane via their hydrophobic regions, forming a bilayer with a hydrophobic core and hydrophilic borders (Figure 1). However, this model did not consider further interactions which may segregate the proteins, supposed to diffuse freely in a sea of lipids. In 1974, it was discovered that in model membranes, temperature changes can produce clusters of lipids (Lee, Birdsall, Metcalfe, Toon, & Warren, 1974), an idea that was further developed to the concept of lipids in an ordered state (Wunderlich, Kreutz, Mahler, Ronai, & Heppeler, 1978). A functional role of such dynamically forming lipid platforms, the so-called lipid rafts, was proposed in signal transduction. According to the raft hypothesis, rafts form from the tight packing of sphingolipids and cholesterol, thus providing a nucleus for further association with specific proteins (Simons & Ikonen, 1997). Rafts were operationally defined as low–density lipid/protein assemblies called detergent-resistant membranes (DRMs), which were obtained after cell lysis with strong detergents followed by buoyant centrifugation. Although the raft hypothesis was initially intended to explain domains formed only

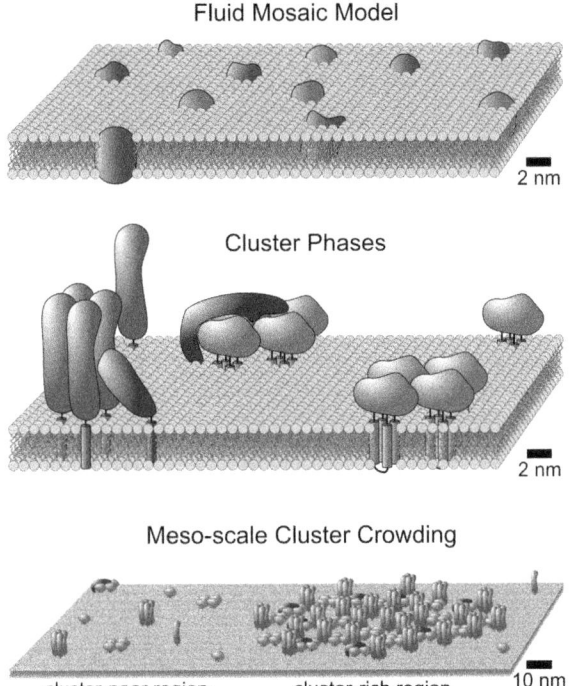

Fluid Mosaic Model

2 nm

Cluster Phases

2 nm

Meso-scale Cluster Crowding

cluster-poor region   cluster-rich region   10 nm

Figure 1 *Top: Fluid-mosaic model; middle: Cluster phases; bottom: Cluster phases underlying mesoscale organization.* When the fluid-mosaic model was proposed in 1972 (Singer & Nicolson, 1972), its main intention was not to explain the micropatterning phenomenon, but to propose a general structural model of a lipid bilayer membrane with embedded proteins therein. At that time no concepts were allowing for proper differentiation between transmembrane segments and globular bulky domains. However, it also made a prediction on the lateral dynamics proposing that membrane proteins, although occasionally interacting with each other, in principle diffuse freely in a sea of lipids. It took more than 40 years until reliable estimates on the ratio between lipids and proteins became available. Two studies, one on synaptic vesicles (Takamori et al., 2006) and one on the red blood cell membrane (Dupuy & Engelman, 2008) suggest that 18−25% of the hydrophobic volume is occupied by the transmembrane segments of proteins. Because membrane proteins often have large globular domains, it is likely that more than 50% of the surface is covered with proteinaceous structures. Hence, compared to the fluid-mosaic model, biological membranes are much more dominated by proteins than initially assumed. The cluster phase model suggests the concentration of only one or a small group of proteins (and in some cases lipids) into a densely packed protein cluster. For clarity, the cartoon shows less crowding than experimentally determined and clusters composed of maximal four proteins. Real numbers are likely much higher, although data is available for only a few proteins (see text). The clusters are separated from each other through a protein-free zone and exchange proteins by dynamically capturing and releasing molecules. The mesoscale organization is an addition to the cluster phase model. Here, clusters from different proteins clump together into a large mesoscale assembly. However, it is important to note that within the assembly, the clusters retain their identity.

by a particular selection of proteins, it was also considered an example for a more general mechanism of membrane organization. It quickly became very popular as more and more proteins were found in rafts, and thus raised interest in the micropatterning phenomenon. Another model for a general micropatterning mechanism is the fence-and-picket hypothesis (Kusumi et al., 2005; Kusumi, Sako, & Yamamoto, 1993). It suggests that subplasmalemmal protein networks represent barriers for diffusion and are the underlying mechanism for the widely observed corralled and hop diffusion (when a particle exhibiting corralled diffusion jumps from one neighboring corral into another).

Over the years it became more and more clear that DRMs not only present structures already present in live cells but also postsolubilization artifacts (Lai, 2003; Munro, 2003; Pierce, 2004). It is difficult to say at what point the DRM criteria has been abandoned for defining rafts (some still use it today) but several attempts for redefining rafts have since been undertaken. During a keystone symposium in 2006 it was suggested that "Membrane rafts are small (10–200 nm), heterogeneous, highly dynamic, sterol- and sphingolipid-enriched domains that compartmentalize cellular processes" (Pike, 2006). Notably, detergent resistance and liquid ordering were left out from the definition, because it has been agreed upon that no evidence exists for the presence of such ordered phases in complex cellular systems. A few years later, lipid rafts were supposed to represent "fluctuating nanoscale assemblies of sphingolipid, cholesterol, and proteins that can be stabilized to coalesce" (Lingwood & Simons, 2010).

Until 10 years ago, a description of the basic physical features of a lipid raft or any other domain was almost impossible because microdomains are abundant structures present at high densities. In electron microscopy, the labeling efficiency is too poor to describe the size and density of domains, and diffraction-limited microscopy (with a resolution of 250–300 nm) generally visualizes blurry, diffraction limited patterns from nonresolved domains, although in cases of particularly abundant domains a uniform appearance may be observed, casting doubts that microdomains exist. This limitation was overcome by the introduction of super-resolution microscopy (Lang & Rizzoli, 2010) allowing one to appreciate the high density of membrane protein domains and to measure their physical dimensions. Most domains have a diameter in the range of 100 nm and look round or ellipsoid. Little is known about how densely proteins are packed within the domains and whether they are preferentially formed of only one or several types of proteins. Live imaging approaches such as fluorescence recovery after photobleaching (FRAP) or single particle tracking suggest that domains are dynamic structures, constantly releasing and capturing molecules (Barg, Knowles, Chen, Midorikawa, & Almers, 2010; Espenel et al., 2008; Saka et al., 2014; Sieber et al., 2007). Hence, domains are not stable structures but rather exchange components with other domains over large distances.

More recently, an additional mesoscale organization principle was proposed. In this model, membrane protein clusters do not randomly scatter across the plasma membrane but instead locally crowd together to form mesoscale assemblies into which the individual clusters may retain their unique identities (Figure 1). Such assemblies are rather variable and irregular in shape and size, with physical dimensions of several hundred nanometers in width and length and separated by areas of low-protein density (Lillemeier, Pfeiffer, Surviladze, Wilson, & Davis, 2006; Saka et al., 2014). Finally, possible roles of ankyrin and spectrin in the long-range organization of plasma membranes via proto-domains have also been proposed (Bennett & Lorenzo, 2013).

At present, it can safely be concluded that most, if not all, membrane proteins segregate into different domains or clusters via specific mechanisms. The biological functions of clustering are diverse (see Section 2.2), or in some cases may not even exist. One difficulty in unraveling their functions is that most studies are limited in correlative conclusions, showing that functions are disturbed or lost when domains are disintegrated (in contrast to a gain of function concept). In the early days, this was mostly achieved by means of cholesterol depletion, which has been widely used to disintegrate cholesterol-rich rafts. However, this approach does not elucidate the molecular mechanism by which a domain influences a process. Moreover, it turns out that cholesterol is a much more general membrane structure organizer than previously envisioned, potentially affecting the organization and diffusional behavior of nearly any protein (Saka et al., 2014). Thus, this opens the possibility that observed inhibition following cholesterol depletion may in some cases be a secondary effect and not directly related to the loss of function of the microdomain being studied. In any case, the high density of proteins necessarily will have a regulatory effect in at least two scenarios. In some cases, a high local concentration might be required to generate sufficiently strong signals (Gurry, Kahramanogullari, & Endres, 2009), while in other cases, active protein domains might be embedded inside of the cluster to regulate or conserve their reactivity (Lang & Rizzoli, 2010).

2. ZOOMING INTO A MEMBRANE DOMAIN—USING SNARE PROTEIN CLUSTERS AS A PARADIGM

What level of insight can be achieved when experimentally studying membrane proteins? For a small group of membrane proteins, a large amount of quantitative data is available. These proteins belong to the Soluble N-ethylmaleimide-sensitive-factor Attachment protein REceptor (SNARE) family and are small membrane proteins essential for all intracellular fusion steps, with the exception of mitochondrial fusion (Jahn & Fasshauer, 2012). They share a conserved 60–70 amino acid stretch, the SNARE-motif, which is typically attached to a C-terminal transmembrane segment (Figure 2). Most

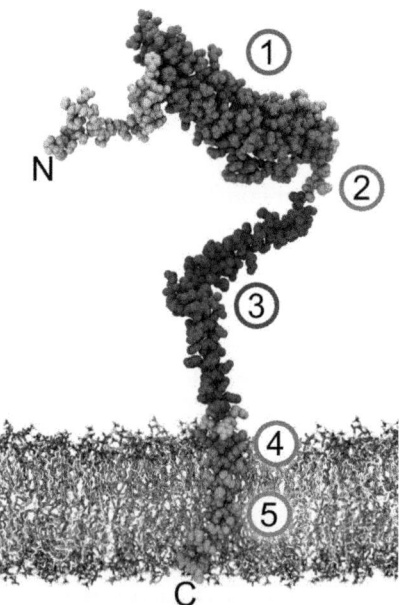

Figure 2 *Defined segments of syntaxin 1 and how they are related to clustering.* Structural model of rat syntaxin 1A, a 288-amino acid-long protein. The protein has a short N-terminal peptide attached to an autonomously folded N-terminal domain (1; blue (dark gray in print versions)) which is linked via a long linker region (2) to the SNARE-motif (3; red (darker gray in print versions)). Between the SNARE-motif and a relatively short transmembrane segment (TMS) (5; yellow (light gray in print versions)) exists a short stretch harboring five positively charged amino acids, the polybasic linker region (4; orange (gray in print versions)). For definition of segments see van den Bogaart et al. (2013). Discussed clustering mechanisms involve the TMS (5) via mechanisms involving cholesterol or hydrophobic mismatch (see text). The polybasic linker regions (4) are clustered together through charge bridges involving PI(4,5)P$_2$ or PI(4,5)P$_3$ (van den Bogaart et al., 2011; Khuong et al., 2013) and the SNARE-motif (3) clusters syntaxin molecules through homo-oligomerization involving the N-terminal part of the SNARE-motif (Sieber et al., 2006, 2007).

SNAREs also have an autonomously folded N-terminal domain attached via a linker region to the SNARE-motif (Figure 2).

Different types of SNAREs, occupying opposed membranes, can form complexes via their SNARE-motifs, thereby mediating membrane fusion. Such heterooligomeric SNARE complexes are formed from four SNARE-motifs provided from three to four different SNAREs (please note that a few SNAREs have two SNARE-motifs).

In 2001, both microscopic and biochemical evidence suggested that members of the SNAREs involved in neuronal exocytosis, syntaxin 1 and SNAP25, are concentrated in cholesterol-dependent domains (Chamberlain, Burgoyne, & Gould, 2001; Lang et al., 2001), also called syntaxin and SNAP25 clusters. Following this observation, a large number of studies have focused on unraveling the mechanisms of clustering and its putatively associated functions (van den Bogaart, Lang, & Jahn, 2013). Although all SNAREs form domains, the SNARE syntaxin 1 has the most available information and thus allows for a more detailed and quantitative discussion. The second most studied SNARE is SNAP25, which has two SNARE-motifs connected by a linker region attached to the plasma membrane via palmitoylation.

In the following sections, we discuss three aspects of SNARE domains, (1) their physical parameters, (2) the function of clustering, and (3) what drives and limits clustering. During the course of this discussion, questions arise which refer in general to all membrane protein clusters, therefore we consider SNARE clustering as a paradigm of membrane protein clustering which illustrates the fundamental questions of the field.

2.1 Physical parameters of SNARE clusters

Syntaxin 1 and SNAP25 are highly abundant both in the presynaptic membrane (Wilhelm et al., 2014) as well as in the plasma membrane of neuroendocrine cells. For syntaxin 1, quantitative estimates from the neuroendocrine cell line PC12 yield values from 540 molecules/μm^2 (Knowles et al., 2010) to approximately 2000 molecules/μm^2 (Sieber et al., 2007; James, Kowalchyk, Daily, Petrie, & Martin, 2009), while for SNAP25 7500 molecule/μm^2 were determined (Knowles et al., 2010). For syntaxin, the almost fourfold difference might be due in part to different definitions of the membrane area. The value of 2000 molecules was obtained by relating total molecules to imaged cell surface area, or rather the 2D projection of an imperfectly flat membrane, while the value of 540 molecules is related to membranes measured by capacitance measurements. Because these average values do not provide any information about local heterogeneities, for clarification the molecule number per cluster was determined for syntaxin 1 by three different approaches. In the first approach, syntaxin 1 molecules were counted by quantitative western blot while clusters were counted by super-resolution Stimulated Emission Depletion (STED) microscopy, which provided an upper limit of 90 molecules per cluster. A Brownian dynamics simulation of fluorescence recovery after photobleaching (FRAP) kinetics yielded an estimate on the fraction of freely diffusing, unclustered syntaxin 1. Both allowed for the conclusion that the average cluster contains 75 syntaxin 1 molecules, which, according to the cluster size determined by STED microscopy, populate an area with a diameter of 50–60 nm (Sieber et al., 2007). Overexpression generates more rather than larger clusters (Sieber, Willig, Heintzmann, Hell, & Lang, 2006), suggesting that cluster size depends not much or not at all on syntaxin concentration. In the second approach, when overexpressing fluorescent protein labeled syntaxin 1, photoactivated localization microscopy (PALM) images display 30 to 40 molecules in a 50 nm large spot (Rickman et al., 2010). Considering the fact that PALM underestimates molecules and that endogenous syntaxin was not taken into account, the value of 30–40 molecules agrees well with the model suggesting 75 molecules. In the third and final approach, again using fluorescent protein-labeling and total internal reflection fluorescence (TIRF) microscopy, signals arising from single clusters were divided by the intensity of a single GFP molecule. Correcting for the presence of endogenous syntaxin, a value of 50 syntaxin molecules per cluster was determined (Knowles et al., 2010). Likewise, a value of 70 molecules per cluster was predicted for SNAP25 (Knowles et al., 2010).

Antibody labeling was applied for the visualization of SNARE clusters by STED microscopy. Due to space limitations caused by the high density of epitopes, each syntaxin 1 or SNAP25 molecule cannot be decorated with a primary antibody, resulting in substochiometric labeling. A secondary antibody was then bound to the primary antibody. Because the chain "epitope"–"primary antibody"–"secondary antibody" is flexible, fluorescence may not arise exactly from the epitope site. Both effects, substochiometric labeling and fluorophore flexibility, may allow for determination of the average size but not for the resolution of small irregularities in cluster shape. Perhaps for this reason, STED micrographs of syntaxin 1 (Sieber et al., 2006, 2007) and SNAP25 clusters (Willig, Keller, Bossi, & Hell, 2006) generally have a circular shape while imperfectly round clusters are detected imaging ensembles of single molecules by PALM-/STORM-based approaches (Pertsinidis et al., 2013; Rickman et al., 2010). Another study applying dSTORM microscopy (Bar-On et al., 2012) showed more elliptical clusters, though antibody labeling was also used in this approach. In this study, evidence was presented that clusters are not uniformly dense but rather become less crowded towards the periphery of the cluster.

Regarding the mobility of the entire syntaxin 1 cluster in live PC12 (Barg et al., 2010) or MIN6 cells (Ohara-Imaizumi et al., 2004), the clusters do not migrate over larger distances and in plasma membrane sheets they appear rather stationary (Sieber et al., 2007). However, despite the overall immobility, several observations indicate that clusters are highly dynamic because they exchange molecules with each other. First, the kinetics of FRAP is consistent with an exchange of molecules between clusters (see above). Second, under conditions overexpressing only a few fluorescent protein-labeled syntaxin 1 molecules, single molecules change between a fast and a slow diffusion mode (Barg et al., 2010). Additionally, syntaxin molecules reversibly accumulate beneath docked granules (Barg et al., 2010). This suggests that single clusters can strongly vary in their size and can even dissolve completely. Similar observations were made when single SNAP25 molecules were tracked (Knowles et al., 2010). Moreover, imaging localized fluorescence spots of fluorescent protein-labeled SNAP25, it was observed that clusters are highly dynamic, staying or disappearing on a timescale of only a few minutes (Antoku, Dedecker, Pinheiro, Vosch, & Sørensen, 2015).

These data, although variable within a range, give a good formal description of a syntaxin cluster and a SNAP25 cluster.

2.2 Putative functions of SNARE clustering

As outlined above, syntaxin 1 and SNAP25 are involved in Ca^{2+}-triggered fusion of vesicles with the plasma membrane, as occurs in the synapse or in neuroendocrine cells. Neuronal fusion also requires the SNARE synaptobrevin 2, localized on the vesicle surface. After fusion, a ternary complex composed of all three SNAREs remains in the plasma membrane. This ternary complex is disassembled into the individual SNAREs by NSF (along with its cofactor αSNAP and with requisite ATP consumption), fueling SNAREs with energy for another fusion cycle.

Most likely, the SNARE-cycle begins with an interaction between syntaxin 1 and SNAP25 which forms an acceptor complex for synaptobrevin 2. However, a number of regulatory factors (e.g., Munc18) are required for coordination upon reaching a late primed state from which fusion is triggered within milliseconds by Ca^{2+} sensors from the synaptotagmin family (Jahn & Fasshauer, 2012).

As SNAREs are key players in fusion, their local concentrations pose the basic question of whether functionally relevant SNARE complexes form inside or outside of the clusters. Should clusters be functionally relevant, is their role up- or downstream in the SNARE cycle, and do they influence the cycle at more than one step?

Some insights are provided by comparing complex formation between syntaxin 1 and SNAP25 in solution (in the absence of the short transmembrane segment (TMS) and at roughly equimolar concentrations) and in native membranes. In solution, instead of a 1:1 syntaxin:SNAP25 complex, a 2:1 (syntaxin 1)$_2$:SNAP25 complex forms to which binding of synaptobrevin 2 is precluded (Fasshauer & Margittai, 2004). This kinetic dead-end complex can be avoided by increasing the ratio between SNAP25 and syntaxin 1 in liposomes (Pobbati, Stein, & Fasshauer, 2006). In native membranes there is no evidence for the 2:1 complex. First, soluble syntaxin 1 (lacking the TMS) readily binds to endogenous SNAP25 in membrane sheets, indicating that endogenous syntaxin 1 does not compete efficiently with the exogenous syntaxin 1 (Lang, Margittai, Hölzler, & Jahn, 2002), although it may be locally available at a much higher concentration. Even the elimination of endogenous syntaxin 1 does not increase the binding capacity of SNAP25 for exogenously added syntaxin 1 (Lang et al., 2002). This suggests that the interaction between syntaxin 1 and SNAP25 is tightly controlled in native membranes, or rather in steady state only a small fraction of SNAP25 and syntaxin 1 molecules are interacting. Second, in FRAP experiments in PC12 cells, overexpression of syntaxin 1 slowed down the mobility of SNAP25, which was used for characterizing the interaction between SNAP25 and syntaxin 1. The data showed that the second SNARE-motif of SNAP25 is not involved in any interactions (Halemani, Bethani, Rizzoli, & Lang, 2010). As the second SNARE-motif is required for the 2:1 complex, the finding argues against its presence in native membranes. Additionally, Förster's resonance energy transfer (FRET) lifetime imaging was applied to probe the state of SNARE complexes (Rickman et al., 2010). This study also suggested that within the clusters syntaxin 1 and SNAP25 do interact but two types of complexes were detected. One, previously mentioned above, and another into which the second SNARE-motif of SNAP25 is involved (called the three helical SNARE-complexes). Interestingly, the two types of complexes were unmixed on the level of clusters. Upon cholesterol depletion, all complexes adopted the three helical SNARE-complex conformations.

SNARE clustering provides an explanation for the phenomenon of highly abundant SNAREs' incapacity to interact efficiently with each other.

Clustering not only separates syntaxin 1 and SNAP25 spatially, but involves syntaxin 1 self-oligomerization via the SNARE-motif (see Section 2.3) which directly competes with the interaction of syntaxin 1 with SNAP25. Along with the excess of SNAP25 (see above), by these means the ratio between reactive SNAP25 and syntaxin 1 is kept very high, promoting not only the 1:1 (or the three helical SNARE-complex) over the 2:1 complex but also allowing only a small fraction of syntaxin 1 to be involved in any complex formation.

How clustering influences SNARE complex formation kinetics was also studied from another angle. Probing ternary SNARE-complexes in native membranes, two phases of fusion-independent SNARE-complex formation were observed (Bar-On et al., 2008). The reconstruction of the experimental data in a kinetic model agreed well with the idea of a small pool of unclustered syntaxin 1 forming complexes rapidly, while the large majority residing in clusters form complexes at a rate of at least fourfold slower (Bar-On et al., 2009). Similarly, the model predicts that the disassembly of clustered *cis*-SNARE-complexes by NSF is 20-fold faster than that of unclustered complexes.

Clustering also likely has other effects, such as allowing the assembly of complex molecular architectures for the coordination of the SNARE-cycle in space and time. Accumulating evidence for this idea is based on the early suggestion that syntaxin 1 clusters define sites for vesicle docking and fusion (Lang et al., 2001; Ohara-Imaizumi et al., 2004). In following studies, factors involved in the regulation of exocytosis were localized to syntaxin clusters such as Munc18 (Pertsinidis et al., 2013), the calcium sensor synaptotagmin 1 (Honigmann et al., 2013), or $PI(4,5)P_2$ (Aoyagi et al., 2005; James, Khodthong, Kowalchyk, & Martin, 2008).

In this context, factors come together and complexes are stabilized at clusters, perhaps by avidity effects, in which case clusters would act as templates for molecular scaffolds creating fusion sites. At the same time, this clustering may alleviate inhibitory effects. Regarding $PI(4,5)P_2$, a liposome fusion assay involving reconstituted SNAREs and natural concentrations of $PI(4,5)P_2$ has shown that its inhibitory effect is increased when two of the positively charged amino acids in the polybasic linker region of syntaxin 1 (Figure 2) are replaced by alanine residues (James et al., 2008). This effect is likely mediated via $PI(4,5)P_2$ action on membrane curvature. This suggests that syntaxin 1 sequesters $PI(4,5)P_2$ via its polybasic linker region and thus diminishes its inhibitory effect (James et al., 2008; see Section 2.3 for the putative roles of $PI(4,5)P_2$ in syntaxin clustering).

However, a few studies also propose other views. For instance, one study analyzed the distribution of SNAREs by PALM and the location and mobility of secretory granules. Combining the data in a model suggests that vesicles dock and fuse in areas of low SNARE density, where the number of SNAREs beneath a granule ranges from zero to seven (Yang et al., 2012). On the other hand, when applying single molecule imaging it is observed that syntaxin clusters form beneath already docked granules (Barg et al., 2010). Here, syntaxin clustering and docking are also related, though in this view docking precedes clustering.

Taken together, there is still ongoing research in this direction, as even more roles of clusters, e.g., in endocytosis during vesicle recycling (Milovanovic & Jahn, 2015), may be discovered in the future.

2.3 Mechanisms promoting clustering

A complete description of the underlying attractive forces is not sufficient to explain the phenomenon of clustering. Further, a size limiting mechanism is required, as otherwise clusters would grow to one cluster per cell. This will be discussed in a more general framework in Section 4.

It is known that syntaxin clustering is highly specific, because isoforms of syntaxin 1 and 4 or syntaxin 3 and 4 segregate into different clusters when present in the same cell membrane (Low et al., 2006; Milovanovic et al., 2015; Sieber et al., 2006). Because syntaxin 1 and SNAP25 functionally interact with each other during membrane fusion, it is possible that some SNAP25 molecules are embedded in the syntaxin 1 clusters or that syntaxin 1 and SNAP25 clusters do partially mix or rather overlap. Here we encounter largely contradictory findings in literature (van den Bogaart et al., 2013), perhaps because syntaxin 1 and SNAP25 clusters are certainly in close proximity to each other (Halemani et al., 2010; Pertsinidis et al., 2013) and in most cases an overlap has been determined with diffraction-limited resolution, which cannot discriminate between close proximity and concentric overlap.

Several factors influence SNARE clustering. The first clustering mechanism was suggested to depend on cholesterol, based on a type of partitioning in cholesterol-enriched domains or rafts (Chamberlain et al., 2001; Lang, 2007; Lang et al., 2001; Salaün, James, & Chamberlain, 2004). However, as outlined above, one has to question the specificity of the cholesterol effect on membrane domain organization (Saka et al., 2014). Moreover, cholesterol alone is not sufficient as cholesterol-mediated clustering can be disrupted by $PI(4,5)P_2$ through electrostatic binding to the polybasic linker region

(Figure 2) of syntaxin 1 (Murray & Tamm, 2009, 2011). The same interaction was proposed to promote clustering in native membranes (van den Bogaart et al., 2011), albeit here in principle $PI(4,5)P_3$ can also be involved (Khuong et al., 2013).

The theory that some type of electrostatic mechanism controls SNARE clusters is also supported by the effect of micromolar concentrations of Ca^{2+}, which leads to stronger clustering of both syntaxin 1 and SNAP25 (Zilly et al., 2011). As these two proteins are highly negatively charged, it was suggested that the positively charged Ca^{2+} diminishes the repulsive forces which develop with the number of SNAREs, and thus allows more SNAREs to cluster within a defined area. Hence, the accumulation of charges during clustering could be one component that limits cluster size.

Finally, hydrophobic mismatch was added as an additional mechanism promoting syntaxin clustering via lipid TMS interactions, a mechanism apparently sufficient for the separation of syntaxin 1 and syntaxin 4, which have TMSs of slightly different lengths (Milovanovic et al., 2015).

When studying the role of the large cytoplasmic domain in a co-clustering assay, it was shown that the SNARE-motif (Figure 2) is essential for targeting syntaxin 1 into a syntaxin 1 cluster (Sieber et al., 2006). Deletion of the N-terminal domain still allowed for proper targeting, while deletion of the entire cytoplasmic domain (without deleting the polybasic linker region) allowed only for the formation of clusters separated from full-length syntaxin clusters. This finding underlines the intrinsic feature of the TMS to form clusters on its own. In line with these observations are FRAP studies showing that the deletion of the SNARE-motif but not the N-terminal domains increases apparent diffusion in PC12 cells (Sieber et al., 2007) and rat spinal cord neurons (Ribrault et al., 2011). Mechanistically, it has been suggested that homo–oligomerization of the SNARE-motif, as observed in solution (Lerman, Robblee, Fairman, & Hughson, 2000), is also responsible for oligomerization or cluster formation in the membrane (Sieber et al., 2006).

While these studies suggest no role for the N-terminal domain in syntaxin clustering per se, the N-terminal domain is required to initiate cluster formation specifically beneath secretory granules (Barg et al., 2010).

More clustering mechanisms are related to the actin cytoskeleton (van den Bogaart et al., 2013) though here it becomes difficult to differentiate direct from indirect effects, similar to effects observed after cholesterol depletion.

So far, the data suggest that the TMS of syntaxin 1 and likely of all syntaxins is capable of clustering via lipid interactions. For more specific or tighter clustering, protein—protein interactions might be involved. At some point, the accumulation of charges and steric hindrance may stop further cluster growth. SNAP25 has no TMS but still forms clusters in the absence of both SNARE-motifs (Halemani et al., 2010). SNAP25 is a negatively charged protein but is similar to syntaxin 1 in that it has polybasic regions close to its membrane-anchored section. It remains to be shown whether the membrane-anchored region alone is sufficient to generate clusters or if other domains in the linker region are required.

3. FUNDAMENTAL OPEN QUESTIONS

The micropatterning field is largely dominated by experimental scientists covering the spectrum of the life sciences, with physicists only occasionally being represented. When conceptualizing their data, biologists draw a model sketch showing the players doing what they are supposed to do, leaving out unknown factors, as well as temporal, spatial, stoichiometric, and thermodynamic aspects. Hence, biological models tend to be nonquantitative and in a way selective. In contrast, physical models are mandatorily quantitative, because arguments are derived from mathematical relationships where orders of magnitudes must be physically relevant. Only in a physical model can the important determinants be adequately described. However, a problem is that experimental evidence is not available for all parameters, resulting in some factors being left out (assuming they are not essential) while others require approximations. Both of the approaches have their limitations when explaining why so far we only partially understand the micropatterning phenomenon.

One fundamental open question refers to the observed specificity of micropatterning. None of the available concepts seems capable of embracing the pivotal specificity question in its global nature. For instance, the fence-and-picket hypothesis predicts sorting by steric effects, but, as discussed in section 4.3, examples are known where two isoforms of a protein with virtually the same structure segregate into different clusters. The lipid raft hypothesis suggests a mechanism that should not be capable of very specifically discriminating membrane proteins. Finally, hydrophobic mismatch, suggesting that variability in local membrane thickness leads to recruitment of adequately matching TMS, may allow for sorting into a few categories but again does not provide a sufficiently large range of interaction variability to account for the high degree of micropatterning. Perhaps it is a combination of all mechanisms and small contributions of specific protein—protein interactions that are sufficient for sorting. In fact, a few studies have shown that the presence of a particular protein domain, which may be just a few amino acids long, is necessary for the sorting of a protein into a respective cluster (Homsi et al., 2014; Schreiber, Fischer, & Lang, 2012; Sieber et al., 2006). This suggests that fine-tuning of interactions of proteins with their neighborhood can segregate them into relevant clusters. However, here another problem comes into play. While the raft, the fence-and-picket, or the hydrophobic mismatch models can readily explain the growth of domains up to several hundred nanometers, protein—protein interactions are commonly considered to form only dimers or low oligomers. Only when we assume the existence of a complex network of binding partners with several binding valences per protein can large-scale aggregation into a 2D platform be explained.

Another open question is why clusters stop growing. Clustering decreases the entropy and must be driven by a release of energy (e.g., binding energy, perhaps also by depletion attraction, see below), resulting in an overall gain of free energy (please note that free energy and free enthalpy are essentially identical in an aqueous solution when the volume is constant). Because such a system would tend to minimize its boundary length, for each protein type only one "giant" cluster per cell should form. Hence, some mechanisms must exist which stop cluster growth at some point, for instance the accumulation of charges or steric restriction development with cluster size.

In the following, we discuss these problems in the context of physical models and physical views on membrane micropatterning. Statistical mechanics has historically been designed in order to tackle large and complex systems of interacting objects in a thermally agitated, fluctuating medium, as well as to derive collective behaviors from the knowledge of microscopic properties of the interacting objects. Statistical mechanics is thus a priori adapted to address the key questions discussed above.

4. THE AVAILABLE PHYSICAL MODELING BACKGROUND: ITS PREDICTIONS AND OPEN QUESTIONS

Physical approaches from statistical mechanics and soft matter have dealt with biomembranes for more than two decades, as introduced in the notable works of Safran (Dan, Pincus, & Safran, 1993), Pincus (Goulian,

Bruinsma, & Pincus, 1993), Lipowsky (Lipowsky, 1993) or Helfrich (Weikl, Kozlov, & Helfrich, 1998), and their collaborators. It has been understood that membrane inclusions feel mutual forces propagated by the elastic, fluctuating membrane, both of mechanical and entropic origin, notably generic and partially nonspecific short-range attractions larger than a few $k_B T$. Such forces, stronger than the thermal energy, are likely to shape the membrane in spite of thermal agitation and molecular diffusion, and thus provide a first organization principle beyond Singer and Nicolson's proposal. If the attraction of the membrane molecules is strong enough or if the density is high enough, a (generally first-order) phase transition occurs: a macroscopic (or "giant") protein-rich phase, the size of which is smaller than but comparable to the size of the system itself (here, the cell size), coexists with a low-density "gas" phase, containing essentially "monomeric" proteins (Safran, 1994). This is a classical liquid–vapor phase transition. This phenomenon of two-dimensional phase separation has very recently been demonstrated/ observed on reconstituted supported cell membranes (Banjade & Rosen, 2014), where large domains of Nephrin, Nck, and N-WASP proteins were observed to form above a critical concentration, without the necessity of lipid segregation. As expected, Banjade and Rosen found that "the critical concentration depends on how strongly the proteins interact." Note, however, that in this work the system is not asserted to be at equilibrium (reaching equilibrium through domain coarsening can be a very long process (Bray, 1994)), thus the large micrometric clusters are likely to coalesce further and to produce "giant" phases.

This prediction of a *macroscopic* protein phase is in evident contradiction with the experimental observation of smaller, *submicrometric* domains in live cells (Fan, Sammalkorpi, & Haataja, 2010a; Lenne & Nicolas, 2009). Therefore, a physical mechanism must exist which explains why phase–condensation stops at a given domain size, or in other words, why domains that are too large tend to fractionate into smaller ones. One speaks of "microphase" separation (sometimes called "mesophase"), in opposition to the macroscopic phase separation described above. Two classes of mechanisms have been discussed in literature: several works have focused on equilibrium thermostatistical mechanisms, whereas more recent publications consider the out-of-equilibrium character of a cell membrane as pivotal (Figure 3).

In this section, following the model proposed by Lenne and Nicolas (2009), the term "domain" is used with a broad meaning and stands for any membrane area where molecules of some kind are more concentrated than in the surrounding membrane.

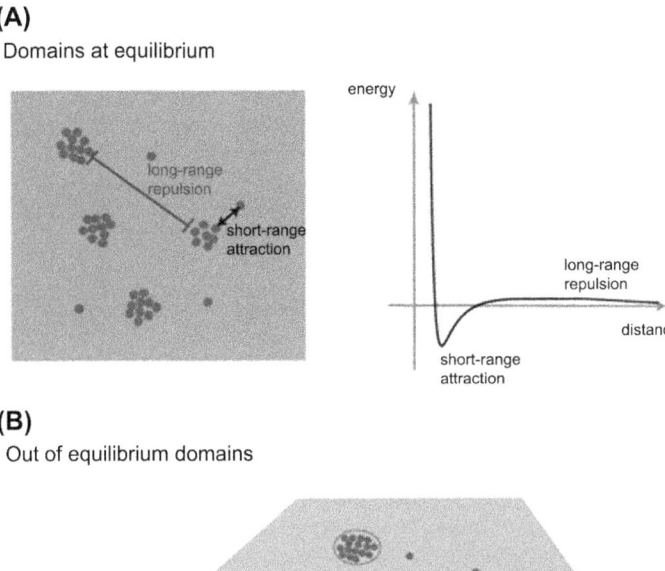

(A)
Domains at equilibrium

(B)
Out of equilibrium domains

Figure 3 *The two main classes of physical mechanisms having been proposed to account for the existence of finite-size clusters.* (A) Long-range repulsion prevents the growth of large domains towards a macroscopic phase and stops the condensation process at a finite cluster size. Clusters coexist with a "gas" of monomeric proteins. (B) Out-of-equilibrium processes, such as active membrane "recycling," break large domains into smaller ones and also prevent the condensation into a macroscopic phase. *Reproduced from Lenne and Nicolas (2009) with permission of The Royal Society of Chemistry.*

4.1 Equilibrium models, cluster-phase hypothesis

There exist several nonspecific, short–range, attractive forces mediated (or "propagated") by the membrane which can act on membrane inclusions such as proteins at a range of a few nanometers and with the binding energy on the order of the thermal energy $k_B T$ (Figure 4).

Most of them have been evoked in Section 2.3 at a qualitative level. Starting from elementary principles, physicists have worked on these quantifications for about two decades.

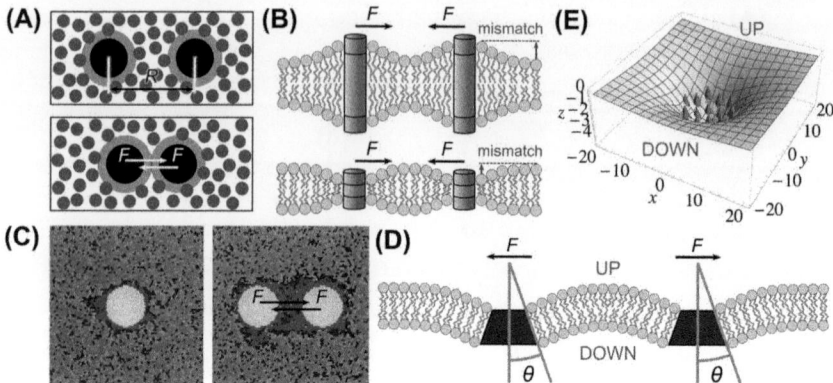

Figure 4 *Different origins of membrane-propagated forces between membrane proteins.* The arrows indicate the orientation of the force *F* exerted on proteins separated by a distance *R*. Short-range attractions: (A) Depletion force due to the lateral osmotic pressure of lipids (schematized as red (dark gray in print versions) discs) on the proteins (in black). The gray annuli represent the minimal distance of approach of lipid centers. When the proteins are far away (top panel), the osmotic pressure is isotropic and the net force vanishes. When the gray annuli start overlapping (bottom panel), the lipids are depleted in-between the proteins and the osmotic pressure becomes anisotropic. A nonvanishing force ensues. (B) Hydrophobic mismatch forces due to the mismatch between the width of the membrane hydrophobic layer and the height of the protein hydrophobic cores (in gray). (C) In the case of complex lipidic composition, a single protein (left panel) is surrounded by an annulus of "wetting" lipids (in red (dark gray in print versions)). When proteins get closer (right panel), sharing their lipid annuli becomes advantageous in terms of interface energy between the red (dark gray in print versions) and purple (black in print versions) lipids *(Reprinted from Gil et al. (1997) with permission from Elsevier (the original figure was modified by adding labelled arrows)).* Longer-range repulsion: (D) Origin of curvature-mediated long-range repulsion. Each inclusion, modeled in first approximation as a cone of half-aperture angle θ, elastically deforms the membrane and breaks the up/down symmetry. (E) Example of collective membrane bending (or budding) induced by a cluster of 19 identical up/down asymmetric inclusions (pink (gray in print versions) cones). Such a deformation is not the simple sum of individual deformations as in panel D, evidencing the many-body character of the interaction energy *(Adapted from Weitz and Destainville (2013) with permission of The Royal Society of Chemistry.).*

Even though they are generally not mentioned, depletion forces due to the 2D osmotic pressure laterally exerted by lipids on proteins, which tend to bring them closer when they are about a nanometer away, have been quantified and shown to lead to protein–protein attraction (Borodich, Rojdestvenski, & Cottam, 2003) (see Figure 4(A)). As previously discussed concerning SNARE proteins, hydrophobic mismatch forces exist which act when the hydrophobic belt of a protein does not match the width of the membrane hydrophobic layer (Contreras, Ernst, Wieland, & Brügger, 2011; Mouritsen, 2005). The energy cost of the subsequent membrane elastic deformation grows with the distance between two identical proteins, thus resulting in an attractive effective force, the energy scale of which is also on the order of $k_B T$ (Neder, West, Nielaba, & Schmidt, 2011; Nielsen, Goulian, & Andersen, 1998) (Figure 4(B)). Furthermore, biomembranes contain several lipid species, and an additional mechanism then potentially leads to attractive forces: proteins, either integral or peripheral (e.g., GPI-anchored proteins), can recruit in their neighborhood "wetting" lipids having a better affinity for them (van den Bogaart et al., 2011; Lee, 2003; Vereb et al., 2003; Hancock, 2006). The recruited lipids form "annuli" (or "shells") (Anderson & Jacobson, 2002; Contreras et al., 2011) surrounding the proteins, which favor configurations where proteins are not far apart because sharing their annuli reduces the interfacial energy between different lipid species (Figure 4(C)). Ensuing effective binding energies are also of order $k_B T$ or even larger (Gil, Ipsen, Mouritsen, et al., 1998; Gil, Ipsen, Tejero, 1998; Gil, Sabra, Ipsen, & Mouritsen, 1997; Mouritsen, 2005; Reynwar & Deserno, 2009). Such forces are more specific than those previously mentioned. For example, it is reasonable to anticipate that mechanisms involving cholesterol and sphingolipids belong to this category, but other lipids, such as $PI(4,5)P_2$, may also play a role (see Section 2).

Additionally, the hydrophobic layer of the membrane is a low dielectric medium, with a relative permittivity $\varepsilon_r < 5$ (Buyukdagli, Manghi, & Palmeri, 2010) and width d (few nanometers), which sets the range of electrostatic interactions between charged residues (Abney & Scalettar, 1993). This allows for electrostatic interactions which are capable of acting beyond a nanometer. Their role has been observed in the formation of K-Ras (Plowman, Ariotti, Goodall, Parton, & Hancock, 2008) and syntaxin-1A (see Section 2.3) clusters.

Put together, this indicates that energies at close contact between membrane proteins—which we call "binding energies" in the following—are a few times larger than the thermal energy $k_B T$, typically $E_{att} = 3-6\ k_B T$, as observed in refined numerical simulations (de Meyer, Venturoli, & Smit, 2008; Periole, Knepp, Sakmar, Marrink, & Hubert, 2012; Schmidt, Guigas, & Weiss, 2008), or even measured experimentally by high-speed Atomic Force Microscopy (AFM) (Casuso, Sens, Rico, & Scheuring, 2010). The exact value of E_{att} is modulated by the nature of the proteins (the amino acids exposed to the lipids and their ternary structure) as well as their lipidic environment.

In 1995, Abney and Scalettar were, to our knowledge, the first authors to quantify the effect of such short-range forces on collective protein organization (Abney & Scalettar, 1995). In the disordered phase—that is to say at protein density ϕ sufficiently low to avoid macroscopic phase separation as discussed above, or if the transition temperature has been exceeded—thermodynamics indicate that there nevertheless exist density fluctuations, the amplitude of which increases when getting closer to the critical point (situated at the end of the liquid—vapor coexistence line in the (P,T) phase diagram, where the phase transition becomes second order) (Chaikin & Lubensky, 1995; Onuki, 2002). Abney and Scalettar defined domains as membrane patches where the protein density is larger by a given amount than the mean membrane density. Attractive forces tend to increase domain sizes further, up to ~ 100 nm for density fluctuations of 30%, for example. Essentially, experiments and theories on phase transitions show that the correlation length ξ becomes very large when getting close to the critical point in the homogenous phase (Onuki, 2002) (a celebrated phenomenon such as critical opalescence is related to this fact (Chaikin & Lubensky, 1995)). This correlation length can also be seen as a typical domain size. However, ξ only depends on the protein concentration and the temperature, whereas experimental data tend to prove it as a quantity that is not significantly concentration dependent when a protein is overexpressed in a live cell (see Section 4.4).

By nature, such domains are transient, with lifetimes estimated to be at most on the order of 1 s due to rapid destruction by diffusion. These domains could explain density fluctuations in snapshots obtained by electron microscopy, but are less likely to explain the existence of long-lived domains (Fan et al., 2010a; Homsi et al., 2014), as observed by live fluorescence microscopy. Furthermore, they suppose the membrane composition to be permanently maintained close to the critical point (Veatch et al., 2008), which is a very strong biological constraint. These different observations indicate that an additional ingredient is certainly lacking to explain the existence of finitely sized domains, in case they were to be accounted for by equilibrium arguments.

More recent ideas originate from colloid science (Stradner et al., 2004) and belong to a vast field of phenomena where a competition between short-range attraction, as above, and longer-range repulsions leads to modulated phases or micropatterning (Seul & Andelman, 1995). Before going into further details below, we can already say that the mechanism is as follows: when entering the phase-separation region in the (T, ϕ) phase diagram, domains start growing, but domains which are too large eventually repel each other due to long-range repulsion and the condensation process "stalls." In other words, clusters which are too large are also unstable (and fragment into smaller ones, as illustrated (e.g., Meilhac & Destainville, 2011), because the ensuing energetic cost would overcome the gain in terms of line tension (the line tension is the excess interfacial energy per unit length, of proteins lying on the domain boundary) if they condensed into a giant, macroscopic phase. Starting a decade ago, a series of papers explored this idea.

When studying the clustering of syntaxin 1 and 4 through super-resolution STED fluorescence microscopy, Sieber et al. (2006) demonstrated that lipids alone were not sufficient for correct syntaxin clustering, but that protein—protein attractions (with binding energies of ≈ 3 $k_B T$ and a range of ≈ 1 nm (Sieber et al., 2007)), primarily involving the SNARE-motif, were required to account for experimental observations. To account for the observed finite size of syntaxin clusters (Sieber et al., 2006), it was proposed that repulsive forces might balance short-range attractive forces (Sieber et al., 2007). Steric repulsion between the large cytoplasmic domains of syntaxin proteins was proposed to cause the instability of too large oligomers, and thus to be at the origin of finite-size domains. However, this effect was not quantified and an *ad hoc* term was added in simulations in an empirical way. Very recently, a comparable approach was applied to the sporulation of the protein SpoVM, the membrane-inserted helix of which acts as a wedge-shaped inclusion (Wasnik, Wingreen, & Mukhopadhyay, 2015).

Ras GTPases are a superfamily of lipid-anchored peripheral membrane proteins, the clustering of which has been characterized by immunogold labeling and electron microscopy (Gurry et al., 2009). Small Ras clusters contain dozens of proteins, independent of the level of overexpression, which also led the authors to conclude that, without assuming the existence of lipid rafts which were incompatible with the observed gold point patterns, a competition between short-range attractions (with binding energies of five $k_B T$ and a range of 2 nm) and longer-range repulsion might be at the origin of clusters. Indeed, the obtained simulated data were shown to account quantitatively for the experimental observations. In this work, the Ras—Ras repulsion was supposed to result from membrane curvature induced by the insertion of the Ras lipid anchor in the inner lipid leaflet. This hypothesis was not quantified further on a physical basis.

In 2008, in order to further refine an anterior proposition illuminating the role of protein—protein long-range interactions (Daumas et al., 2003), it was proposed to explore the capabilities of curvature-mediated long-range repulsive forces on physical grounds (Destainville, 2008), using anterior quantitative findings on interactions between biomembrane inclusions such as proteins. Indeed, the way in which elastic deformations and thermal fluctuations of a biomembrane, seen as an elastic medium, are at the origin of specific interactions between objects adsorbed on or inserted into the membrane has been thoroughly explored by physicists since the early 1990s (Dan et al., 1993; Goulian et al., 1993; Weikl et al., 1998; Lenne & Nicolas, 2009). Such interactions are said to be *propagated* by the membrane (in analogy, for example, with electromagnetic interactions that are *propagated* by the electromagnetic field). Kinetic Monte Carlo simulations (Destainville, 2008), as well as statistical mechanics calculations (Destainville & Foret, 2008), indicate that such forces, the intensity of which is reasonable in regard to known lipidic membrane properties, are able to account for the existence of domains containing tens to hundreds of proteins, as observed experimentally.

Membrane proteins both peripheral and integral impose a local curvature to the membrane, one potential cause being their crystallographic shape (Bass, Strop, Barclay, & Rees, 2002; Blood & Voth, 2006; Doyle et al., 1998; Manglik et al., 2012; Park, Scheerer, Hofmann, Choe, & Ernst, 2008). In Section 4.4 below, we shall discuss further the various origins of protein-induced membrane curvature, generically related to local up/down symmetry breaking ("up" or "down" refers to a membrane that would be globally horizontal). For two (identical) inclusions at a distance r (Figure 4(D)), the ensuing repulsion falls off exponentially with r, and its range is set by $\lambda_{\text{rep}} = (\kappa/\sigma)^{1/2}$, where $\kappa \approx 20-100\ k_B T$ is the (intrinsic) lipid bilayer elastic bending modulus, and σ is the membrane tension (due to external constraints) (Weikl et al., 1998; Weitz & Destainville, 2013). For live cells, these authors used measured values $\sigma \approx 10^{-5}$ to $10^{-4}\ \text{J/m}^2$, which leads to $\lambda_{\text{rep}} = 30-200$ nm. The intensity of the repulsion depends on the curvature imposed onto the membrane by each protein, which can itself be quantified by the half-aperture angle θ of the protein conical shape, in a first and rough approximation (see Figure 4(D)). For two identical inclusions, the interaction energy is proportional to θ^2.

Beyond the two-particle case, a practical way to tackle the statistical mechanics of protein assemblies is to use tools from micellization theory (Safran, 1994; Turner, Sens, & Socci, 2005; Destainville & Foret, 2008) and to write the free energy of a cluster constituted of k proteins (or k-cluster, or k-mer) as:

$$F(k) = -f_0 k + \gamma\, k^{1/2} + \nu\, k^{\alpha}. \tag{1}$$

This is the so-called *liquid droplet approximation*. The first right-hand-side term of this equation accounts for the free-energy gain when a protein enters a cluster, due to the short-range attraction plus some entropic contribution. The free energy per protein f_0 is related to the binding energy E_{att} as defined above (Destainville & Foret, 2008). The second term quantifies the energy penalty for proteins belonging to the cluster boundary, where γ is proportional to the line tension between the condensed and gas phases. It is responsible for the collapse of the condensed phase into a giant phase in absence of the third term, which accounts for the long-range repulsion, and where the exponent $\alpha > 1$ depends on the nature of the repulsion and ν measures its intensity. As a simple example, if the repulsion is pairwise additive and if its range λ_{rep} is much larger than the typical cluster diameter, then all protein pairs repel each other equally with intensity E_{rep}, and the total repulsive energy in a k-cluster is $E_{rep}\, k(k-1)/2 \approx E_{rep}\, k^2/2$ (Wasnik et al., 2015), which suggests that $\alpha \approx 2$. In reality, the situation is more complex because of the many-bodied character of the repulsion propagated by the elastic membrane (Weitz & Destainville, 2013). Indeed, it is demonstrated in this work that, since proteins in a cluster bend the membrane in a collective manner, they tend to form a small spherical bud (Figure 4(E)), which yields an exponent $\alpha > 1$, equal to or close to 2 for a large range of parameters even if the half-aperture angle θ imposed individually by each protein is as weak as a few degrees (α tends to 2 in the low-tension limit).

Once the free energy of each k-cluster is set as in Eqn (1), the system can be tackled with the usual tools of thermodynamics (Destainville & Foret, 2008; Safran, 1994) as an ideal mixture of monomers, dimers, ..., k-mers. At equilibrium, a "gas" of monomers (and rare dimers or trimers) is then shown to coexist with finite-sized clusters, the typical size $k^\star$ of which is not simply set by λ_{rep}. It depends on the exponent α, on the protein concentration ϕ, and on the attraction-to-repulsion ratio, f_0/ν: the stronger the attraction, the larger the clusters; the stronger the repulsion, the smaller the clusters. The size distribution is generically bimodal (monomers, rare dimers or trimers, and k-clusters with $k \approx k^\star$) even though this is probably not systematic in the very low-tension limit (Meilhac & Destainville, 2011). However, a striking feature of such cluster phases is that once clusters have nucleated above the critical concentration ϕ_c, their typical size $k^\star$ grows very slowly with ϕ, making it essentially insensitive to concentration.

Note that such a statistical mechanics model assumes that clusters are highly dynamic at the molecular length scale. They presumably behave as liquid droplets of roundish shape (in order to minimize the interfacial energy) and appear in simulations to be nonrigid, deformable, and fluctuating, with their own internal diffusive dynamics where neighborhoods are continuously reshuffled (Destainville, 2008). Although they are long-lived, even when at thermodynamic equilibrium, clusters continuously exchange material with the gas phase of monomers. This conclusion could also be drawn from experimental evidence (see Hancock, 2006 and references therein). Consistently, in Sieber et al. (2007), fluorescence recovery after photobleaching (FRAP) was observed on experimental timescales where syntaxin clusters were essentially immobile (they probably contribute more generally to the immobile fraction observed in many FRAP experiments). A similar conclusion was made based on reversible accumulations of single syntaxin molecules and transient changes between fast and slow diffusion modes of SNAREs (Barg et al., 2010; Knowles et al., 2010). For CD9, a member of the tetraspanin family, real-time dual color fluorescence microscopy provided similar evidence when examining their larger micrometric domains (Espenel et al., 2008). In Banjade and Rosen (2014), even though the large, micrometric clusters persisted for hours, it was observed that individual proteins were not retained in the clusters for very long. A highly dynamic, back-and-forth exchange between the clusters and their surroundings was also observed.

4.2 Out-of-equilibrium models

Alternatively, some authors have argued that active processes might explain finite domain size (Hancock, 2006; Jacobson, Mouritsen, & Anderson, 2007), arising from the competition between the coalescence of domains by diffusion, and the traffic of lipids to and from the membrane, either by exchange of monomers or by endocytosis and exocytosis of membrane patches (also called membrane "recycling"). Active membrane recycling breaks large assemblies, the growth of which is thus stopped at a given typical size (Fan, Sammalkorpi, & Haataja, 2010b; Foret, 2005; Gómez et al., 2009; Turner et al., 2005). Even though the latter models have primarily been designed to account for the finite size of *lipid* domains ("rafts"), they are sufficiently general to be applicable to protein domains. As in Abney and Scalettar's work discussed above (Abney & Scalettar, 1995), Gheber and Edidin had already anticipated such a mechanism in 1999 in the *disordered* phase (Gheber & Edidin, 1999). In their model, an additional meshgrid of obstacles to diffusion, mimicking the presence of the cortical cytoskeleton, ensured a sufficient lifetime of domains by preventing the rapid diffusion of molecules. Note that if the transport between the membrane and the cytosol is *passive* (i.e., *not active*, without any energy consumption), one recovers the equilibrium case discussed in the previous section (Foret, 2012).

A simple dynamic scaling argument was proposed by Foret in 2005: if starting from a random configuration, the typical domain size grows with time as $L(t) = C (Dat)^{1/3}$, where C is a constant; D is the protein diffusion coefficient and a is the protein lateral size. This phenomenon is known as coarsening in the literature from the field of physics (Bray, 1994). Additionally, protein traffic to and from the membrane sets a typical recycling time τ, itself depending on off- and on-rates. Recycling prevents equilibration beyond the timescale τ, and the coarsening stops at a domain size $L = C (Da\tau)^{1/3}$. Monte-Carlo simulations (Foret, 2005), linear stability analysis (Gómez et al., 2009), and numerical or analytical resolutions of stochastic equations (Fan et al., 2010b; Foret, 2012; Turner et al., 2005) have completed this simple argument. Note that even though these different models agree on the prediction of finite-size domains, they lead to quantitatively different predictions about the steady-state distributions of cluster sizes k, depending on how the traffic to and from the membrane is precisely modeled (see Fan et al., 2010a; Foret, 2012 for thorough discussions) and whether spatial correlations are taken into account or not. In particular, Foret argues that the traffic should be modeled differently for peripheral and transmembrane proteins (Foret, 2012), because the latter preferentially escape and join the membrane by endo- and exocytosis, respectively. Consequently, depending on the model details, some authors predict bimodal size distributions (Foret, 2005, 2012; Gómez et al., 2009), where monomers coexist with clusters of typical size $k^\star$ as above, whereas others predict large, power-law distributions, with a large-size cutoff (Fan et al., 2010b; Foret, 2012; Turner et al., 2005).

Intermittently active proteins (undergoing conformational transitions driven by external energy sources of any form, e.g., light or ATP/GTP) have also been predicted to break macroscopic proteolipidic domains into finite-size domains, when their affinity for a given lipidic phase depends on their conformation (Sabra & Mouritsen, 1998).

4.3 Protein segregation and domain specialization

Several membrane protein isoforms or structurally similar family members have been observed to segregate in separate domains (Hancock, 2003; Henis, Hancock, & Prior, 2009; Kai et al., 2006; Low et al., 2006; Milovanovic et al., 2015; Sieber et al., 2006; Uhles, Moede, Leibiger, Berggren, & Leibiger, 2003; Zuidscherwoude et al., 2015). More generally, clusters are thought to be specialized, in the sense that they gather one or a few protein species

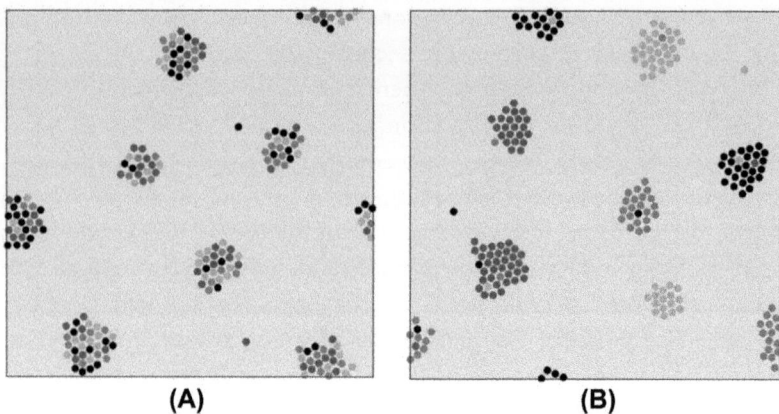

Figure 5 *Cluster-specificity, illustrated with four protein families (four different colors;* ***q*** *= 4).* (A) If proteins of different colors attract each other sufficiently with respect to proteins of the same color, they are mixed in clusters. (B) In the opposite case, clusters are essentially mono-colored, i.e., they are highly concentrated in proteins of a single color (they have "demixed"). This demixing process ensures nanodomain specialization. *Reprinted and adapted from Destainville, N. (2010) with permission from EPL.* (See color plate)

out of the hundreds of different populations that a cell membrane usually contains, in order to perform a determined biological task (see Section 2.2, Sharma et al., 2004; Lang & Rizzoli, 2010 for discussions on the biological functions of clusters). The more natural way to account for such observations in the scenarios discussed above is to suppose that fine-tuning of protein–protein attractions E_{att}, which we have observed to be modulated by the nature of the interacting proteins, refines the segregation process.

In this respect, a simple mechanism at equilibrium has been proposed in Destainville (2010) as follows: even though mixing entropy would stabilize multiple-species clusters, the Flory–Huggins theory (de Gennes, 1979) shows that if like particles have a favorable binding energy E_{att} compared to different ones, they tend to assemble together (to "demix," see Figure 5) provided that the difference of affinity $\Delta E_{att} = E_{att,like} - E_{att,different}$ between identical and different particles is sufficiently large, as the ensuing energy gain is larger than the entropic cost related to the gained order. This is reminiscent of the phase transition discussed above except that there is now a potentially large number of protein species, which we denote by q. Note that in this scenario, identical *particles* are not necessarily identical *proteins*, but proteins that need to be segregated in the same clusters in order to perform their biological function together; In Figure 5, the same particle color can represent a few different protein species.

Merging the Flory—Huggins and cluster phase theories, it has been shown by analytical calculations (Destainville, 2010) and later confirmed by numerical simulations (Meilhac & Destainville, 2011) that the critical value of ΔE_{att} grows moderately with q, more precisely $\Delta E_{att,c} \approx k_B T \ln(q)/3$. Even for a number of distinct functional clusters as large as $q = 10^4$, $\Delta E_{att} > 3\ k_B T$ ensures demixing and cluster specialization as desired. Such a value of ΔE_{att} is perfectly realistic in terms of modulations of the binding energy E_{att} depending on the protein nature (Reynwar & Deserno, 2009; Schmidt et al., 2008).

The example mentioned in Section 2 of syntaxin 1 and 4, which presumably segregate in different clusters because their TMSs have slightly different lengths (Milovanovic et al., 2015), falls within this framework. But hydrophobic mismatch is now seen as one of the many mechanisms which is likely to modulate attraction. Recently, by the help of super-resolution microscopy, several members of the tetraspanin family have also been found to segregate in distinct membrane nanoclusters in spite of their strong structural similarities (Zuidscherwoude et al., 2015).

4.4 Discussion with respect to experimental questions

We have discussed above the promising propositions that are likely to account on a physical basis for many experimental observations concerning clusters and to answer the pivotal questions raised in Section 3 such as the existence of finite-size clusters (as opposed to a "giant" macrophase), their typical size (on the order of $10-100$ proteins), their highly dynamic character, and their specificity. We have also been able to rationalize the respective roles of the different forces evoked in the biological literature, and to incorporate them in a global theoretical perspective. In spite of these successes, several points remain to be clarified before a consensus can definitively emerge. We now tackle some of them.

We have written that, in addition to their globally "conical" crystallographic shape, proteins can impose a local spontaneous curvature to lipidic membranes by various mechanisms. Zimmerberg and Kozlov have reviewed some of them (Zimmerberg & Kozlov, 2006). However, the authors mostly insisted on the most extreme cases, such as clathrins, BAR domains, or dynamins, "scaffold" proteins that are known to shape the membrane on rather large scales for biological purposes. We are mostly interested in proteins that *locally* bend the membrane, a phenomenon that has never been explored systematically. Generally speaking, local curvature, from which large cluster instability derives, occurs whenever the up/down symmetry is locally broken due to the presence of a protein. This is the case for peripheral proteins (Reynwar & Deserno, 2009; Saric & Cacciuto, 2013) or proteins (or more simply α-helices) partially inserted in a single monolayer (Campelo & Kozlov, 2014). Even in the case where a transmembrane protein has no global intrinsic spontaneous curvature, it is still likely to interact with different lipids in the two monolayers because of the different in vivo lipid compositions (van Bogaart et al., 2011; Lenne & Nicolas, 2009; Zimmerberg & Kozlov, 2006). If the consequent local spontaneous curvature induced by these lipids differs in both monolayers, the bilayer will itself experience a local spontaneous curvature, independently of the protein crystallographic shape. It thus appears that a large amount of proteins are likely sufficient to bend the membrane and to contribute to the cluster phase scenario by promoting protein–protein repulsion. We recall that an effective half-aperture angle θ as small as a few degrees is sufficient to destabilize macro-phases because of repulsion.

However, two main objections can be stated: (1) some proteins will impose an inward curvature, whereas the remaining ones will curve the bilayer outward. *A priori*, both categories should be grossly equally present in a cell. By contrast to identical protein pairs, "head-to-tail" ones experience an *attraction* due to curvature effects, of range λ_{rep} (Weitz & Destainville, 2013), which seems to contradict the previous arguments. However mixing head-to-tail proteins in the same cluster imposes a high energetic penalty. Consequently, inward- and upward-oriented proteins tend to demix in distinct clusters, which solves this apparent contradiction (Weitz & Destainville, 2013). (2) Even though it is certainly difficult to ascertain, there probably exist some transmembrane proteins that impose a negligible curvature to the membrane (a very small θ) and do not partake to the above scenario. This point has been explored and quantified (Meilhac & Destainville, 2011): if not too numerous, up/down symmetric inclusions mix into the clusters of asymmetric proteins discussed so far, and their effect is simply to slightly reduce the intensity of the repulsion and thus to increase the typical cluster size.

It is also worth mentioning that similar equilibrium mechanisms have recently been explored for the formation of *lipidic* finite size domains, through numerical simulations (Goh, Amazon, & Feigenson, 2013), theoretical investigations (Gueguen, Destainville, & Manghi, 2014), and experimentation (Goh et al., 2013; Shimobayashi, Ichikawa, & Taniguchi, 2015). The key ingredient here is again the previously evoked up/down symmetry-breaking mechanism, which provokes instability of too large lipidic domains. This can be induced through the lipid composition imposing an intrinsic curvature, which differs between both monolayers in cell membranes. Alternatively, spontaneous symmetry breaking (namely buckling in the present case) can be due to a competition between bending energy and line tension in artificial vesicles where both leaflets have the same composition (Lipowsky, 1993; Ursell, Klug, & Phillips, 2009). Beyond this, it is generally accepted that membrane domains are maintained by lipid–lipid, lipid–protein, and protein–protein interactions working together in the membrane (Anderson & Jacobson, 2002; van den Bogaart et al., 2011; Hancock, 2006; Harder, 2003; Lingwood & Simons, 2010; Vereb et al., 2003, and references therein), and that neither lipids nor proteins alone can account for the variety of domains observed in cell membranes. For instance, a functional network connecting the palmitoyltransferases DHHC5/8 with ankyrin-G, ankyrin-G with βII-spectrin, and eventually βII-spectrin with phosphoinositides has recently been shown to be required for the correct morphology of MDCK epithelial cells (He, Abdi, & Bennett, 2014). A scenario based on protein–protein interactions does not exclude a concomitant recruitment by clusters of specific lipids having a higher affinity for the clustered proteins (especially as such lipids potentially participate to protein–protein short-range attraction, as discussed previously). A general theoretical development embracing both proteins and lipids in the same curvature-induced repulsion mechanism is still lacking.

4.4.1 Some additional experimental evidence for the cluster-phase hypothesis

A pioneering study by freeze-fracture electron microscopy of monomer–oligomer equilibrium of bacteriorhodopsin in reconstituted proteoliposomes (Gulik-Krzywicki, Seigneuret, & Rigaud, 1987) demonstrated that bacteriorhodopsin resides in small clusters following a bimodal distribution, behaving as anticipated in Destainville (2008) and Destainville and Foret (2008) (notably the typical cluster size $k^\star$ is essentially insensitive to the concentration ϕ). In Sharma et al. (2004), using a combination of FRAP and FRET, the authors concluded that GPI-anchored proteins were divided into small nanoscale clusters and monomers. In agreement with our predictions at equilibrium, the cluster size $k^\star$ also seemed more or less constant (at most four GPI-anchored proteins) over the explored concentration range, even though the indirect method used did not enable a precise measure of their size. In Sieber et al. (2006), using a much more precise super-resolution microscopy technique, the number of syntaxin 1 clusters was found to be proportional to the protein density (increased above the endogenous level by transfection, immunostained by antibody labeling and visualized by STED microscopy), also revealing that clusters have an essentially con-centration-independent size. In Homsi et al. (2014), the same conclusion was drawn for CD81 tetraspanins. In Gurry et al. (2009), similar conclusions were shown to hold for Ras proteins, even though they were observed through a different experimental technique (immunogold labeling and elec-tron microscopy). Finally, note that activation of membrane proteins can lead to conformational changes (and thus for example changes of the hydro-phobic width or of the angle θ), resulting in modifications of clustering ca-pabilities. This might explain why KcsA potassium channels reconstituted into liposomes are observed to form nano-clusters when closed, and to be in a monomeric phase when open-gated (Sumino, Yamamoto, Iwamoto, Dewa, & Oiki, 2014). More experiments are needed to ascertain the exis-tence of cluster phases in both model and cellular systems.

4.4.2 What is the relevant mechanism?

In conclusion, we tackle an important open question in the context of cell membranes: How do we discriminate the relevant mechanism between the various out-of-equilibrium and equilibrium mechanisms discussed above? We propose several strategies for future investigations. (1) Experiments (Gulik-Krzywicki et al., 1987; Sharma et al., 2004) seem to demonstrate evidence for bimodal cluster-size distributions, which might rule out some out-of-equilibrium models. However, the experimental finite reso-lution length-scale, even of super-resolution microscopy, does not permit a definitive conclusion because very small clusters are not reliably detectable. (2) Different models propose different behaviors of the average (or typical) cluster size $k^\star$ in functions of the concentration ϕ, whereas experiments tend to prove that cluster sizes are rather insensitive to concentration when proteins are overexpressed (Gulik-Krzywicki et al., 1987; Gurry et al., 2009; Homsi et al., 2014; Sieber et al., 2006). Notwithstanding the resolution issue addressed above, this might be a way to distinguish be-tween the available models. However, even though some authors propose scaling laws quantifying how $k^\star$ depends on ϕ (Destainville & Foret, 2008; Fan et al., 2010b; Foret, 2012; Turner et al., 2005), this point should be clarified further for the different models and compared to experiments on model systems before being thoroughly compared to experiments on live cells. (3) A partial classification based upon structure factors and corre-lation is also proposed by Fan et al. (2010c) and might be compared to experiments. (4) Finally, in the context of real cell membranes, we have shown that equilibrium approaches can account for the specialization of nanodomains. Are out-of-equilibrium approaches also able to account for protein sorting, as in Figure 5 (right panel)? This would require the typical equilibrium sorting time to be smaller than the recycling time τ such that recycling does not destroy protein sorting. Further investigations are also required in this case.

5. CONCLUSION

Deciphering functional cell membrane micropatterning is a long-standing question that is likely to find a consensual answer in the near future, thanks to the convergence of both super-resolution microscopy techniques and quantitative (equilibrium and far-from-equilibrium) statistical mechanics approaches. Starting from experimental observations, different mechanisms have been proposed that are likely to explain the existence and lifetime of specialized membrane microdomains, and physicists aim at proposing arguments and/or new experiments that would definitively discriminate between them.

Until the beginning of the past decade, though highly debated, two main paradigms were pushed forward, namely the raft-hypothesis and the fence-and-picket model. The development of high-resolution (both in space and time) experimental techniques indicated that even though they are perfectly adapted to numerous experimental contexts, they are not able to address all theoretical predictions. A unifying paradigm is thus required that would extend the previous ones. The cluster phase scenario under review here is likely to be a step forward in this direction, because in its ultimate form its purpose is to embrace lipid–lipid, lipid–protein, and protein–protein interaction in a single, unifying theory. It has already been shown to be able to account for the existence of finite-sized domains—even though there is no consensus on the dominant mechanism yet, either of equilibrium nature or of out-of-equilibrium origin—as well as their biological specialization. It has not been designed to address the interactions with the cytoskeleton yet, but including essentially immobile or quite slowly diffusing proteins in the model does not present any conceptual issue. These "pinning defects," by interacting favorably with the molecules of some clusters, might anchor the clusters to fixed positions and hinder their diffusion, but ultimately should not modify their lifetime and specificity. Such an approach can easily reconcile the cluster phase model and the observations that had led to fence-and-picket-like scenarios. Trapping both proteins and lipids in one cluster might also explain why lipids observed by single-particle tracking techniques exhibit very slow or confined diffusion modes (Kusumi et al., 2005).

Recently, an additional concept, on top of cluster phases (Figure 1), has been proposed: a mesoscale organization of the plasma membrane has been deciphered, where membrane protein clusters are not randomly distributed across the plasma membrane but seem to have a nonuniform density distribution (Figure 1). This could be related to nested-compartment models previously proposed to account for high-speed tracking data (Murase et al., 2004). Explaining the existence of this hierarchy of structures will be another challenge for physicists and a way to test the robustness of their previous theoretical developments. By the help of super-resolution imaging, it has also recently been discovered that actin, spectrin, and associated molecules form a spectacular periodic, lattice-like arrangement of protein clusters in neuronal axons (Xu, Zhong, & Zhuang, 2013). This is another instance of mesoscale organization that will deserve investigation on a physical basis in the near future. In particular, it has been shown to be dependent on the concentration of available βII spectrin, suggesting that a phase transition might also be at play at some stage in neuronal development (Zhong et al., 2014).

REFERENCES

Abney, J. R., & Scalettar, B. A. (1993). Molecular crowding and protein organization. In M. B. Jackson (Ed.), *Thermodynamics of membrane receptors and channels* (pp. 183–225). Boca Raton: CRC Press.

Abney, J. R., & Scalettar, B. A. (1995). Fluctuations and membrane heterogeneity. *Biophysical Chemistry, 57*, 27–36.

Anderson, R. G., & Jacobson, K. (2002). A role for lipid shells in targeting proteins to caveolae, rafts, and other lipid domains. *Science, 296*, 1821–1825.

Antoku, Y., Dedecker, P., Pinheiro, P. S., Vosch, T., & Sørensen, J. B. (2015). Spatial distribution and temporal evolution of DRONPA-fused SNAP25 clusters in adrenal chromaffin cells. *Photochemical and Photobiological Sciences, 14*, 1005–1012.

Aoyagi, K., Sugaya, T., Umeda, M., Yamamoto, S., Terakawa, S., & Takahashi, M. (2005). The activation of exocytotic sites by the formation of phosphatidylinositol 4,5-bisphosphate microdomains at syntaxin clusters. *The Journal of Biological Chemistry, 280*, 17346–17352.

Banjade, S., & Rosen, M. K. (2014). Phase transitions of multivalent proteins can promote clustering of membrane receptors. *eLife, 3*, e04123.

Barg, S., Knowles, M. K., Chen, X., Midorikawa, M., & Almers, W. (2010). Syntaxin clusters assemble reversibly at sites of secretory granules in live cells. *Proceedings of the National Academy of Sciences of the United States of America, 107*, 20804–20809.

Bar-On, D., Gutman, M., Mezer, A., Ashery, U., Lang, T., & Nachliel, E. (2009). Evaluation of the heterogeneous reactivity of the syntaxin molecules on the inner leaflet of the plasma membrane. *Journal of Neuroscience, 29*, 12292—12301.

Bar-On, D., Winter, U., Nachliel, E., Gutman, M., Fasshauer, D., Lang, T., et al. (2008). Imaging the assembly and disassembly kinetics of *cis*-SNARE complexes on native plasma membranes. *FEBS Letters, 582*, 3563—3568.

Bar-On, D., Wolter, S., van de Linde, S., Heilemann, M., Nudelman, G., Nachliel, E., et al. (2012). Super-resolution imaging reveals the internal architecture of nano-sized syntaxin clusters. *The Journal of Biological Chemistry, 287*, 27158—27167.

Bass, R. B., Strop, P., Barclay, M., & Rees, D. C. (2002). Crystal structure of *Escherichia coli* MscS, a voltage-modulated and mechanosensitive channel. *Science, 298*, 1582—1587.

Bennett, V., & Lorenzo, D. N. (2013). Spectrin- and ankyrin-based membrane domains and the evolution of vertebrates. *Current Topics in Membranes, 72*, 1—37.

Blood, P. D., & Voth, G. A. (2006). Direct observation of Bin/amphiphysin/Rvs (BAR) domain-induced membrane curvature by means of molecular dynamics simulations. *Proceedings of the National Academy of Sciences of the United States of America, 103*, 15068—15072.

van den Bogaart, G., Lang, T., & Jahn, R. (2013). Microdomains of SNARE proteins in the plasma membrane. *Current Topics in Membranes, 72*, 193—230.

van den Bogaart, G., Meyenberg, K., Risselada, H. J., Amin, H., Willig, K. I., Hubrich, B. E., et al. (2011). Membrane protein sequestering by ionic protein-lipid interactions. *Nature, 479*, 552—555.

Borodich, A., Rojdestvenski, I., & Cottam, M. (2003). Lateral heterogeneity of photosystems in thylakoid membranes studied by brownian dynamics simulations. *Biophysical Journal, 85*, 774—789.

Bray, A. J. (1994). Theory of phase ordering kinetics. *Advances in Physics, 43*, 357—459.

Buyukdagli, S., Manghi, M., & Palmeri, J. (2010). Ionic capillary evaporation in weakly charged nanopores. *Physical Review Letters, 105*, 158103.

Campelo, F., & Kozlov, M. M. (2014). Sensing membrane stresses by protein insertions. *PLoS Computational Biology, 10*, e1003556.

Casuso, I., Sens, P., Rico, F., & Scheuring, S. (2010). Experimental evidence for membrane-mediated protein—protein interaction. *Biophysical Journal, 99*, L47—L49.

Chaikin, P. M., & Lubensky, T. C. (1995). *Principles of condensed matter physics*. Cambridge: Cambridge University Press.

Chamberlain, L. H., Burgoyne, R. D., & Gould, G. W. (2001). SNARE proteins are highly enriched in lipid rafts in PC12 cells: implications for the spatial control of exocytosis. *Proceedings of the National Academy of Sciences of the United States of America, 98*, 5619—5624.

Contreras, F. X., Ernst, A. M., Wieland, F., & Brügger, B. (2011). Specificity of intramembrane protein—lipid interactions. *Cold Spring Harbor Perspectives in Biology, 3*, a004705.

Dan, N., Pincus, P., & Safran, S. A. (1993). Membrane-induced interactions between inclusions. *Langmuir, 9*, 2768—2771.

Daumas, F., Destainville, N., Millot, C., Lopez, A., Dean, D., & Salome, L. (2003). Confined diffusion without fences of a G-protein-coupled receptor as revealed by single particle tracking. *Biophysical Journal, 84*, 356—366.

Destainville, N. (2008). Cluster phases of membrane proteins. *Physical Review E: Statistical, Nonlinear, and Soft Matter Physics, 77*, 011905.

Destainville, N. (2010). An alternative scenario for the formation of specialized protein nanodomains (cluster phases) in biomembranes. *Europhysics Letters, 91*, 58001.

Destainville, N., & Foret, L. (2008). Thermodynamics of nano-cluster phases: a unifying theory. *Physical Review E: Statistical, Nonlinear, and Soft Matter Physics, 77*, 051403.

Doyle, D. A., Morais Cabral, J., Pfuetzner, R. A., Kuo, A., Gulbis, J. M., Cohen, S. L., et al. (1998). The structure of the potassium channel: molecular basis of K^+ conduction and selectivity. *Science, 280*, 69—77.

Dupuy, A. D., & Engelman, D. M. (2008). Protein area occupancy at the center of the red blood cell membrane. *Proceedings of the National Academy of Sciences of the United States of America, 105,* 2848−2852.

Espenel, C., Margeat, E., Dosset, P., Arduise, C., Le Grimellec, C., Royer, C. A., et al. (2008). Single-molecule analysis of CD9 dynamics and partitioning reveals multiple modes of interaction in the tetraspanin web. *Journal of Cell Biology, 182,* 765−776.

Fan, J., Sammalkorpi, M., & Haataja, M. (2010a). Formation and regulation of lipid microdomains in cell membranes: theory, modeling, and speculation. *FEBS Letters, 584,* 1678−1684.

Fan, J., Sammalkorpi, M., & Haataja, M. (2010b). Influence of nonequilibrium lipid transport, membrane compartmentalization, and membrane proteins on the lateral organization of the plasma membrane. *Physical Review E: Statistical, Nonlinear, and Soft Matter Physics, 81,* 011908.

Fan, J., Sammalkorpi, M., & Haataja, M. (2010c). Lipid microdomains: structural correlations, fluctuations, and formation mechanisms. *Physical Review Letters, 104,* 118101.

Fasshauer, D., & Margittai, M. (2004). A transient N-terminal interaction of SNAP-25 and syntaxin nucleates SNARE assembly. *The Journal of Biological Chemistry, 279,* 7613−7621.

Foret, L. (2005). A simple mechanism of raft formation in two-component fluid membranes. *Europhysics Letters, 71,* 508−514.

Foret, L. (2012). Aggregation on a membrane of particles undergoing active exchange with a reservoir. *European Physical Journal E Soft Matter, 35,* 12.

de Gennes, P. G. (1979). *Scaling concepts in polymer physics.* Ithaca: Cornell University Press.

Gheber, L. A., & Edidin, M. (1999). A model for membrane patchiness: lateral diffusion in the presence of barriers and vesicle traffic. *Biophysical Journal, 77,* 3163−3175.

Gil, T., Ipsen, J. H., Mouritsen, O. G., Sabra, M. C., Sperotto, M. M., & Zuckermann, M. J. (1998a). Theoretical analysis of protein organization in lipid membranes. *Biochimica et Biophysica Acta Biomembranes, 1376,* 245−266.

Gil, T., Ipsen, J. H., & Tejero, C. F. (1998b). Wetting controlled phase transitions in two-dimensional systems of colloids. *Physical Review E: Statistical, Nonlinear, and Soft Matter Physics, 57,* 3123−3133.

Gil, T., Sabra, M. C., Ipsen, J. H., & Mouritsen, O. G. (1997). Wetting and capillary condensation as means of protein organization in membranes. *Biophysical Journal, 73,* 1728−1741.

Goh, S. L., Amazon, J. J., & Feigenson, G. W. (2013). Toward a better raft model: modulated phases in the four-component bilayer, DSPC/DOPC/POPC/CHOL. *Biophysical Journal, 104,* 853−862.

Gómez, J., Sagués, F., & Reigada, R. (2009). Nonequilibrium patterns in phase-separating ternary membranes. *Physical Review E: Statistical, Nonlinear, and Soft Matter Physics, 80,* 011920.

Goulian, M., Bruinsma, R., & Pincus, P. (1993). Long-range forces in heterogeneous fluid membranes. *Europhysics Letters, 22,* 145−150.

Gueguen, G., Destainville, N., & Manghi, M. (2014). Mixed lipid bilayers with locally varying spontaneous curvature and bending. *European Physical Journal E Soft matter, 37,* 76.

Gulik-Krzywicki, T., Seigneuret, M., & Rigaud, J. L. (1987). Monomer-oligomer equilibrium of bacteriorhodopsin in reconstituted proteoliposomes. *The Journal of Biological Chemistry, 262,* 15580−15588.

Gurry, T., Kahramanogullari, O., & Endres, R. G. (2009). Biophysical mechanism for Ras-nanocluster formation and signaling in plasma membrane. *PLoS One, 4,* e6148.

Halemani, N. D., Bethani, I., Rizzoli, S. O., & Lang, T. (2010). Structure and dynamics of a two-helix SNARE complex in live cells. *Traffic, 11,* 394−404.

Hancock, J. F. (2003). Ras proteins: different signals from different locations. *Nature Review Molecular Cell Biology, 4*, 373—384.

Hancock, J. F. (2006). Lipid rafts: contentious only from simplistic standpoints. *Nature Review Molecular Cell Biology, 7*, 456—462.

Harder, T. (2003). Formation of functional cell membrane domains: the interplay of lipid- and protein-mediated interactions. *Philosophical Transactions of the Royal Society of London B, 358*, 863—868.

He, M., Abdi, K. M., & Bennett, V. J. (2014). Ankyrin-G palmitoylation and βII-spectrin binding to phosphoinositide lipids drive lateral membrane assembly. *Journal of Cell Biology, 206*, 273—288.

Henis, Y. I., Hancock, J. F., & Prior, I. A. (2009). Ras acylation, compartmentalization and signaling nanoclusters (Review). *Molecular Membrane Biology, 26*, 80—92.

Homsi, Y., Schloetel, J., Scheffer, K. D., Schmidt, T. H., Destainville, N., Florin, L., et al. (2014). The extracellular δ-domain is essential for the formation of CD81 tetraspanin webs. *Biophysical Journal, 107*, 100—113.

Honigmann, A., van den Bogaart, G., Iraheta, E., Risselada, H. J., Milovanovic, D., Mueller, V., et al. (2013). Phosphatidylinositol 4,5-bisphosphate clusters act as molecular beacons for vesicle recruitment. *Nature Structural and Molecular Biology, 20*, 679—686.

Jacobson, K., Mouritsen, O. G., & Anderson, R. G. W. (2007). Lipid rafts: at a crossroad between cell biology and physics. *Nature Cell Biology, 9*, 7—14.

Jahn, R., & Fasshauer, D. (2012). Molecular machines governing exocytosis of synaptic vesicles. *Nature, 490*, 201—207.

James, D. J., Khodthong, C., Kowalchyk, J. A., & Martin, T. F. J. (2008). Phosphatidylino-sitol 4,5-bisphosphate regulates SNARE-dependent membrane fusion. *Journal of Cell Biology, 182*, 355—366.

James, D. J., Kowalchyk, J., Daily, N., Petrie, M., & Martin, T. F. J. (2009). CAPS drives trans-SNARE complex formation and membrane fusion through syntaxin interactions. *PNAS, 106*, 17308—17313. http://dx.doi.org/10.1073/pnas.0900755106.

Kai, M., Sakane, F., Jia, Y. F., Imai, S., Yasuda, S., & Hanoh, H. (2006). Lipid phosphate phosphatases 1 and 3 locate in distinct lipid rafts. *Journal of Biochemistry, 140*, 677—686.

Khuong, T. M., Habets, R. L., Kuenen, S., Witkowska, A., Kasprowicz, J., Swerts, J., et al. (2013). Synaptic PI(3,4,5)P3 is required for Syntaxin1A clustering and neurotransmitter release. *Neuron, 77*, 1097—10108.

Knowles, M. K., Barg, S., Wan, L., Midorikawa, M., Chen, X., & Almers, W. (2010). Single secretory granules of live cells recruit syntaxin-1 and synaptosomal associated protein 25 (SNAP-25) in large copy numbers. *Proceedings of the National Academy of Sciences of the United States of America, 107*, 20810—20815.

Kusumi, A., Nakada, C., Ritchie, K., Murase, K., Suzuki, K., Murakoshi, H., et al. (2005). Paradigm shift of the plasma membrane concept from the two-dimensional continuum fluid to the partitioned fluid: high-speed single-molecule tracking of membrane molecules. *Annual Review of Biophysics and Biomolecular Structure, 34*, 351—378.

Kusumi, A., Sako, Y., & Yamamoto, M. (1993). Confined lateral diffusion of membrane receptors as studied by single particle tracking (nanovid microscopy). Effects of cal-cium-induced differentiation in cultured epithelial cells. *Biophysical Journal, 65*, 2021—2040.

Lai, E. C. (2003). Lipid rafts make for slippery platforms. *Journal of Cell Biology, 162*, 365—370.

Lang, T. (2007). SNARE proteins and 'membrane rafts'. *The Journal of Physiology, 585*, 693—698.

Lang, T., Bruns, D., Wenzel, D., Riedel, D., Holroyd, P., Thiele, C., et al. (2001). SNAREs are concentrated in cholesterol-dependent clusters that define docking and fusion sites for exocytosis. *The EMBO Journal, 20*, 2202—2213.

Lang, T., Margittai, M., Hölzler, H., & Jahn, R. (2002). SNAREs in native plasma membranes are active and readily form core complexes with endogenous and exogenous SNAREs. *Journal of Cell Biology, 158,* 751–760.

Lang, T., & Rizzoli, S. O. (2010). Membrane protein clusters at nanoscale resolution: more than pretty pictures. *Physiology, 25,* 116–124.

Lee, A. G. (2003). Lipid–protein interactions in biological membranes: a structural perspective. *Biochimica et Biophysica Acta Biomembranes, 1612,* 1–40.

Lee, A. G., Birdsall, N. J., Metcalfe, J. C., Toon, P. A., & Warren, G. B. (1974). Clusters in lipid bilayers and the interpretation of thermal effects in biological membranes. *Biochemistry, 13,* 3699–3705.

Lenne, P. F., & Nicolas, A. (2009). Physics puzzles on membrane domains posed by cell biology. *Soft Matter, 5,* 2841–2848.

Lerman, J. C., Robblee, J., Fairman, R., & Hughson, F. M. (2000). Structural analysis of the neuronal SNARE protein syntaxin-1A. *Biochemistry, 39,* 8470–8479.

Lillemeier, B. F., Pfeiffer, J. R., Surviladze, Z., Wilson, B. S., & Davis, M. M. (2006). Plasma membrane-associated proteins are clustered into islands attached to the cytoskeleton. *Proceedings of the National Academy of Sciences of the United States of America, 103,* 18992–18997.

Lingwood, D., & Simons, K. (2010). Lipid rafts as a membrane-organizing principle. *Science, 327,* 46–50.

Lipowsky, R. (1993). Domain-induced budding of fluid membranes. *Biophysical Journal, 64,* 1133–1138.

Low, S. H., Vasanji, A., Nanduri, J., He, M., Sharma, N., Koo, M., et al. (2006). Syntaxins 3 and 4 are concentrated in separate clusters on the plasma membrane before the establishment of cell polarity. *Molecular Biology of the Cell, 17,* 977–989.

Manglik, A., Kruse, A. C., Kobilka, T. S., Thian, F. S., Mathiesen, J. M., Sunahara, R. K., et al. (2012). Crystal structure of the μ-opioid receptor bound to a morphinan antagonist. *Nature, 485,* 321–326.

Meilhac, N., & Destainville, N. (2011). Clusters of proteins in bio-membranes: insights into the roles of interaction potential shapes and of protein diversity. *Journal of Physical Chemistry B, 115,* 7190–7199.

de Meyer, F. J. M., Venturoli, M., & Smit, B. (2008). Molecular simulations of lipid-mediated protein–protein interactions. *Biophysical Journal, 95,* 1851–1865.

Milovanovic, D., Honigmann, A., Koike, S., Göttfert, F., Pähler, G., Junius, M., et al. (2015). Hydrophobic mismatch sorts SNARE proteins into distinct membrane domains. *Nature Communications, 6,* 5984.

Milovanovic, D., & Jahn, R. (2015). Organization and dynamics of SNARE proteins in the presynaptic membrane. *Frontiers in Physiology, 6,* 89.

Mouritsen, O. G. (2005). *Life—As a matter of fat.* Berlin: Springer.

Munro, S. (2003). Lipid rafts: elusive or illusive? *Cell, 115,* 377–388.

Murase, K., Fujiwara, T., Umemura, Y., Suzuki, K., Iino, R., Yamashita, H., et al. (2004). Ultrafine membrane compartments for molecular diffusion as revealed by single molecule techniques. *Biophysical Journal, 86,* 4075–4093.

Murray, D. H., & Tamm, L. K. (2009). Clustering of syntaxin-1A in model membranes is modulated by phosphatidylinositol 4,5-bisphosphate and cholesterol. *Biochemistry, 48,* 4617–4625.

Murray, D. H., & Tamm, L. K. (2011). Molecular mechanism of cholesterol- and polyphosphoinositide-mediated syntaxin clustering. *Biochemistry, 50,* 9014–9022.

Neder, J., West, B., Nielaba, P., & Schmidt, F. (2011). Membrane-mediated protein-protein interaction: a Monte Carlo study. *Current Nanoscience, 7,* 656–666.

Nielsen, C., Goulian, M., & Andersen, O. S. (1998). Energetics of inclusion-induced bilayer deformations. *Biophysical Journal, 74,* 1966–1983.

Ohara-Imaizumi, M., Nishiwaki, C., Nakamichi, Y., Kikuta, T., Nagai, S., & Nagamatsu, S. (2004). Correlation of syntaxin-1 and SNAP-25 clusters with docking and fusion of insulin granules analysed by total internal reflection fluorescence microscopy. *Diabetologia, 47,* 2200−2207.

Onuki, A. (2002). *Phase transition dynamics.* Cambridge: Cambridge University Press.

Park, J. H., Scheerer, P., Hofmann, K. P., Choe, H. W., & Ernst, O. P. (2008). Crystal structure of the ligand-free G-protein-coupled receptor opsin. *Nature, 454,* 183−187.

Periole, X., Knepp, A. M., Sakmar, T. P., Marrink, S. J., & Hubert, T. (2012). Structural determinants of the supramolecular organization of G protein-coupled receptors in bilayers. *Journal of the American Chemical Society, 134,* 10959−10965.

Pertsinidis, A., Mukherjee, K., Sharma, M., Pang, Z. P., Park, S. R., Zhang, Y., et al. (2013). Ultrahigh-resolution imaging reveals formation of neuronal SNARE/Munc18 complexes in situ. *Proceedings of the National Academy of Sciences of the United States of America, 110,* E2812−E2820.

Pierce, S. K. (2004). To cluster or not to cluster: FRETting over rafts. *Nature Cell Biology, 6,* 180−181.

Pike, L. J. (2006). Rafts defined: a report on the Keystone Symposium on Lipid Rafts and Cell Function. *Journal of Lipid Research, 47,* 1597−1598.

Plowman, S. J., Ariotti, N., Goodall, A., Parton, R. G., & Hancock, J. F. (2008). Electrostatic interactions positively regulate K-Ras nanocluster formation and function. *Molecular and Cellular Biology, 28,* 4377−4385.

Pobbati, A. V., Stein, A., & Fasshauer, D. (2006). N- to C-terminal SNARE complex assembly promotes rapid membrane fusion. *Science, 313,* 673−676.

Reynwar, B. J., & Deserno, M. (2009). Membrane composition-mediated protein-protein interactions. *Biointerphases, 3,* FA117−FA124.

Ribrault, C., Reingruber, J., Petković, M., Galli, T., Ziv, N. E., Holcman, D., et al. (2011). Syntaxin1A lateral diffusion reveals transient and local SNARE interactions. *Journal of Neuroscience, 31,* 17590−17602.

Rickman, C., Medine, C. N., Dun, A. R., Moulton, D. J., Mandula, O., Halemani, N. D., et al. (2010). t-SNARE protein conformations patterned by the lipid microenvironment. *The Journal of Biological Chemistry, 285,* 13535−13541.

Sabra, M. C., & Mouritsen, O. G. (1998). Steady-state compartmentalization of lipid membranes by active proteins. *Biophysical Journal, 74,* 745−752.

Safran, S. A. (1994). *Statistical thermodynamics of surfaces, interfaces, and membranes.* Cambridge: Perseus.

Saka, S. K., Honigmann, A., Eggeling, C., Hell, S. W., Lang, T., & Rizzoli, S. O. (2014). Multi-protein assemblies underlie the mesoscale organization of the plasma membrane. *Nature Communications, 5,* 4509.

Salaün, C., James, D. J., & Chamberlain, L. H. (2004). Lipid rafts and the regulation of exocytosis. *Traffic, 5,* 255−264.

Saric, A., & Cacciuto, A. (2013). Self-assembly of nanoparticles adsorbed on fluid and elastic membranes. *Soft Matter, 9,* 6677−6695.

Schmidt, U., Guigas, G., & Weiss, M. (2008). Cluster formation of transmembrane proteins due to hydrophobic mismatching. *Physical Review Letters, 101,* 128104.

Schreiber, A., Fischer, S., & Lang, T. (2012). The amyloid precursor protein forms plasmalemmal clusters via its pathogenic amyloid-β domain. *Biophysical Journal, 102,* 1411−1417.

Seul, M., & Andelman, D. (1995). Domain shapes and patterns: the phenomenology of modulated phases. *Science, 267,* 476−483.

Sharma, P., Varma, R., Sarasij, R. C., Ira, Gousset, K., Krishnamoorthy, G., et al. (2004). Nanoscale organization of multiple GPI-anchored proteins in living cell membranes. *Cell, 116,* 577−589.

Shimobayashi, S. F., Ichikawa, M., & Taniguchi, T. (2015). *Phase-separation transitions in asymmetric lipid bilayers.* Preprint http://arxiv.org/abs/1504.03038.

Sieber, J. J., Willig, K. I., Heintzmann, R., Hell, S. W., & Lang, T. (2006). The SNARE motif is essential for the formation of syntaxin clusters in the plasma membrane. *Biophysical Journal, 90,* 2843–2851.

Sieber, J. J., Willig, K. I., Kutzner, C., Gerding-Reimers, C., Harke, B., Donnert, G., et al. (2007). Anatomy and dynamics of a supramolecular membrane protein cluster. *Science, 317,* 1072–1076.

Simons, K., & Ikonen, E. (1997). Functional rafts in cell membranes. *Nature, 387,* 569–572.

Singer, S. J., & Nicolson, G. L. (1972). The fluid mosaic model of the structure of cell membranes. *Science, 175,* 720–731.

Stradner, A., Sedgwick, H., Cardinaux, F., Poon, W. C. K., Egelhaaf, S. U., & Schurtenberger, P. (2004). Equilibrium cluster formation in concentrated protein solutions and colloids. *Nature, 432,* 492–495.

Sumino, A., Yamamoto, D., Iwamoto, M., Dewa, T., & Oiki, S. (2014). Gating-associated clustering-dispersion dynamics of the KcsA potassium channel in a lipid membrane. *Physical Chemistry Letters, 5,* 578–584.

Takamori, S., Holt, M., Stenius, K., Lemke, E. A., Grønborg, M., Riedel, D., et al. (2006). Molecular anatomy of a trafficking organelle. *Cell, 127,* 831–846.

Turner, M. S., Sens, P., & Socci, N. D. (2005). Nonequilibrium raftlike membrane domains under continuous recycling. *Physical Review Letters, 95,* 168301.

Uhles, S., Moede, T., Leibiger, B., Berggren, P. O., & Leibiger, I. B. (2003). Isoform-specific insulin receptor signaling involves different plasma membrane domains. *Journal of Cell Biology, 163,* 1327–1337.

Ursell, T. S., Klug, W. S., & Phillips, R. (2009). Morphology and interaction between lipid domains. *Proceedings of the National Academy of Sciences of the United States of America, 106,* 13301–13306.

Veatch, S. L., Cicuta, P., Sengupta, P., Honerkamp-Smith, A., Holowka, D., & Baird, B. (2008). Critical fluctuations in plasma membrane vesicles. *ACS Chemical Biology, 3,* 287–293.

Vereb, G., Szollosi, J., Matko, J., Nagy, P., Farkas, T., Vigh, L., et al. (2003). Dynamic, yet structured: the cell membrane three decades after the Singer–Nicolson model. *Proceedings of the National Academy of Sciences of the United States of America, 100,* 8053–8058.

Wasnik, V., Wingreen, N. S., & Mukhopadhyay, R. (2015). Modeling curvature-dependent subcellular localization of the small sporulation protein SpoVM in bacillus subtilis. *PLoS One, 10,* e0111971.

Weikl, T. R., Kozlov, M. M., & Helfrich, W. (1998). Interaction of conical membrane inclusions: effect of lateral tension. *Physical Review E: Statistical, Nonlinear, and Soft Matter Physics, 57,* 6988–6995.

Weitz, S., & Destainville, N. (2013). Attractive asymmetric inclusions in elastic membranes under tension: cluster phases and membrane invaginations. *Soft Matter, 9,* 7804–7816.

Wilhelm, B. G., Mandad, S., Truckenbrodt, S., Kröhnert, K., Schäfer, C., Rammner, B., et al. (2014). Composition of isolated synaptic boutons reveals the amounts of vesicle trafficking proteins. *Science, 344,* 1023–1028.

Willig, K., Keller, J., Bossi, M., & Hell, S. W. (2006). STED microscopy resolves nanoparticle assemblies. *New Journal of Physics, 8,* 106.

Wunderlich, F., Kreutz, W., Mahler, P., Ronai, A., & Heppeler, G. (1978). Thermotropic fluid goes to ordered "discontinuous" phase separation in microsomal lipids of Tetrahymena. An X-ray diffraction study. *Biochemistry, 17,* 2005–2010.

Xu, K., Zhong, G., & Zhuang, X. (2013). Actin, spectrin, and associated proteins form a periodic cytoskeletal structure in axons. *Science, 339,* 452–456.

Yang, L., Dun, A. R., Martin, K. J., Qiu, Z., Dunn, A., Lord, G. J., et al. (2012). Secretory vesicles are preferentially targeted to areas of low molecular SNARE density. *PLoS One*, 7, e49514.

Zhong, G., He, J., Zhou, R., Lorenzo, D., Babcock, H. P., Bennett, V., et al. (2014). Developmental mechanism of the periodic membrane skeleton in axons. *eLife*, 3, 04581.

Zilly, F. E., Halemani, N. D., Walrafen, D., Spitta, L., Schreiber, A., Jahn, R., et al. (2011). Ca^{2+} induces clustering of membrane proteins in the plasma membrane via electrostatic interactions. *The EMBO Journal*, 30, 1209—1220.

Zimmerberg, J., & Kozlov, M. M. (2006). How proteins produce cellular membrane curvature. *Nature Review Molecular Cell Biology*, 7, 9—19.

Zuidscherwoude, M., Gottfert, F., Dunlock, V. M. E., Fidgor, C. G., van den Bogaart, G., & van Spriel, A. B. (2015). The tetraspanin web revisited by super-resolution microscopy. *Scientific Reports*, 5, 12201.

CHAPTER THREE

Plasma Membrane Repair in Health and Disease

Alexis R. Demonbreun and Elizabeth M. McNally*

Center for Genetic Medicine, Northwestern University, Chicago, IL, USA
*Corresponding author: E-mail:elizabeth.mcnally@northwestern.edu

Contents

Abstract

Since an intact membrane is required for normal cellular homeostasis, membrane repair is essential for cell survival. Human genetic studies, combined with the development of novel animal models and refinement of techniques to study cellular injury, have now uncovered series of repair proteins highly relevant for human health. Many of the deficient repair pathways manifest in skeletal muscle, where defective repair processes result in myopathies or other forms of muscle disease. Dysferlin is a membrane-associated protein implicated in sarcolemmal repair and also linked to other membrane functions including the maintenance of transverse tubules in muscle. MG53, annexins, and Eps15 homology domain-containing proteins interact with dysferlin to form a

Current Topics in Membranes, Volume 77
ISSN 1063-5823
http://dx.doi.org/10.1016/bs.ctm.2015.10.006

membrane repair complex and similarly have roles in membrane trafficking in muscle. These molecular features of membrane repair are not unique to skeletal muscle, but rather skeletal muscle, due to its high demands, is more dependent on an efficient repair process. Phosphatidylserine and phosphatidylinositol 4,5-bisphosphate, as well as Ca^{2+}, are central regulators of membrane organization during repair. Given the importance of muscle health in disease and in aging, these pathways are targets to enhance muscle function and recovery from injury.

List of Abbreviations

CK   Creatine Kinase
EHD   Eps15 Homology Domain
PIP2   Phosphatidylinositol 4,5-bisphosphate

1. MEMBRANE INTEGRITY ESSENTIAL FOR CELL SURVIVAL

To preserve cellular integrity and structure, maintenance of the plasma membrane is required to sustain proper osmotic homeostasis and cell survival. This review will focus on the sarcolemma, the plasma membrane in striated muscle cells of cardiac and skeletal muscle. The unique shape and function of striated muscle is designed to accommodate marked cell shape changes that occur concomitant with contraction and to transmit force. Like other membranes, the sarcolemma is a lipid bilayer composed of individual phospholipids oriented in an impermeable sheet that relies on the cytoskeleton for structure and stability. Disruptions altering the lipid bilayer or the cytoskeleton result in membrane dysfunction and disease. Dystrophin is a cytoskeletal protein essential for sarcolemmal stability; loss of dystrophin leads to a fragile sarcolemma that is highly susceptible to contraction-induced damage. Dysferlin, another protein whose loss is linked to muscle disease, impairs muscle membrane repair as well as other membrane trafficking functions. Many of the proteins that interact with dystrophin and dysferlin are also regulators of sarcolemmal membrane integrity and repair.

Plasma membrane disruption, due to mechanical forces, and chemical or toxin onslaughts are common forms of cellular injury for many different mammalian cell types under normal physiological conditions (McNeil & Khakee, 1992; McNeil & Kirchhausen, 2005). Skeletal and cardiac muscle undergo extreme physiological daily stress and continually undergo bouts of membrane damage, requiring quick and efficient means of membrane repair. Additionally, epidermal cells, epithelial, and endothelial cells undergo membrane disruption and have evolved related mechanisms for cellular repair (Yu & McNeil, 1992). Lesions occurring in the plasma membrane are a major threat to the survival of cells as extracellular Ca^{2+} flows freely through the damaged membrane. Normally, Ca^{2+} levels are found at low levels within the cell. In muscle, increased intracellular Ca^{2+} triggers not only muscle contraction but also sustained elevated intracellular Ca^{2+} elicits a number of downstream signaling cascades, including the activation of Ca^{2+}-dependent proteases causing degradation of structural and mechanical proteins of the cell, ultimately leading to cell death. The *mdx* mouse model of muscular dystrophy lacks dystrophin, a multispectrin repeat containing protein, that stabilizes the sarcolemma (Bulfield, Siller, Wight, & Moore, 1984). Dystrophin-deficient myofibers and cardiomyocytes are fragile; they display increased sarcolemmal disruption and leak, typically monitored with vital tracers and fluorescent dyes (Straub, Rafael, Chamberlain, & Campbell, 1997). The absence of dystrophin produces sarcolemmal fragility in a muscle that is thought to have normal repair mechanisms. Dystrophin binds to cytoskeletal actin and also to a complex of transmembrane (TMEM) proteins linked to extracellular matrix proteins (Constantin, 2014). Mutations in the membrane-associated components of the dystrophin complex similarly lead to a fragile sarcolemma indicating that the entire complex is required for sarcolemmal stability (Hack et al., 2000).

2. MODELS OF CELL MEMBRANE REPAIR

Efficient membrane resealing in mammalian cell membranes generally occurs within $10-30$ s (McNeil & Steinhardt, 1997). Healthy cells, specifically myofibers, trigger a Ca^{2+}-induced, phospholipid-dependent repair mechanism, originally thought to involve the fusion of intracellular vesicles at the site of damage (Bansal et al., 2003; Davis, Doherty, Delmonte, & McNally, 2002). In this model, elevated levels of Ca^{2+} at the site of injury induce rapid vesicular fusion at the site of damage. Through vesicle cycling, a repair patch forms at the site of membrane disruption and described as the "patch repair" hypothesis (McNeil & Khakee, 1992; McNeil & Kirchhausen, 2005). Piccolo, Moore, Ford, and Campbell (2000) provided further evidence for the patch repair model using electron microscopy to document increased vesicle accumulation at the sarcolemma in muscle lacking dysferlin, a protein implicated in membrane repair. The authors suggested that the enrichment of vesicles on the cytoplasmic face of the sarcolemma, which were not seen in healthy cells, represented vesicular trapping due to delayed fusion (Piccolo et al., 2000). However, unlike in nerve terminal synaptic vesicle transmission, there is no direct evidence that muscle contains a reservoir of "predocked" vesicles beneath the plasma membrane, poised for fusion, in the event of a membrane lesion. Thus, the patch hypothesis has been challenged recently (see below). Notably, the activation and translocation of nonlocal vesicles to the site of damage is a slow process that would not prevent the rapid influx of intracellular Ca^{2+}.

Another repair model also maintains that membranes are repaired through a patching mechanism, but postulates that the vesicles responsible are not generic cellular vesicles, but instead are lysosomes. Experiments in cells including 3T3 fibroblasts, CHO cells, and NRK cells have shown that during mechanical damage, a subset of lysosomes are activated by the influx of Ca^{2+}, translocate to the site of damage, and fuse to the membrane lesion forming a repair patch (Reddy, Caler, & Andrews, 2001; Rodriguez, Webster, Ortego, & Andrews, 1997). However, live-cell imaging of damaged zebra fish muscle with fluorescently labeled Lamp1, a lysosome-associated membrane protein, did not find that Lamp1-positive vesicles translocated to the site of injury suggesting that lysosome-mediated repair may not be conserved among all cell types or types of damage (Roostalu & Strahle, 2012).

These data are consistent with a model where lysosomes may not play a direct role in membrane repair, but instead may serve a secondary role in maintaining muscle health during damage. The rise in intracellular Ca^{2+} during injury also activates lysosomal removal of potential toxins and cellular debris that, if insufficiently scavenged, may enhance membrane damage. Where injury was induced by viruses, bacteria, or chemical compounds, lysosomal scavenging is a critical element for mitigating injury. The lysosome model has been further refined by demonstrating that lysosomes secrete acid sphingomyelin, triggering vesicle endocytosis, to aid in the removal of toxins and likely facilitating membrane repair (Corrotte et al., 2013). Additionally, toxins are expelled along with other cytoplasmic contents in the form of macrovesicles or blebs bolstering membrane recovery and reducing intracellular Ca^{2+} levels (Babiychuk, Monastyrskaya, Potez, & Draeger, 2011). However, these two models do not address the main mechanism of rapid membrane repair in response to contraction-induced damage in muscle or disruption of the plasma membrane in epithelial cell damage.

A more recent model suggests that repair of the disrupted sarcolemma relies on lateral recruitment of membrane to reseal the site of injury. McDade and colleagues used confocal imaging of fluorescently labeled membrane to visualize membrane movement during laser damage. This method demonstrated that sarcolemma adjacent to the site of flowed toward the lesion to plug the site of injury (McDade, Archambeau, & Michele, 2014). In this model, the plasma membrane is already positioned to diffuse into the lesion upon disruption, making the repair process and blockade of Ca^{2+} influx quick and efficient. This lateral recruitment model does not exclude the possibility that the patch hypothesis and/or the lysosome repair mechanism act in conjunction with lateral diffusion to repair the plasma membrane. However, the lateral recruitment model is attractive in that it does provide evidence for a model that correlates with the timing of repair required in cells to maintain Ca^{2+} homeostasis.

3. MODELING MEMBRANE REPAIR

Several *in vitro* methods have been designed in multiple cell types to understand the process of membrane repair. These include mechanical injury methods, such as cell scraping, in which cells are injured by scraping a razor blade or pipette across a field of cells (Reddy et al., 2001). Cultured cells are also injured by the addition of microscopic glass beads rolled over cells to induce mechanical damage (Reddy et al., 2001). Both glass bead rolling and scraping elicit membrane damage that lacks precision with regard to position of injury. Damage between cell cultures and within cultures can vary considerably. These methods are thought to resemble a mechanical-type injury creating large lesions and high levels of Ca^{2+} influx. As an alternative, laser-induced damage, a method by which a confocal laser burns a hole into the cell membrane and microneedle puncturing, the act of inserting a fine point into the cell have been utilized (Bansal et al., 2003; Cai, Weisleder, et al., 2009; Swaggart et al., 2014). Laser-induced injury and cell puncturing allow for the location and dimension of damage to be precisely tuned, providing more reproducible results. The use of high heat during laser-induced damage may denature local proteins if not done appropriately, and cell puncturing highly disorganizes the underlying cytoskeleton by pushing the membrane into the interior of the cell making real-time analysis of the damaged area more difficult. Despite these issues, laser injury and microneedle damage have emerged as two methods that are utilized in combination with live-cell fluorescent imaging. Combining these methods allows the study of fluorescently tagged protein trafficking in real time to assess timing, localization, and response to a variety of experimental manipulations (Cai, Masumiya, et al., 2009; Swaggart et al., 2014).

In muscle, electroporation can be used to introduce plasmids to overexpress fluorescently tagged proteins of interest in live muscle. Following electroporation, muscle cells are isolated and available for damage assays (DiFranco, Quinonez, Capote, & Vergara, 2009). Tagged proteins of interest can then be observed before, during, and after repair. The development of high-resolution confocal microscopy including structured illumination microscopy, stochastic optical reconstruction microscopy, and stimulated emission depletion allows for high-resolution imaging of repair structures (Jaiswal et al., 2014; Swaggart et al., 2014). The utility of the method is its relative speed. However, the use of electroporation introduces injury to the recipient cell and this can be accompanied by a repair response (Roche et al., 2011). Most investigators allow 7–14 days between electroporation and laser-induced injury, but even this interval may be insufficient to allow for full recovery. Like other methods, electroporation results in varying levels of plasmid transduction and therefore protein expression. Protein overexpression itself can induce toxic effects on cells from endoplasmic reticulum (ER) stress and other methods. Thus, while this is a highly useful method, electroporation, overexpression, and laser injury all have experimental caveats that must be considered when interpreting results.

4. MEMBRANE REPAIR AND HUMAN MYOPATHY

4.1 Ferlins

Muscular dystrophy occurs when muscle degeneration exceeds muscle regeneration. Loss-of-function mutation in the gene encoding dysferlin, also known as *Fer1L1*, results in limb-girdle muscular dystrophy 2B (LGMD2B), Miyoshi myopathy (MM), or distal anterior compartment myopathy. LGMD2B is an autosomal recessive form of muscular dystrophy that is most often associated with muscle weakening of distal muscles presenting in humans in the second decade of life (Bashir et al., 1998; Liu et al., 1998). Cardiac function is generally preserved with dysferlin mutations (Kuru et al., 2004). Creatine kinase (CK) is an enzyme localized in the interior of cells that can leak from the sarcolemma into the circulation during membrane damage and is a commonly used marker of muscle degeneration. With loss of dysferlin, serum CK is extremely elevated

(10–40,000 U/l compared to normal 100 U/l) consistent with increased leak of muscle contents into the serum. Muscle biopsies often reveal inflammatory infiltrate, and they can lead to the misdiagnosis of polymyositis, an autoimmune disorder (Nguyen et al., 2007). Dysferlin is the best-studied member of the ferlin family, and multiple mouse models have been generated that recapitulate human disease (Bansal et al., 2003; Bittner et al., 1999; Demonbreun, Fahrenbach, et al., 2010; Demonbreun et al., 2014; Ho et al., 2004; Lostal et al., 2010).

Dysferlin is 230 kDa protein that belongs to the ferlin family of proteins. These proteins are highly similar in structure with each containing a carboxyl-terminal TMEM domain, which anchors the protein, and up to seven C2 domains. At the time of its discovery, the only other known ferlin protein was the *Caenorhabditis elegans* protein *fer1*, and *fer1* was linked to membrane fusion defects that impaired fertilization (Achanzar & Ward, 1997). Fer1 contains multiple C2 domains like dysferlin. C2 domains are Ca^{2+}-sensitive phospholipid binding domains of about 130 amino acids in length. The C2 domains in dysferlin have high relationship to those in the fusion protein synaptotagmin (Bansal & Campbell, 2004; Bansal et al., 2003; Davis et al., 2002). Synaptotagmins generally contain two C2 domains and mice-null for synaptotagmin 7, which develop autoimmune myositis with defective membrane resealing properties (Chakrabarti et al., 2003). The amino-terminal C2 domain, C2A, mediates Ca^{2+}-sensitive lipid binding, and this topology positions the C2A domain to mediate membrane fusion within the cell (Davis et al., 2002; Therrien, Di Fulvio, Pickles, & Sinnreich, 2009). In addition to the C2 domains, ferlins contain Ferlin-motifs, FerI, FerA, and FerB, although these 60 amino acid structures while conserved have no yet defined function. A DysF domain, situated as a nested repeat, is also present. The DysF domain also has no-known function, but a similar domain is found in the peroxisomal proteins (Lek, Evesson, Sutton, North, & Cooper, 2012).

Dysferlin is highly expressed in skeletal and cardiac tissue, despite the lack of cardiac involvement in the majority of patients and mouse models (Bashir et al., 1998). Dysferlin expression is comparatively low during muscle development with progressive upregulation as muscle differentiates into mature myotubes (Davis et al., 2002). Previously, dysferlin was thought to localize primarily to the sarcolemma in muscle cells (Bansal et al., 2003; Piccolo et al., 2000). However, using a pH-sensitive tag fused to the carboxy-terminus of dysferlin, dysferlin was found to enrich in the transverse (T)-tubule structure of mature muscle fibers Kerr et al. (2013).

T-tubules are specialized membranous invagination, continuous with the sarcolemma, and these invaginations are responsible for promoting Ca^{2+} propagation throughout the myofiber (Flucher, 1992; Porter & Palade, 1957). The localization of dysferlin at the T-tubule correlates developmentally with the onset of dysferlin expression, as T-tubules form as muscle differentiation proceeds (Klinge et al., 2010; Lee et al., 2002). Additionally, dysferlin has been found in tubular aggregates derived from sarcoplasmic reticular, another membranous compartment critically important in the regulation of calcium (Ikezoe et al., 2003). Dysferlin is expressed in many other tissue types including monocytes, lung, skin, brain, testis, and placenta (Bashir et al., 1998). The role of dysferlin in these tissues is not known.

Due to the high homology of dysferlin to synaptotagmin and the muscle pathology found in dysferlinopathy patients, it was hypothesized that dysferlin may play a role in vesicle fusion-induced repair within skeletal muscle. Bansal et al. (2003) first described dysferlin's role in repair by generating mice lacking dysferlin and using laser-induced injury on isolated myofibers. Laser injury was conducted in the presence of FM1-43, a lipophilic dye that normally is nonfluorescent but upon binding negatively charged phospholipids exhibits increased fluorescence intensity. Dysferlin–null myofibers had much slower membrane resealing than control myofibers (Bansal et al., 2003). Control myofibers damaged in the absence of Ca^{2+} displayed repair defects similar to that of dysferlin-null muscle, demonstrating the requirement of extracellular Ca^{2+} for membrane repair. Dysferlin has roles beyond membrane repair, extending to general trafficking. For example, dysferlin-regulated vesicle trafficking in both fibroblasts and myoblasts was shown using pulse-chase experiments with Alexa-488 labeled transferrin (Demonbreun, Fahrenbach, et al., 2010). In dysferlin-null cells, labeled transferrin accumulated within cells indicative of delayed recycling. The accumulation of transferrin in dysferlin-null cells is consistent with the role of dysferlin as a participant in the endocytic pathway regulating membrane fusion. Dysferlin-null mice exhibit muscle wasting and weakness, similar to humans with LGMD2B. To determine the muscle-specific role of dysferlin in muscle disease, dysferlin was overexpressed in the skeletal muscle of dysferlin-null mice. Overexpression of dysferlin rescued the pathological phenotype and functional deficits in dysferlin deficient mice. These data suggest that functional deficits due to dysferlin are muscle-intrinsic and not caused by a more general role dysferlin may play in recycling (Lostal et al., 2010; Millay et al., 2009).

Myoferlin, another ferlin family member, is highly homologous to dysferlin, 57% at the amino acid level. Like dysferlin, myoferlin contains seven C2 domains, and like dysferlin, myoferlin's C2A domain binds phospholipids in a Ca^{2+}-sensitive manner (Davis et al., 2002). Unlike dysferlin, myoferlin is highly expressed during early muscle development in the skeletal muscle precursor cell, the myoblast, and is downregulated as the muscle fuses into mature myotubes (Doherty et al., 2005). During the repair process when the embryonic programs are reinitiated, myoferlin expression is upregulated (Demonbreun, Lapidos, et al., 2010). Upregulation of myoferlin is thought to promote myoblast fusion enhancing regeneration during repair. Myoblasts from a mouse model lacking myoferlin display-delayed myoblast fusion resulting in smaller myotubes in vitro. Concomitantly, myoferlin-null mice have a mild dystrophic phenotype characterized by smaller myofibers and delayed regeneration, characteristic of fusion defects (Doherty et al., 2005). Like dysferlin, myoferlin regulates endocytic recycling and receptor trafficking, and these effects were seen in both fibroblasts and myoblasts (Demonbreun, Posey, et al., 2010; Doherty et al., 2008). Mice lacking myoferlin not only have delayed trafficking of transferrin but also of receptors that are translocated like the IGF1-receptor. Delayed recycling of the IGF1-receptor was also correlated with decreased downstream signaling cascades. Vesicle trafficking and signaling are critical components required for proper muscle growth and membrane repair (Demonbreun, Posey, et al., 2010; Doherty et al., 2008). This role for myoferlin in regulating vesicle trafficking explains both growth and repair defects observed in myoferlin-null muscle.

Myoferlin-null mice are born in Mendelian ratios and live a normal life span. Human diseases linked to mutations in the gene encoding myoferlin have not been reported. However, recent reports have linked decreased levels of myoferlin to modulate the invasive potential of cancerous cells, where decreased proliferation and migration of myoferlin-depleted cells was seen *in vitro*. Additionally, myoferlin reduction through siRNA of Lewis lung carcinoma cells resulted in reduced membrane repair measured through dye uptake after laser damage (Leung, Yu, Lin, Tognon, & Bernatchez, 2013). These data provide evidence myoferlin directs membrane repair and play a role in efficient proliferation and growth in tumor cells. Overall, the roles of dysferlin and myoferlin in nonmuscle cells are expected to be similar to what has been seen in muscle cells.

In silico prediction, algorithms have identified three other potential ferlinlike proteins—Fer1L4, Fer1L5, and Fer1L6. Fer1L5 is most similar to

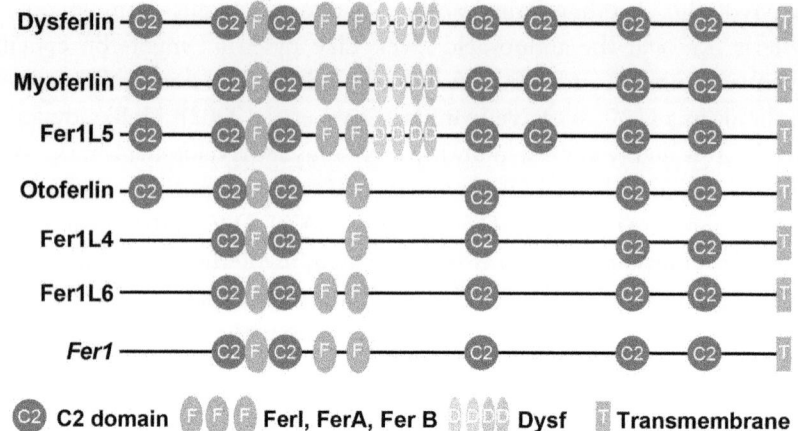

Figure 1 *Schematic of ferlin family members.* The ferlins share similar structure. Each contains multiple C2 domains (circles), which bind phospholipids in a calcium-dependent manner. Fer domains (dark ovals) and DysF domains (light ovals) have unknown function and are present within only some ferlin family members. A carboxy-terminal transmembrane domain (rectangle) anchors the protein.

myoferlin and has been studied in vitro. Similar to myoferlin, Fer1L5 is expressed during skeletal muscle differentiation and has been shown to interact with the Eps15 homology domain (EHD) containing protein EHD1 and EHD2 through the interaction of the C2 domains (Posey et al., 2011). The interaction with the EHD proteins is discussed below. In total, in humans there are six ferlin family members, including otoferlin, which has been linked to genetically mediated deafness (Roux et al., 2006). The six ferlin proteins all share their multi-C2 domain nature and the presence of additional conserved regions (Figure 1).

4.2 Anoctamin 5

Anoctamin 5 (ANO5/TMEM16E) belongs to a large family of TMEM proteins. ANO5, specifically belongs to the TMEM16 family, consisting of 10 highly homologous members. TMEM16 proteins contain eight hydrophobic helices predicted to be TMEM domains and are described as Ca^{2+}-activated chloride channels (CaCCs), phospholipid scramblases or important for protein–protein interactions (Pedemonte & Galietta, 2014). ANO1 and ANO2 were shown to function as CaCCs, and the remaining family members are thought to have this role by analogy (Ousingsawat et al., 2009; Stephan et al., 2009). Loss-of-function mutations *ANO5* result in LGMD2L and MM dystrophy-3 (MMD-3). The muscle manifestations in LGMD2L and MMD-3 resemble the clinical presentation of dysferlinopathies, LGMD2B, and MM (Bolduc et al., 2010). Specifically, individuals with *ANO5* mutations have lower limb weakness and progressive muscle atrophy. Like dysferlin gene mutations, *ANO5* mutations are associated with very elevated serum CK levels and a similar phenotypic range of muscle disease that typically does not associate with cardiac defects (Mahjneh et al., 2010). The ANO5 protein is localized to the ER and intracellular vesicles, and ANO5 is upregulated during muscle differentiation similar to dysferlin, suggesting an equivalent role in muscle (Mizuta et al., 2007; Tsutsumi et al., 2005). Although not specifically localized to the sarcolemma, ANO5 was analyzed for a role in membrane repair. In HEK293 cells, ANO5 was seen to translocate to the cell periphery after saponin treatment suggesting ANO5 is participating in membrane repair (Tian, Wright, Cebotaru, Wang, & Guggino, 2015). Using adeno-associated virus (AAV), ANO5 was introduced into dysferlin-null mice as a potential membrane repair therapy (Monjaret et al., 2013). Laser injury was used to assess damage, and dysferlin-null myofibers with and without ANO5 expression were damaged (Monjaret et al., 2013). The presence of ANO5 did not significantly improve the membrane repair capacities of dysferlin-null myofibers, nor did ANO5 improve the histological pathology of skeletal muscle in mice, suggesting distinct roles for ANO5 and dysferlin in membrane repair (Monjaret et al., 2013). Members of the TMEM16 family have also been implicated in distributing phospholipids within membranes, referred to as phospholipid scramblase activity (Suzuki et al., 2013). Further studies are required to elucidate the role of ANO5 in membrane repair and determine how ANO5 functions within the membrane repair complex. Dominant mutations in *ANO5* lead to gnathodiaphyseal dysplasia, a disorder of bone dystrophy that results in bone thinning, bowing, and occasionally bony necrosis (Marconi et al., 2013). The role of *ANO5* gene mutations in both bone and muscle disease suggests that these pathways are used by multiple tissues and cell types.

5. REGULATORS OF MEMBRANE REPAIR

5.1 Annexins

The annexin protein family is characterized by the ability to bind phospholipids and actin in a Ca^{2+}-dependent manner. Annexins preferentially bind phosphatidylserine (PS), phosphatidylinositols (PtdIns), and cholesterol

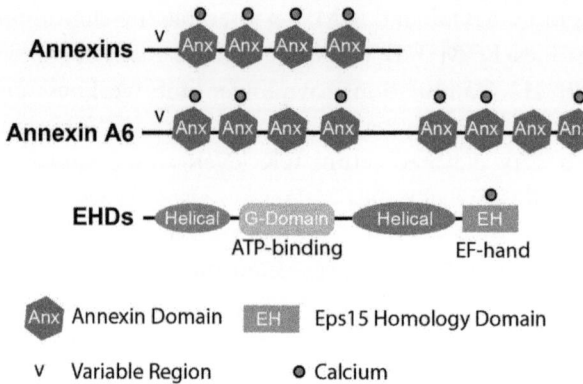

Figure 2 *Schematic of annexin and Eps15 homology domain (EHD) proteins.* The annexin family contains 12 mammalian proteins each containing four annexin repeats (hexagon) and an amino-terminal variable (v) region. Annexin A6 is an atypical annexin containing eight annexin repeats and variable (v) amino-terminal region. The annexin repeats coordinate calcium binding (green (dark gray in print versions) circle) despite lacking EF hands. The EHD containing family of proteins contains four highly homologous family members, EHD1—4. Each members is composed of an amino-terminal helical region, an ATP-binding G-domain, a helical region, and an EH domain containing a calcium-binding EF hand.

(Gerke, Creutz, & Moss, 2005). Dominant or recessive mutations in annexin genes have not been associated with muscle disease; however, annexin A5 genetic variants associate with pregnancy loss (de Laat et al., 2006). The annexin family is known to comprise over 160 distinct proteins that are present in more than 65 unique species (Gerke & Moss, 2002). Humans have 12 different annexin genes, characterized by distinct tissue expression and localization. Annexins are involved in a variety of cellular processes including membrane permeability, mobility, and vesicle fusion. These properties are Ca^{2+}-dependent. Although annexins do not contain EF hand domains, calcium ions bind to the individual annexin repeat domains (Figure 2). Differential Ca^{2+} affinity allows each annexin protein to respond to changes in intracellular Ca^{2+} levels under unique spatiotemporal conditions (Blackwood & Ernst, 1990).

Structurally, the annexin family of proteins contains a conserved carboxy-terminal core domain composed of multiple annexin repeats and a variable amino-terminal head. The amino-terminus differs in length and amino acid sequence among the annexin family members. Additionally, posttranslational modifications alter protein function and protein localization (Goulet, Moore, & Sartorelli, 1992; Kaetzel et al., 2001). Annexin proteins have the potential to self-oligomerize and interact with membrane surfaces in the presence of Ca^{2+} (Zaks & Creutz, 1991). The amino-terminal region is thought to bind one lipid membrane in a Ca^{2+}-dependent manner, while the annexin core region binds an additional lipid membrane. *In vitro* studies have shown that annexin domains lacking the amino-terminus can aggregate membrane through dimerization and annexin complex formation (Wang & Creutz, 1994).

Because of this plasma membrane binding capacity, annexins have broad membrane trafficking and actin organization roles. Specifically, annexins have been shown to interact with actin in a Ca^{2+}-dependent manner coordinating the assembly and organization of the cytoskeleton, another critical component required during the repair process. Annexin A2 is linked to the secretory protein transport pathway and facilitates formation of filamentous (F)-actin (Hayes, Shao, Bailly, & Moss, 2006). The loss of annexin A2 decreases actin formation and vesicle trafficking. Additionally, annexin A2 binds phosphatidylinositol 4,5-bisphosphate (PIP2) and cholesterol with high affinity, making the lipid bilayer a target for annexin A2 binding (Hayes et al., 2004). The binding of annexin A2 to both actin and lipids is Ca^{2+}-dependent allowing for tight control of the repair process.

Jaiswal et al. (2014) showed that the influx of Ca^{2+} upon membrane damage triggers annexin A2 to the site of membrane injury, facilitating actin reorganization near the injury lesion. It is hypothesized that the reorganization of actin at the site of damage facilitates aggregation of vesicles and phospholipids at the site of damage as well as providing changes in membrane tension required that aid repair. Notably, these studies were performed in MCF7 and HeLa cancer cells, although similar mechanisms may occur in other cells types, as actin reorganization at membrane lesions is observed in both Xenopus and Drosophila (Bement, Mandato, & Kirsch, 1999; Clark et al., 2009).

Annexins do not contain a predicted hydrophobic signal sequence targeting the annexins for classical secretion through the ER, yet annexins are found both on the interior and exterior of the cell (Christmas, Callaway, Fallon, Jones, & Haigler, 1991; Deora, Kreitzer, Jacovina, & Hajjar, 2004; Wallner et al., 1986). The process by which the annexins are externalized remains unknown. It is hypothesized that annexins may be released through granular exocytosis or cell lysis; however, the method of externalization may also vary by cell type. Functionally, localization both inside and outside the cell adds to the complexity of the roles which annexins play within tissues and cell types. Annexins have been shown to have anti-inflammatory, profibrinolytic, and antithrombotic effects. The annexin A1–deleted mouse model exhibits an exacerbated inflammatory response when challenged and is resistant to the anti-inflammatory effects of glucocorticoids (Hannon et al., 2003). The annexin A2-null mouse develops fibrin accumulation in the microvasculature and is defective in clearance of arterial thrombi (Ling et al., 2004).

Annexins have been shown to directly regulate membrane repair (Babbin et al., 2008; Lennon et al., 2003; McNeil, Rescher, Gerke, & McNeil, 2006). Annexins A1, A2, A5, and A6 localize to the site of muscle membrane repair in zebra fish muscle (Roostalu & Strahle, 2012). This localization at the sarcolemma occurs in a sequential manner with annexin A6 arriving at the site first (Roostalu & Strahle, 2012). Marg et al. (2012) also identified annexin A1 at the site of sarcolemmal damage in cultured human muscle cells. Jaiswal et al. (2014) found annexin A1 and annexin A2 at the site of membrane injury in two human cancer cell lines, MCF and HeLa cells. Although little is known about the precise function of annexin A1 and annexin A2 in membrane repair, the expression level of both proteins may function as a diagnostic marker for a number of muscle diseases due to the strong correlation between high expression levels of annexin A1 and A2 and the clinical severity of such forms of muscular dystrophies (Cagliani et al., 2005). Annexins A1 and A2 directly bind dysferlin and are hypothesized to play a role in a larger membrane repair complex (Lennon et al., 2003; Roostalu & Strahle, 2012).

Annexin A6 is involved in the membrane repair process in zebrafish and mouse skeletal muscle. During muscle membrane damage, annexin A6 translocates to the site of injury (Roostalu & Strahle, 2012; Swaggart et al., 2014). Annexin A6 is unique among the annexin family members, as it is the only annexin protein that contains two core domains and eight annexin repeat domains; all other annexin family members contain one core domain and four annexin repeats (Benz et al., 1996) (Figure 2). In vitro studies have shown that annexin A6 is capable of membrane binding through both the amino- and carboxy-terminal annexin core domains facilitating membrane coalescence of two opposing membranes, a requirement when coordinating membrane repair (Buzhynskyy et al., 2009). The annexin A6-null mouse does not display any overt phenotype (Hawkins, Roes, Rees, Monkhouse, & Moss, 1999). However, annexin A6 was found to modify a mouse model of muscular dystrophy, the *Sgcg* mouse that lacks γ-sarcoglycan, a dystrophin-associated protein (Swaggart et al., 2014). The *Sgcg* mouse was bred onto a two murine genetic backgrounds, DBA/2J (D2) and 129T2/SvEmsJ (129), and F4 progeny analyzed. These two models display distinct differences in the severity of muscle disease (Heydemann, Huber, Demonbreun, Hadhazy, & McNally, 2005), and the DBA/2J strain has similarly been shown to enhance the *mdx* model of muscular dystrophy (Fukada et al., 2010). Quantitative trait loci mapping of membrane leak combined with RNAseq pointed identified *Anxa6*, the gene encoding annexin A6, as a modifier of membrane fragility. In the severe strain, a truncated form of the annexin A6 protein, representing the first 32 kDa of annexin A6, was expressed at low levels. This truncated annexin protein lacks the last four carboxy-terminal annexin repeats and is the result of a splice site variant located in the middle of exon 11 causing a premature stop codon in exon 16. This alternate splice form of annexin A6 was also found in the C57BL/6 (B6) mouse strain. The "N32" annexin A6 truncated protein was found to impair trafficking of full-length annexin A6 to the sarcolemma after injury (Swaggart et al., 2014). Evidence for the dominant negative effect annexin A6's amino-terminus was also seen in zebrafish and in human fibroblasts (Kamal, Ying, & Anderson, 1998; Roostalu & Strahle, 2012). Taken together, these data confirm a critical role for annexins in regulating membrane repair.

Although loss of annexin A6 in the mouse does not yield an overt phenotype, overexpression of annexin A6 is associated with a pathogenic phenotype (Gunteski-Hamblin et al., 1996). Using the α-myosin heavy chain promoter, annexin A6 was overexpressed more than 10-fold in the heart. Annexin A6 overexpression resulted in dilated hearts with increased fibrosis. Cardiomyocytes from transgenic positive mice overexpressing A6 had reduced function resulting from altered levels of basal and free Ca^{2+}. These data suggest that although annexin A6 is upregulated during muscle differentiation, coordinated overexpression is required to participate in efficient membrane repair and regeneration.

Annexin A5 is the smallest annexin and one of the most studied of the annexin family. A5 is a Ca^{2+}-responsive protein that oligomerizes within membranes, and annexin A5 assembly is dependent on Ca^{2+} and PS concentrations (Mosser, Ravanat, Freyssinet, & Brisson, 1991). Annexin A5 is also used commonly as a marker for apoptosis due to its binding with PS as PS is known to reverse membrane orientation during cell death from the inner to outer leaflet. Annexin A5 also oligomerizes and binds PS at the site of damage. Annexin A5 may provide stability to the site of injury by reducing the movement of PS-containing membrane, and thereby preventing the torn membrane from extending (Saurel, Cezanne, Milon,

Tocanne, & Demange, 1998). Annexin A5-null perivascular cells were assessed for membrane repair capacity using laser-induced injury (Bouter et al., 2011). Lipophilic dye influx was increased in annexin A5-null cells after injury, compared to wild-type controls, suggesting that annexin A5 regulates the membrane resealing process. The introduction of extracellular recombinant annexin A5 prior to laser-induced damage was sufficient to improve the membrane repair capacity of annexin A5-null cells to near wild-type levels upon injury, suggesting that annexin A5 can act from the exterior of the cell (Bouter et al., 2011). This same process was also seen in human placental trophoblasts, which express both high levels of dysferlin and annexin A5 (Carmeille et al., 2015). In neuroblastoma cells, annexin A5 was shown to assemble into complexes that also contained annexins A1 and A2 in a time-dependent manner at the plasma membrane upon increased Ca^{2+} levels (Skrahina, Piljic, & Schultz, 2008). Bouter et al. showed that preventing annexin A5 from forming two-dimensional membrane-associated arrays resulted in defective repair (Bouter et al., 2011). These studies support a role for annexin A5 in membrane repair and suggest that annexin A5 functions as a molecular repair protein from both the interior and exterior of the cell.

6. TRAFFICKING AND MEMBRANE REPAIR

The EHD-containing family of proteins is comprised of four family members, EHD1−4, and this family has been implicated in membrane trafficking and repair (Grant & Caplan, 2008; Naslavsky & Caplan, 2010; Posey et al., 2014; Rapaport et al., 2006). EHD proteins regulate cytoskeletal rearrangements, specifically actin, and also have been shown to interact with the ferlin family of proteins (Doherty et al., 2008; Posey et al., 2011, 2014). EHD proteins have an amino-terminal nucleotide binding domain, a central coiled−coiled domain, and a carboxy-terminal epsin homology (EH) domain, that harbor a Ca^{2+}-binding EF hand domain. The EH domain is known to coordinate binding proteins that contain an asparagine-proline-phenylalanine (NPF) motif (Paoluzi et al., 1998; Salcini et al., 1997). EHDs are evolutionarily conserved and have been implicated in vesicle trafficking and recycling of many signaling molecules including IGF1, GLUT4, EGFR, and the transferrin receptor, in muscle, fibroblasts, and HeLa cells (Naslavsky & Caplan, 2010; Posey et al., 2014; Rapaport et al.,

2006). While EHDs family members within a species are highly homologous, their roles are nonredundant (George et al., 2007).

EHD1 is the best characterized of the EHD family of proteins and bares the highest homology to the *C. elegans* RME-1. *Ehd1*-null mice have features of muscular dystrophy and also developmental abnormalities (Posey et al., 2014; Rainey et al., 2010). *Ehd1*-null muscle has elongated T-tubules similar to what is seen in dysferlin-null muscle, and this observation is consistent with the role of EHD1 regulating endocytic recycling to the plasma membrane (Posey et al., 2014; Rapaport et al., 2006). Similar to other membrane-affiliated proteins, EHD1 binds to PtdIns, a primarily membrane-associated lipid. EHD1 does not bind sphingolipids, cholesterol, ceramides, phosphatidylcholine or PS, suggesting a preferential interaction with and localization to the membrane (Naslavsky, Rahajeng, Chenavas, Sorgen, & Caplan, 2007). The preferred binding of EHD1 with PtdIns was shown to be Ca^{2+}-dependent in vitro and was mediated by the second half of the EH domain (Naslavsky et al., 2007). Additional data suggest that the orientation of the EHD protein allows for simultaneous binding of both NPF motif-binding partners and PtdIns lipids. This dual binding would facilitate the involvement of EHDs in both endocytic recycling of membrane and membrane repair.

To investigate EHD1's role in membrane repair, Marg et al. analyzed EHD1 localization after laser damage in cultured human myotubes (Marg et al., 2012). EHD1 did not traffic to the site of damage and instead remained in the cytoplasm. However, these data do not negate a role for EHD1 in repair. EHD1 is expressed in muscle during differentiation and localizes to the T-tubule in skeletal muscle similar to dysferlin (Posey et al., 2011). Cultured human myotubes do not form mature T-tubule structures and therefore the lack of EHD1 translocation in cultured myotubes may indicate that EHD1 requires its T-tubule location in order to participate in membrane repair. Unlike EHD1, EHD2 is expressed early in development, potentially before the formation of T-tubules and is downregulated as differentiation proceeds in a pattern similar to myoferlin. In cultured human cultured myotubes, EHD2 was observed to translocate to the site of damage, indicating that EHD2 is not dependent on a T-tubule-dependent trafficking mechanism (Marg et al., 2012).

6.1 EHD3 and EHD4

EHD3 and EHD4 are highly homologous to EHD1 and EHD2, although their roles in skeletal muscle membrane repair have not been extensively tested (Pohl et al., 2000). EHD4 expression during muscle differentiation shares the same pattern as dysferlin with low-level expression in the singly nucleated myoblast and increasing levels of expression as myoblast differentiate into multinucleated myotubes (Posey et al., 2011). Gudmundsson et al. (2012) showed that in rat and mouse models of heart failure, EHD3 expression levels were upregulated in the left ventricle compared to nonischemic controls suggesting a role for EHD3 in cardiac repair. Additionally, upregulation of EHD3 was shown in human heart failure samples validating these studies. Due to the high homology of EHD3 and EHD4 to other EHD proteins, further analysis of these proteins during membrane repair is needed.

6.2 GTPase Regulator Associated with Focal Adhesion Kinase-1

GTPase regulator associated with focal adhesion kinase-1 (GRAF1) is a protein that contains an amino-terminal BAR domain; BAR domains (named for their presence in Bin1, amphiphysin, and Rvs) are implicated in membrane bending (Lundmark et al., 2008). In addition, GRAF1 contains a carboxy-terminal pleckstrin homology-domain that binds PIP2 and PS, and a carboxy-terminal src homology domain. GRAF1 interacts with dynamin suggesting a role for GRAF1 in membrane remodeling and scission, similar to EHD1's interaction with the BAR domain containing protein amphiphysin 2/Bin1 protein (Posey et al., 2014). GRAF1 is expressed in punctate structures in fibroblasts, and it has been shown that the presence of GRAF1 can induce the formation of small intracellular tubules in vitro, a process referred to as tubulation. Unlike many other repair proteins, GRAF1 does not bind actin directly but can associate with actin through Cdc42. GRAF1 expression recapitulates that of myoferlin and EHD2, in that expression decreases with differentiation and is upregulated during muscle regeneration (Doherty et al., 2005; Doherty et al., 2011; Lenhart et al., 2014). A GRAF1 null mouse was found to have defects in myoblast fusion and decreased myofiber size, similar to the phenotype observed in myoferlin-deficient mice (Doherty et al., 2005; Lenhart et al., 2014). Consistent with this coexpression, Lenhart showed GRAF1 associated with the ferlin proteins myoferlin and Fer1L5 during muscle differentiation at prefusion complexes along with the EHD1 and EHD2. The interaction of GRAF1 with ferlin and EHD members suggests that GRAF1 is a member of the repair complex.

6.3 Mitsugumin 53 (Trim72)

Mitsugumin 53, MG53, also known as Trim72 is part of the larger family of E3 ubiquitin ligases, tripartite motif (TRIM) proteins (Meroni & Diez-Roux, 2005). MG53 consists of an amino-terminal TRIM domain and a carboxy-terminal SPRY domain, a common structure of most TRIM family proteins. RNA analysis revealed MG53 is highly expressed in striated muscle and to a lower extent in lung and kidney epithelia (Cai, Masumiya, et al., 2009; Duann et al., 2015; Jia et al., 2014). Because of its high expression in skeletal muscle, MG53 was investigated as a potential mediator of membrane repair at the site of injury (Cai, Masumiya, et al., 2009). MG53-null mice develop progressive myopathy consistent with decreased fiber repair. MG53-deficient muscle fibers exhibit decreased repair after laser–induced damage (Cai, Masumiya, et al., 2009). Mechanical electrode studies in both lung and kidney epithelia were used to show that MG53 translocates to the site of membrane disruption, indicative of a role in membrane repair in multiple cell types (Duann et al., 2015; Jia et al., 2014). MG53 preferentially binds PS at the site of damage, similar to the annexins (Cai, Masumiya, et al., 2009). Mechanical damage of the muscle cell line C2C12 resulted in MG53 translocation to the site of damage. MG53 was also found to colocalize with annexin A5, a surrogate marker of PS and another protein involved in membrane repair (Cai, Weisleder, et al., 2009). MG53 also interacts with dysferlin through the C2A domain in a Ca^{2+}-sensitive manner (Flix et al., 2013; Matsuda et al., 2012). Upon membrane damage, dysferlin was found to travel to the site of injury, followed by MG53. MG53 contains several cysteine residues that upon reduction prevent MG53 oligomerization and the capacity to facilitate membrane repair (Cai, Masumiya, et al., 2009). Mutations in MG53 that inhibit MG53 oligomerization inhibited dysferlin localization at the sarcolemma (Matsuda et al., 2012). This suggests that MG53 acts in conjunction with other proteins, including dysferlin and annexins, in a larger repair complex to facilitate membrane repair (Figure 3).

7. THERAPIES USED TO PROMOTE REPAIR

Maintaining the integrity of membranes is critical in the prevention of disease. The sarcolemma membrane of muscle is a unique model to understand membrane repair processes. The sarcolemma is well-organized, and its primary components are well-studied. The sarcolemma is also under heavy

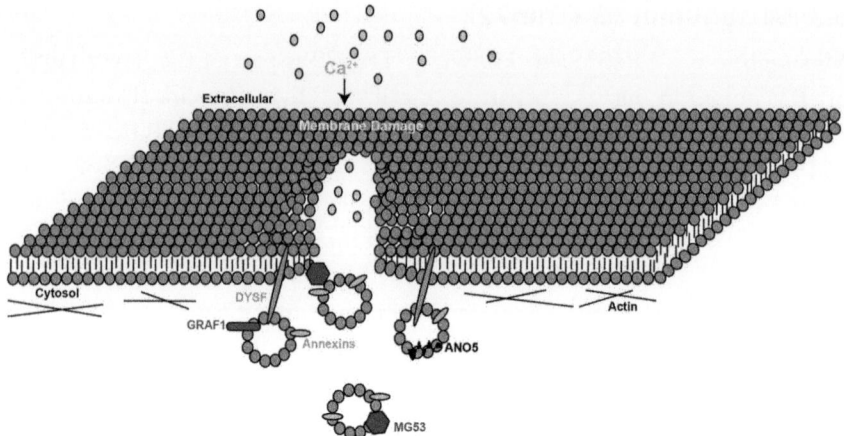

Figure 3 *Model for plasma membrane repair.* A membrane lesion is formed within the sarcolemma creating a microhole within the membrane. The influx of calcium triggers annexin A6 (A6) localization to the repair cap. Other repair complex proteins including annexin A1, A2, A5, dysferlin (DYSF), and MG53 are recruited to the lesion aiding in the resealing of the sarcolemma. (See color plate)

mechanical load compared to other tissues providing a "testing ground" to better understand the repair process as muscle membranes are continually damaged and repaired in the course of normal activity. In the muscular dystrophies and myopathies, this process is often perturbed. In some forms of muscle disease, damage is accelerated, outstripping repair. In other forms of disease, repair is defective. Cells and animal models of these genetic defects provide unique avenues to study the repair process and, in turn, membrane repair treatments. Many other tissue types, including heart and lung, also rely on intact membranes to prevent disease. Thus, methods and/or reagents that increase the efficiency of membrane repair across multiple tissue types could potentially impact multiple disease processes.

7.1 Delivering Mitsugumin 53

Extensive work by Weisleder and colleagues has shown the TRIM family member, MG53 is useful after multiple forms of injury to many different cell types. In muscle fibers, addition of recombinant MG53, rhMG53, to the exterior of the cell localizes to the site of injury, preventing the influx of impermeable dye normally seen in damaged fibers (Weisleder et al., 2012). It is hypothesized that MG53 acts as a molecular "band-aid" on injured cells allowing for the recovery of a membrane that otherwise is unable to repair. These findings are reminiscent of what was seen when purified annexin A5 was introduced into cellular models of injury (Bouter et al., 2011). The observation that MG53 colocalizes with annexin A5 supports that this complex can perhaps be assembled by adding a number of its protein components.

Recombinant MG53 was also tested in vivo in the *mdx* animal model (Weisleder et al., 2012). Initial short-term in vivo studies showed that the addition of rhMG53 improved the membrane integrity of the myofibers preventing dye uptake into the muscle after eccentric running in the *mdx* mouse model. rhMG53 also reduced areas of myonecrosis within these animals suggesting rhMG53 stabilizes the membrane in vivo (Weisleder et al., 2012). Similar protective effects are seen in heart, lung, and kidney cells treated with rhMG53 or AAV-MG53 (Duann et al., 2015; He et al., 2012; Jia et al., 2014; Liu et al., 2015). These data are consistent with the idea that MG53 contributes to normal membrane repair and improves the reparative capacity in multiple organs in vitro and in vivo. The results from long-term rhMG53 studies will prove useful to determine if MG53 repair can maintain improved membrane integrity after cyclic rounds of cellular damage.

7.2 Poloxamer 188

Poloxamers are nonionic copolymers composed of a hydrophilic poly(ethylene oxide) block and a hydrophobic poly(propylene oxide) block (Marks, Pan, Bushell, Cromie, & Lee, 2001). *Poloxamer* 188 (P188) also known as pluronic F68, flocor, or rheothRx and one of the most commonly used copolymers. P188 is used in surfactants and emulsifying agents and has been supported as an *in vivo* for therapies improving the membrane instability that occurs in the absence of dystrophin (Yasuda et al., 2005). In humans, P188 has a half-life of 18 h and is nontoxic. Using X-ray diffraction of a lipid model in vitro, it was shown P188 likely forces lipid compaction restoring the membrane barrier through insertion into the lipid bilayer (Maskarinec, Hannig, Lee, & Lee, 2002). It has been shown that P188 can induce sealing of damaged membranes in vivo following electrical injury, which creates microholes within the membrane, a process termed electroporation now utilized in experimentation for plasmid DNA uptake (Weaver, 1995). Electrical burns produce extensive local trauma to the skin and skeletal muscle, and although this is not a disease in itself, produces injury similar to epidermal and dystrophic diseases. Addition of P188 prevented fluorescent dextran from entering damaged muscle fibers as well as reduced post-trauma inflammatory infiltration (Lee, River, Pan, Ji, & Wollmann, 1992).

Most notably, P188 stabilizes membrane in animal models with Duchenne muscular dystrophy-related cardiomyopathy (Townsend et al., 2010; Yasuda et al., 2005). Addition of P188 was able to stabilize these membranes preventing the continual rounds of degeneration from occurring (Spurney et al., 2011). However, the effects of P188 on skeletal muscle have not been studied as intensely. Terry, Kaneb, and Wells (2014) found the administration of P188 through intraperitoneal injection in *mdx* mice resulted improved histopathology; however, P188 was not able to prevent damage in response to eccentric contraction suggesting that P188 may be able to only partially protect against membrane damage. Evaluating how P188 induces protein relocalization may provide more insight into the mechanism by which P188 exerts its effect. Understanding if P188 only improves membrane stability or whether it also recruits repair proteins or alters phospholipid composition may improve the therapeutic indications for this and related compounds.

7.3 Myoferlin Upregulation

Due to the high level of expression of myoferlin during growth and regeneration, myoferlin is a prime candidate for improving defects in membrane repair. Transgenic mice overexpressing myoferlin under the broadly expressed chicken beta-actin promoter were used to examine the effect of myoferlin (Lostal et al., 2012). These mice express myoferlin mRNA at 200-fold greater levels than wild-type controls at 6 months, but only express myoferlin protein fourfold higher at 3 weeks of age when myoferlin is expressed during the growth phase and 100-fold higher than controls at 4 months when myoferlin is downregulated in adulthood (Lostal et al., 2012). Myoferlin transgenic mice were crossed to dysferlin-null mice to determine if myoferlin overexpression could rescue in this model. Laser repair assays on myoferlin-TG + myofibers showed that the increased expression of myoferlin in dysferlin-null muscle was sufficient to improve repair to wild-type levels as measured by FM dye influx. Despite a correction in the membrane repair assay, a lack of histological improvement was noted, and the authors suggested that myoferlin and dysferlin have different roles during the process of muscle membrane repair. However, because overexpression of membrane-associated proteins can exert stress on many cell types, including muscle, the presence of histopathology from myoferlin overexpression should be cautiously interpreted. Mice lacking both myoferlin and dysferlin do display a significantly enhanced muscular dystrophy phenotype suggesting that both these proteins are implicated in pathogenesis (Demonbreun et al.,

2014). The dysferlin/myoferlin double-null model developed a more progressive dystrophy with more severe T-tubule defects than that of either single mutant. Given that ferlin proteins regulate membrane trafficking beyond membrane repair with complex roles at the T-tubule, further studies are required to better understand the role of myoferlin in muscle repair.

8. CONCLUSIONS

The available genetic models in humans, mice, and other genetic systems have produced a better understanding of the membrane repair complex. Notably, many of these proteins are important for muscle membrane repair, and this likely relates to the physiological roles of skeletal muscle in producing contraction and the manner under which muscle faces stress and strain. The patterns of membrane repair are also seen in nonmuscle cell types, suggesting that the basic molecular machinery is shared in many different types of cells and organisms. The process of membrane repair is critical to cell and organism survival and requires a complex orchestration of Ca^{2+} and phospholipid signaling, as well as the aggregation of protein-repair complexes. A better understanding of this process, including approaches to improving repair, remains central to health and well-being.

ACKNOWLEDGMENTS

This work was supported by National Institutes of Health NS047726, NS072027, AR052646.

REFERENCES

Achanzar, W. E., & Ward, S. (1997). A nematode gene required for sperm vesicle fusion. *Journal of Cell Science, 110*(Pt 9), 1073–1081.
Babiychuk, E. B., Monastyrskaya, K., Potez, S., & Draeger, A. (2011). Blebbing confers resistance against cell lysis. *Cell Death & Differentiation, 18*, 80–89.
Babbin, B. A., Laukoetter, M. G., Nava, P., Koch, S., Lee, W. Y., Capaldo, C. T., Peatman, E., Severson, E. A., Flower, R. J., Perretti, M., Parkos, C. A., & Nusrat, A. (2008). *J Immunol, 181*, 5035–5044.
Bansal, D., & Campbell, K. P. (2004). Dysferlin and the plasma membrane repair in muscular dystrophy. *Trends in Cell Biology, 14*, 206–213.
Bansal, D., Miyake, K., Vogel, S. S., Groh, S., Chen, C. C., Williamson, R., et al. (2003). Defective membrane repair in dysferlin-deficient muscular dystrophy. *Nature, 423*, 168–172.
Bashir, R., Britton, S., Strachan, T., Keers, S., Vafiadaki, E., Lako, M., et al. (1998). A gene related to *Caenorhabditis elegans* spermatogenesis factor fer-1 is mutated in limb-girdle muscular dystrophy type 2B. *Nature Genetics, 20*, 37–42.
Bement, W. M., Mandato, C. A., & Kirsch, M. N. (1999). Wound-induced assembly and closure of an actomyosin purse string in *Xenopus* oocytes. *Current Biology: CB, 9*, 579–587.

Benz, J., Bergner, A., Hofmann, A., Demange, P., Gottig, P., Liemann, S., et al. (1996). The structure of recombinant human annexin VI in crystals and membrane-bound. *Journal of Molecular Biology, 260*, 638–643.

Bittner, R. E., Anderson, L. V., Burkhardt, E., Bashir, R., Vafiadaki, E., Ivanova, S., et al. (1999). Dysferlin deletion in SJL mice (SJL-Dysf) defines a natural model for limb girdle muscular dystrophy 2B. *Nature Genetics, 23*, 141–142.

Blackwood, R. A., & Ernst, J. D. (1990). Characterization of Ca2(+)-dependent phospholipid binding, vesicle aggregation and membrane fusion by annexins. *The Biochemical Journal, 266*, 195–200.

Bolduc, V., Marlow, G., Boycott, K. M., Saleki, K., Inoue, H., Kroon, J., et al. (2010). Recessive mutations in the putative calcium-activated chloride channel Anoctamin 5 cause proximal LGMD2L and distal MMD3 muscular dystrophies. *American Journal of Human Genetics, 86*, 213–221.

Bouter, A., Gounou, C., Berat, R., Tan, S., Gallois, B., Granier, T., et al. (2011). Annexin-A5 assembled into two-dimensional arrays promotes cell membrane repair. *Nature Communications, 2*, 270.

Bulfield, G., Siller, W. G., Wight, P. A., & Moore, K. J. (1984). X chromosome-linked muscular dystrophy (*mdx*) in the mouse. *Proceedings of the National Academy of Sciences of the United States of America, 81*, 1189–1192.

Buzhynskyy, N., Golczak, M., Lai-Kee-Him, J., Lambert, O., Tessier, B., Gounou, C., et al. (2009). Annexin-A6 presents two modes of association with phospholipid membranes. A combined QCM-D, AFM and cryo-TEM study. *Journal of Structural Biology, 168*, 107–116.

Cagliani, R., Magri, F., Toscano, A., Merlini, L., Fortunato, F., Lamperti, C., et al. (2005). Mutation finding in patients with dysferlin deficiency and role of the dysferlin interacting proteins annexin A1 and A2 in muscular dystrophies. *Human Mutation, 26*, 283.

Cai, C., Masumiya, H., Weisleder, N., Matsuda, N., Nishi, M., Hwang, M., et al. (2009). MG53 nucleates assembly of cell membrane repair machinery. *Nature Cell Biology, 11*, 56–64.

Cai, C., Weisleder, N., Ko, J. K., Komazaki, S., Sunada, Y., Nishi, M., et al. (2009). Membrane repair defects in muscular dystrophy are linked to altered interaction between MG53, caveolin-3, and dysferlin. *The Journal of Biological Chemistry, 284*, 15894–15902.

Carmeille, R., Degrelle, S. A., Plawinski, L., Bouvet, F., Gounou, C., Evain-Brion, D., et al. (2015). Annexin-A5 promotes membrane resealing in human trophoblasts. *Biochimica et Biophysica Acta, 1853*(9), 2033–2044.

Chakrabarti, S., Kobayashi, K. S., Flavell, R. A., Marks, C. B., Miyake, K., Liston, D. R., et al. (2003). Impaired membrane resealing and autoimmune myositis in synaptotagmin VII-deficient mice. *The Journal of Cell Biology, 162*, 543–549.

Christmas, P., Callaway, J., Fallon, J., Jones, J., & Haigler, H. T. (1991). Selective secretion of annexin 1, a protein without a signal sequence, by the human prostate gland. *The Journal of Biological Chemistry, 266*, 2499–2507.

Clark, A. G., Miller, A. L., Vaughan, E., Yu, H. Y., Penkert, R., & Bement, W. M. (2009). Integration of single and multicellular wound responses. *Current Biology: CB, 19*, 1389–1395.

Constantin, B. (2014). Dystrophin complex functions as a scaffold for signalling proteins. *Biochimica et Biophysica Acta, 1838*, 635–642.

Corrotte, M., Almeida, P. E., Tam, C., Castro-Gomes, T., Fernandes, M. C., Millis, B. A., et al. (2013). Caveolae internalization repairs wounded cells and muscle fibers. *eLife, 2*, e00926.

Davis, D. B., Doherty, K. R., Delmonte, A. J., & McNally, E. M. (2002). Calcium-sensitive phospholipid binding properties of normal and mutant ferlin C2 domains. *The Journal of Biological Chemistry, 277*, 22883−22888.

Demonbreun, A. R., Fahrenbach, J. P., Deveaux, K., Earley, J. U., Pytel, P., & McNally, E. M. (2010). Impaired muscle growth and response to insulin-like growth factor 1 in dysferlin-mediated muscular dystrophy. *Human Molecular Genetics, 20*(4), 779−789.

Demonbreun, A. R., Lapidos, K. A., Heretis, K., Levin, S., Dale, R., Pytel, P., et al. (2010). Myoferlin regulation by NFAT in muscle injury, regeneration and repair. *Journal of Cell Science, 123*, 2413−2422.

Demonbreun, A. R., Posey, A. D., Heretis, K., Swaggart, K. A., Earley, J. U., Pytel, P., et al. (2010). Myoferlin is required for insulin-like growth factor response and muscle growth. *FASEB Journal: Official Publication of the Federation of American Societies for Experimental Biology, 24*, 1284−1295.

Demonbreun, A. R., Rossi, A. E., Alvarez, M. G., Swanson, K. E., Deveaux, H. K., Earley, J. U., et al. (2014). Dysferlin and myoferlin regulate transverse tubule formation and glycerol sensitivity. *The American Journal of Pathology, 184*, 248−259.

Deora, A. B., Kreitzer, G., Jacovina, A. T., & Hajjar, K. A. (2004). An annexin 2 phosphorylation switch mediates p11-dependent translocation of annexin 2 to the cell surface. *The Journal of Biological Chemistry, 279*, 43411−43418.

DiFranco, M., Quinonez, M., Capote, J., & Vergara, J. (2009). DNA transfection of mammalian skeletal muscles using in vivo electroporation. *Journal of Visualized Experiments: (JoVE)*.

Doherty, J. T., Lenhart, K. C., Cameron, M. V., Mack, C. P., Conlon, F. L., & Taylor, J. M. (2011). Skeletal muscle differentiation and fusion are regulated by the BAR-containing Rho-GTPase-activating protein (Rho-GAP), GRAF1. *The Journal of Biological Chemistry, 286*, 25903−25921.

Doherty, K. R., Cave, A., Davis, D. B., Delmonte, A. J., Posey, A., Earley, J. U., et al. (2005). Normal myoblast fusion requires myoferlin. *Development, 132*, 5565−5575.

Doherty, K. R., Demonbreun, A. R., Wallace, G. Q., Cave, A., Posey, A. D., Heretis, K., et al. (2008). The endocytic recycling protein EHD2 interacts with myoferlin to regulate myoblast fusion. *The Journal of Biological Chemistry, 283*, 20252−20260.

Duann, P., Li, H., Lin, P., Tan, T., Wang, Z., Chen, K., et al. (2015). MG53-mediated cell membrane repair protects against acute kidney injury. *Science Translational Medicine, 7*, 279ra236.

Flix, B., de la Torre, C., Castillo, J., Casal, C., Illa, I., & Gallardo, E. (2013). Dysferlin interacts with calsequestrin-1, myomesin-2 and dynein in human skeletal muscle. *The International Journal of Biochemistry & Cell Biology, 45*, 1927−1938.

Flucher, B. E. (1992). Structural analysis of muscle development: transverse tubules, sarcoplasmic reticulum, and the triad. *Developmental Biology, 154*, 245−260.

Fukada, S., Morikawa, D., Yamamoto, Y., Yoshida, T., Sumie, N., Yamaguchi, M., et al. (2010). Genetic background affects properties of satellite cells and *mdx* phenotypes. *The American Journal of Pathology, 176*, 2414−2424.

George, M., Ying, G., Rainey, M. A., Solomon, A., Parikh, P. T., Gao, Q., et al. (2007). Shared as well as distinct roles of EHD proteins revealed by biochemical and functional comparisons in mammalian cells and *C. elegans*. *BMC Cell Biology, 8*, 3.

Gerke, V., Creutz, C. E., & Moss, S. E. (2005). Annexins: linking Ca^{2+} signalling to membrane dynamics. *Nature Reviews Molecular Cell Biology, 6*, 449−461.

Gerke, V., & Moss, S. E. (2002). Annexins: from structure to function. *Physiological Reviews, 82*, 331−371.

Goulet, F., Moore, K. G., & Sartorelli, A. C. (1992). Glycosylation of annexin I and annexin II. *Biochemical and Biophysical Research Communications, 188*, 554−558.

Grant, B. D., & Caplan, S. (2008). Mechanisms of EHD/RME-1 protein function in endocytic transport. *Traffic, 9,* 2043–2052.

Gudmundsson, H., Curran, J., Kashef, F., Snyder, J. S., Smith, S. A., Vargas-Pinto, P., et al. (2012). Differential regulation of EHD3 in human and mammalian heart failure. *Journal of Molecular and Cellular Cardiology, 52,* 1183–1190.

Gunteski-Hamblin, A. M., Song, G., Walsh, R. A., Frenzke, M., Boivin, G. P., Dorn, G. W., 2nd, et al. (1996). Annexin VI overexpression targeted to heart alters cardiomyocyte function in transgenic mice. *The American Journal of Physiology, 270,* H1091–H1100.

Hack, A. A., Lam, M. Y., Cordier, L., Shoturma, D. I., Ly, C. T., Hadhazy, M. A., et al. (2000). Differential requirement for individual sarcoglycans and dystrophin in the assembly and function of the dystrophin-glycoprotein complex. *Journal of Cell Science, 113*(Pt 14), 2535–2544.

Hannon, R., Croxtall, J. D., Getting, S. J., Roviezzo, F., Yona, S., Paul-Clark, M. J., et al. (2003). Aberrant inflammation and resistance to glucocorticoids in annexin 1-/- mouse. *FASEB Journal: Official Publication of the Federation of American Societies for Experimental Biology, 17,* 253–255.

Hawkins, T. E., Roes, J., Rees, D., Monkhouse, J., & Moss, S. E. (1999). Immunological development and cardiovascular function are normal in annexin VI null mutant mice. *Molecular and Cellular Biology, 19,* 8028–8032.

Hayes, M. J., Merrifield, C. J., Shao, D., Ayala-Sanmartin, J., Schorey, C. D., Levine, T. P., et al. (2004). Annexin 2 binding to phosphatidylinositol 4,5-bisphosphate on endocytic vesicles is regulated by the stress response pathway. *The Journal of Biological Chemistry, 279,* 14157–14164.

Hayes, M. J., Shao, D., Bailly, M., & Moss, S. E. (2006). Regulation of actin dynamics by annexin 2. *The EMBO Journal, 25,* 1816–1826.

He, B., Tang, R. H., Weisleder, N., Xiao, B., Yuan, Z., Cai, C., et al. (2012). Enhancing muscle membrane repair by gene delivery of MG53 ameliorates muscular dystrophy and heart failure in delta-Sarcoglycan-deficient hamsters. *Molecular Therapy: The Journal of the American Society of Gene Therapy, 20,* 727–735.

Heydemann, A., Huber, J. M., Demonbreun, A., Hadhazy, M., & McNally, E. M. (2005). Genetic background influences muscular dystrophy. *Neuromuscular Disorders: NMD, 15,* 601–609.

Ho, M., Post, C. M., Donahue, L. R., Lidov, H. G., Bronson, R. T., Goolsby, H., et al. (2004). Disruption of muscle membrane and phenotype divergence in two novel mouse models of dysferlin deficiency. *Human Molecular Genetics, 13,* 1999–2010.

Ikezoe, K., Furuya, H., Ohyagi, Y., Osoegawa, M., Nishino, I., Nonaka, I., et al. (2003). Dysferlin expression in tubular aggregates: their possible relationship to endoplasmic reticulum stress. *Acta Neuropathologica, 105,* 603–609.

Jaiswal, J. K., Lauritzen, S. P., Scheffer, L., Sakaguchi, M., Bunkenborg, J., Simon, S. M., et al. (2014). S100A11 is required for efficient plasma membrane repair and survival of invasive cancer cells. *Nature Communications, 5,* 3795.

Jia, Y., Chen, K., Lin, P., Lieber, G., Nishi, M., Yan, R., et al. (2014). Treatment of acute lung injury by targeting MG53-mediated cell membrane repair. *Nature Communications, 5,* 4387.

Kaetzel, M. A., Mo, Y. D., Mealy, T. R., Campos, B., Bergsma-Schutter, W., Brisson, A., et al. (2001). Phosphorylation mutants elucidate the mechanism of annexin IV-mediated membrane aggregation. *Biochemistry, 40,* 4192–4199.

Kamal, A., Ying, Y., & Anderson, R. G. (1998). Annexin VI-mediated loss of spectrin during coated pit budding is coupled with delivery of LDL to lysosomes. *The Journal of Cell Biology, 142,* 937–947.

Kerr, J. P., Ziman, A. P., Mueller, A. L., Muriel, J. M., Kleinhans-Welte, E., Gumerson, J. D., et al. (2013). Dysferlin stabilizes stress-induced Ca^{2+} signaling in the transverse tubule membrane. *Proceedings of the National Academy of Sciences of the United States of America, 110,* 20831−20836.

Klinge, L., Harris, J., Sewry, C., Charlton, R., Anderson, L., Laval, S., et al. (2010). Dysferlin associates with the developing T-tubule system in rodent and human skeletal muscle. *Muscle & Nerve, 41,* 166−173.

Kuru, S., Yasuma, F., Wakayama, T., Kimura, S., Konagaya, M., Aoki, M., et al. (2004). [A patient with limb girdle muscular dystrophy type 2B (LGMD2B) manifesting cardiomyopathy]. *Rinsho shinkeigaku = Clinical Neurology, 44,* 375−378.

de Laat, B., Derksen, R. H., Mackie, I. J., Roest, M., Schoormans, S., Woodhams, B. J., et al. (2006). Annexin A5 polymorphism (-1C−>T) and the presence of anti-annexin A5 antibodies in the antiphospholipid syndrome. *Annals of the Rheumatic Diseases, 65,* 1468−1472.

Lee, E., Marcucci, M., Daniell, L., Pypaert, M., Weisz, O. A., Ochoa, G. C., et al. (2002). Amphiphysin 2 (Bin1) and T-tubule biogenesis in muscle. *Science, 297,* 1193−1196.

Lee, R. C., River, L. P., Pan, F. S., Ji, L., & Wollmann, R. L. (1992). Surfactant-induced sealing of electropermeabilized skeletal muscle membranes in vivo. *Proceedings of the National Academy of Sciences of the United States of America, 89,* 4524−4528.

Lek, A., Evesson, F. J., Sutton, R. B., North, K. N., & Cooper, S. T. (2012). Ferlins: regulators of vesicle fusion for auditory neurotransmission, receptor trafficking and membrane repair. *Traffic, 13,* 185−194.

Lenhart, K. C., Becherer, A. L., Li, J., Xiao, X., McNally, E. M., Mack, C. P., et al. (2014). GRAF1 promotes ferlin-dependent myoblast fusion. *Developmental Biology, 393,* 298−311.

Lennon, N. J., Kho, A., Bacskai, B. J., Perlmutter, S. L., Hyman, B. T., & Brown, R. H., Jr. (2003). Dysferlin interacts with annexins A1 and A2 and mediates sarcolemmal wound-healing. *The Journal of Biological Chemistry, 278*(50), 50466−50473.

Leung, C., Yu, C., Lin, M. I., Tognon, C., & Bernatchez, P. (2013). Expression of myoferlin in human and murine carcinoma tumors: role in membrane repair, cell proliferation, and tumorigenesis. *The American Journal of Pathology, 182,* 1900−1909.

Ling, Q., Jacovina, A. T., Deora, A., Febbraio, M., Simantov, R., Silverstein, R. L., et al. (2004). Annexin II regulates fibrin homeostasis and neoangiogenesis in vivo. *The Journal of Clinical Investigation, 113,* 38−48.

Liu, J., Aoki, M., Illa, I., Wu, C., Fardeau, M., Angelini, C., et al. (1998). Dysferlin, a novel skeletal muscle gene, is mutated in Miyoshi myopathy and limb girdle muscular dystrophy. *Nature Genetics, 20,* 31−36.

Liu, J., Zhu, H., Zheng, Y., Xu, Z., Li, L., Tan, T., et al. (2015). Cardioprotection of recombinant human MG53 protein in a porcine model of ischemia and reperfusion injury. *Journal of Molecular and Cellular Cardiology, 80,* 10−19.

Lostal, W., Bartoli, M., Bourg, N., Roudaut, C., Bentaib, A., Miyake, K., et al. (2010). Efficient recovery of dysferlin deficiency by dual adeno-associated vector-mediated gene transfer. *Human Molecular Genetics, 19,* 1897−1907.

Lostal, W., Bartoli, M., Roudaut, C., Bourg, N., Krahn, M., Pryadkina, M., et al. (2012). Lack of correlation between outcomes of membrane repair assay and correction of dystrophic changes in experimental therapeutic strategy in dysferlinopathy. *PLoS One, 7,* e38036.

Lundmark, R., Doherty, G. J., Howes, M. T., Cortese, K., Vallis, Y., Parton, R. G., et al. (2008). The GTPase-activating protein GRAF1 regulates the CLIC/GEEC endocytic pathway. *Current Biology: CB, 18,* 1802−1808.

Mahjneh, I., Jaiswal, J., Lamminen, A., Somer, M., Marlow, G., Kiuru-Enari, S., et al. (2010). A new distal myopathy with mutation in anoctamin 5. *Neuromuscular Disorders: NMD, 20,* 791−795.

Marconi, C., Brunamonti Binello, P., Badiali, G., Caci, E., Cusano, R., Garibaldi, J., et al. (2013). A novel missense mutation in ANO5/TMEM16E is causative for gnathodiaphyseal dyplasia in a large Italian pedigree. *European Journal of Human Genetics: EJHG, 21,* 613—619.

Marg, A., Schoewel, V., Timmel, T., Schulze, A., Shah, C., Daumke, O., et al. (2012). Sarcolemmal repair is a slow process and includes EHD2. *Traffic, 13,* 1286—1294.

Marks, J. D., Pan, C. Y., Bushell, T., Cromie, W., & Lee, R. C. (2001). Amphiphilic, tri-block copolymers provide potent membrane-targeted neuroprotection. *FASEB Journal: Official Publication of the Federation of American Societies for Experimental Biology, 15,* 1107—1109.

Maskarinec, S. A., Hannig, J., Lee, R. C., & Lee, K. Y. (2002). Direct observation of poloxamer 188 insertion into lipid monolayers. *Biophysical Journal, 82,* 1453—1459.

Matsuda, C., Miyake, K., Kameyama, K., Keduka, E., Takeshima, H., Imamura, T., et al. (2012). The C2A domain in dysferlin is important for association with MG53 (TRIM72). *PLoS Currents, 4,* e5035add5038caff5034.

McDade, J. R., Archambeau, A., & Michele, D. E. (2014). Rapid actin-cytoskeleton-dependent recruitment of plasma membrane-derived dysferlin at wounds is critical for muscle membrane repair. *FASEB Journal: Official publication of the Federation of American Societies for Experimental Biology, 28,* 3660—3670.

McNeil, A. K., Rescher, U., Gerke, V., & McNeil, P. L. (2006). *J Biol Chem, 281,* 35202—35207.

McNeil, P. L., & Khakee, R. (1992). Disruptions of muscle fiber plasma membranes. Role in exercise-induced damage. *The American Journal of Pathology, 140,* 1097—1109.

McNeil, P. L., & Kirchhausen, T. (2005). An emergency response team for membrane repair. *Nature Reviews Molecular Cell Biology, 6,* 499—505.

McNeil, P. L., & Steinhardt, R. A. (1997). Loss, restoration, and maintenance of plasma membrane integrity. *The Journal of Cell Biology, 137,* 1—4.

Meroni, G., & Diez-Roux, G. (2005). TRIM/RBCC, a novel class of 'single protein RING finger' E3 ubiquitin ligases. *BioEssays: News and Reviews in Molecular, Cellular and Developmental Biology, 27,* 1147—1157.

Millay, D. P., Maillet, M., Roche, J. A., Sargent, M. A., McNally, E. M., Bloch, R. J., et al. (2009). Genetic manipulation of dysferlin expression in skeletal muscle: novel insights into muscular dystrophy. *The American Journal of Pathology, 175,* 1817—1823.

Mizuta, K., Tsutsumi, S., Inoue, H., Sakamoto, Y., Miyatake, K., Miyawaki, K., et al. (2007). Molecular characterization of GDD1/TMEM16E, the gene product responsible for autosomal dominant gnathodiaphyseal dysplasia. *Biochemical and Biophysical Research Communications, 357,* 126—132.

Monjaret, F., Suel-Petat, L., Bourg-Alibert, N., Vihola, A., Marchand, S., Roudaut, C., et al. (2013). The phenotype of dysferlin-deficient mice is not rescued by adeno-associated virus-mediated transfer of anoctamin 5. *Human Gene Therapy. Clinical Development, 24,* 65—76.

Mosser, G., Ravanat, C., Freyssinet, J. M., & Brisson, A. (1991). Sub-domain structure of lipid-bound annexin-V resolved by electron image analysis. *Journal of Molecular Biology, 217,* 241—245.

Naslavsky, N., & Caplan, S. (2010). EHD proteins: key conductors of endocytic transport. *Trends in Cell Biology, 21*(2), 122—131.

Naslavsky, N., Rahajeng, J., Chenavas, S., Sorgen, P. L., & Caplan, S. (2007). EHD1 and Eps15 interact with phosphatidylinositols via their Eps15 homology domains. *The Journal of Biological Chemistry, 282,* 16612—16622.

Nguyen, K., Bassez, G., Krahn, M., Bernard, R., Laforet, P., Labelle, V., et al. (2007). Phenotypic study in 40 patients with dysferlin gene mutations: high frequency of atypical phenotypes. *Archives of Neurology, 64,* 1176—1182.

Ousingsawat, J., Martins, J. R., Schreiber, R., Rock, J. R., Harfe, B. D., & Kunzelmann, K. (2009). Loss of TMEM16A causes a defect in epithelial Ca^{2+}-dependent chloride transport. *The Journal of Biological Chemistry, 284,* 28698—28703.

Paoluzi, S., Castagnoli, L., Lauro, I., Salcini, A. E., Coda, L., Fre, S., et al. (1998). Recognition specificity of individual EH domains of mammals and yeast. *The EMBO Journal, 17,* 6541—6550.

Pedemonte, N., & Galietta, L. J. (2014). Structure and function of TMEM16 proteins (anoctamins). *Physiological Reviews, 94,* 419—459.

Piccolo, F., Moore, S. A., Ford, G. C., & Campbell, K. P. (2000). Intracellular accumulation and reduced sarcolemmal expression of dysferlin in limb—girdle muscular dystrophies. *Annals of Neurology, 48,* 902—912.

Pohl, U., Smith, J. S., Tachibana, I., Ueki, K., Lee, H. K., Ramaswamy, S., et al. (2000). EHD2, EHD3, and EHD4 encode novel members of a highly conserved family of EH domain-containing proteins. *Genomics, 63,* 255—262.

Porter, K. R., & Palade, G. E. (1957). Studies on the endoplasmic reticulum. III. Its form and distribution in striated muscle cells. *The Journal of Biophysical and Biochemical Cytology, 3,* 269—300.

Posey, A. D., Jr., Pytel, P., Gardikiotes, K., Demonbreun, A. R., Rainey, M., George, M., et al. (2011). Endocytic recycling proteins EHD1 and EHD2 interact with fer-1-like-5 (Fer1L5) and mediate myoblast fusion. *The Journal of Biological Chemistry, 286,* 7379—7388.

Posey, A. D., Jr., Swanson, K. E., Alvarez, M. G., Krishnan, S., Earley, J. U., Band, H., et al. (2014). EHD1 mediates vesicle trafficking required for normal muscle growth and transverse tubule development. *Developmental Biology, 387,* 179—190.

Rainey, M. A., George, M., Ying, G., Akakura, R., Burgess, D. J., Siefker, E., et al. (2010). The endocytic recycling regulator EHD1 is essential for spermatogenesis and male fertility in mice. *BMC Developmental Biology, 10,* 37.

Rapaport, D., Auerbach, W., Naslavsky, N., Pasmanik-Chor, M., Galperin, E., Fein, A., et al. (2006). Recycling to the plasma membrane is delayed in EHD1 knockout mice. *Traffic, 7,* 52—60.

Reddy, A., Caler, E. V., & Andrews, N. W. (2001). Plasma membrane repair is mediated by Ca(2+)-regulated exocytosis of lysosomes. *Cell, 106,* 157—169.

Roche, J. A., Ford-Speelman, D. L., Ru, L. W., Densmore, A. L., Roche, R., Reed, P. W., et al. (2011). Physiological and histological changes in skeletal muscle following in vivo gene transfer by electroporation. *American Journal of Physiology. Cell Physiology, 301,* C1239—C1250.

Rodriguez, A., Webster, P., Ortego, J., & Andrews, N. W. (1997). Lysosomes behave as Ca^{2+}-regulated exocytic vesicles in fibroblasts and epithelial cells. *The Journal of Cell Biology, 137,* 93—104.

Roostalu, U., & Strahle, U. (2012). In vivo imaging of molecular interactions at damaged sarcolemma. *Developmental Cell, 22,* 515—529.

Roux, I., Safieddine, S., Nouvian, R., Grati, M., Simmler, M. C., Bahloul, A., et al. (2006). Otoferlin, defective in a human deafness form, is essential for exocytosis at the auditory ribbon synapse. *Cell, 127,* 277—289.

Salcini, A. E., Confalonieri, S., Doria, M., Santolini, E., Tassi, E., Minenkova, O., et al. (1997). Binding specificity and in vivo targets of the EH domain, a novel protein-protein interaction module. *Genes & Development, 11,* 2239—2249.

Saurel, O., Cezanne, L., Milon, A., Tocanne, J. F., & Demange, P. (1998). Influence of annexin V on the structure and dynamics of phosphatidylcholine/phosphatidylserine bilayers: a fluorescence and NMR study. *Biochemistry, 37,* 1403—1410.

Skrahina, T., Piljic, A., & Schultz, C. (2008). Heterogeneity and timing of translocation and membrane-mediated assembly of different annexins. *Experimental Cell Research, 314,* 1039—1047.

Spurney, C. F., Guerron, A. D., Yu, Q., Sali, A., van der Meulen, J. H., Hoffman, E. P., et al. (2011). Membrane sealant Poloxamer P188 protects against isoproterenol induced cardiomyopathy in dystrophin deficient mice. *BMC Cardiovascular Disorders, 11,* 20.

Stephan, A. B., Shum, E. Y., Hirsh, S., Cygnar, K. D., Reisert, J., & Zhao, H. (2009). ANO2 is the cilial calcium-activated chloride channel that may mediate olfactory amplification. *Proceedings of the National Academy of Sciences of the United States of America, 106,* 11776–11781.

Straub, V., Rafael, J. A., Chamberlain, J. S., & Campbell, K. P. (1997). Animal models for muscular dystrophy show different patterns of sarcolemmal disruption. *The Journal of Cell Biology, 139,* 375–385.

Suzuki, J., Fujii, T., Imao, T., Ishihara, K., Kuba, H., & Nagata, S. (2013). Calcium-dependent phospholipid scramblase activity of TMEM16 protein family members. *The Journal of Biological Chemistry, 288,* 13305–13316.

Swaggart, K. A., Demonbreun, A. R., Vo, A. H., Swanson, K. E., Kim, E. Y., Fahrenbach, J. P., et al. (2014). Annexin A6 modifies muscular dystrophy by mediating sarcolemmal repair. *Proceedings of the National Academy of Sciences of the United States of America, 111,* 6004–6009.

Terry, R. L., Kaneb, H. M., & Wells, D. J. (2014). Poloxomer 188 has a deleterious effect on dystrophic skeletal muscle function. *PLoS One, 9,* e91221.

Therrien, C., Di Fulvio, S., Pickles, S., & Sinnreich, M. (2009). Characterization of lipid binding specificities of dysferlin C2 domains reveals novel interactions with phosphoinositides. *Biochemistry, 48,* 2377–2384.

Tian, Y., Wright, J., Cebotaru, L., Wang, H., & Guggino, W. B. (2015). Anoctamin5 is related to plasma membrane repair. *JSM Biotechnology & Biomedical Engineering, 3,* 6.

Townsend, D., Turner, I., Yasuda, S., Martindale, J., Davis, J., Shillingford, M., et al. (2010). Chronic administration of membrane sealant prevents severe cardiac injury and ventricular dilatation in dystrophic dogs. *The Journal of Clinical Investigation, 120,* 1140–1150.

Tsutsumi, S., Inoue, H., Sakamoto, Y., Mizuta, K., Kamata, N., & Itakura, M. (2005). Molecular cloning and characterization of the murine gnathodiaphyseal dysplasia gene GDD1. *Biochemical and Biophysical Research Communications, 331,* 1099–1106.

Wallner, B. P., Mattaliano, R. J., Hession, C., Cate, R. L., Tizard, R., Sinclair, L. K., et al. (1986). Cloning and expression of human lipocortin, a phospholipase A2 inhibitor with potential anti-inflammatory activity. *Nature, 320,* 77–81.

Wang, W., & Creutz, C. E. (1994). Role of the amino-terminal domain in regulating interactions of annexin I with membranes: effects of amino-terminal truncation and mutagenesis of the phosphorylation sites. *Biochemistry, 33,* 275–282.

Weaver, J. C. (1995). Electroporation theory. Concepts and mechanisms. *Methods in Molecular Biology, 55,* 3–28.

Weisleder, N., Takizawa, N., Lin, P., Wang, X., Cao, C., Zhang, Y., et al. (2012). Recombinant MG53 protein modulates therapeutic cell membrane repair in treatment of muscular dystrophy. *Science Translational Medicine, 4,* 139ra185.

Yasuda, S., Townsend, D., Michele, D. E., Favre, E. G., Day, S. M., & Metzger, J. M. (2005). Dystrophic heart failure blocked by membrane sealant poloxamer. *Nature, 436,* 1025–1029.

Yu, Q. C., & McNeil, P. L. (1992). Transient disruptions of aortic endothelial cell plasma membranes. *The American Journal of Pathology, 141,* 1349–1360.

Zaks, W. J., & Creutz, C. E. (1991). Ca(2+)-dependent annexin self-association on membrane surfaces. *Biochemistry, 30,* 9607–9615.

CHAPTER FOUR

Local Palmitoylation Cycles and Specialized Membrane Domain Organization

Yuko Fukata[1,2], Tatsuro Murakami[1,2], Norihiko Yokoi[1,2] and Masaki Fukata[1,2,*]

[1]Division of Membrane Physiology, Department of Cell Physiology, National Institute for Physiological Sciences, National Institutes of Natural Sciences, Okazaki, Japan
[2]Department of Physiological Sciences, School of Life Science, SOKENDAI (The Graduate University for Advanced Studies), Okazaki, Japan
*Corresponding author: E-mail: mfukata@nips.ac.jp

Contents

Abstract

Palmitoylation is an evolutionally conserved lipid modification of proteins. Dynamic and reversible palmitoylation controls a wide range of molecular and cellular properties of proteins including the protein trafficking, protein function, protein stability, and specialized membrane domain organization. However, technical difficulties in (1) detection of palmitoylated substrate proteins and (2) purification and enzymology of palmitoylating enzymes have prevented the progress in palmitoylation research,

Current Topics in Membranes, Volume 77
ISSN 1063-5823
http://dx.doi.org/10.1016/bs.ctm.2015.10.003

compared with that in phosphorylation research. The recent development of proteomic and chemical biology techniques has unexpectedly expanded the known complement of palmitoylated proteins in various species and tissues/cells, and revealed the unique occurrence of palmitoylated proteins in membrane-bound organelles and specific membrane compartments. Furthermore, identification and characterization of DHHC (Asp-His-His-Cys) palmitoylating enzyme—substrate pairs have contributed to elucidating the regulatory mechanisms and pathophysiological significance of protein palmitoylation. Here, we review the recent progress in protein palmitoylation at the molecular, cellular, and in vivo level and discuss how locally regulated palmitoylation machinery works for dynamic nanoscale organization of membrane domains.

1. INTRODUCTION

Posttranslational modifications of proteins, including phosphorylation, ubiquitinylation, glycosylation, and lipidation, provide a central molecular mechanism for cells to maintain their homeostasis and appropriately respond to extracellular stimuli. Lipidation such as cotranslational myristoylation and posttranslational prenylation and palmitoylation essentially increases the hydrophobicity of proteins and is thought to be responsible for their proper membrane-bound localization in a cell, especially the plasma membrane localization. Palmitoylation occurs on specific cysteine residues either through the stable amide-linkage (*N*-palmitoylation) or labile thioester-linkage (*S*-palmitoylation). Because *S*-palmitoylation is found more frequently in palmitoylated proteins, we use the term of protein palmitoylation as *S*-palmitoylation in this review. Unlike other irreversible lipid modifications, such as myristoylation and prenylation, palmitoylation represents the reversible nature, suggesting that palmitoylation dynamically regulates subcellular localization of proteins.

In 1979, the first palmitoylated proteins were reported for the glycoproteins of Sindbis virus (Schmidt, Bracha, & Schlesinger, 1979) and vesicular stomatitis virus (Schmidt & Schlesinger, 1979; Veit, 2012). In 1987, the reversibility of protein palmitoylation was shown for ankyrin in erythrocytes in which protein synthesis does not occur but ankyrin can be labeled with a [3H]-palmitate (Staufenbiel, 1987). This result that protein synthesis is not coupled with protein palmitoylation in mature erythrocytes led to the discovery of rapid turnover of palmitate on ankyrin protein. Then, a number of proteins were reported to undergo palmitoylation (as palmitoyl–proteins).

Palmitoylation occurs on both soluble and integral membrane proteins. Examples include GTP-binding proteins (G protein α (Gα) subunit, H/N-Ras), SNARE proteins (SNAP-25), postsynaptic scaffolding proteins (PSD-95, GRIP), cell adhesion molecules (integrin, claudin, and NCAM), and G protein-coupled receptors (GPCR) (Chamberlain & Shipston, 2015; El-Husseini Ael & Bredt, 2002; Fukata & Fukata, 2010; Linder & Deschenes, 2007). Importantly, the specific extracellular signals modify the palmitoylation—depalmitoylation balance on specific palmitoyl-proteins, such as Gαs and PSD-95. Thus, protein palmitoylation appears to be one of the key posttranslational modifications that alter localizations and functions of proteins and have a great impact upon the cellular signaling as in the case of protein phosphorylation. However, the technical difficulties in (1) detection and identification of palmitoyl-modified proteins; (2) identification of the enzymes that add palmitate to proteins (palmitoyl acyltransferase, PAT) and those that cleave the thioester bond (palmitoyl-protein thioesterase, PPT); and (3) spatiotemporal visualization of palmitoylated state of proteins in cells have prevented the progress of research in the field of protein palmitoylation, compared with that of protein phosphorylation. In fact, the identification of palmitoyl-proteins completely depended on the deep insights and hypothesis of investigators because the purification method of palmitoylated proteins was not established and the consensus sequence for protein palmitoylation remained undetermined unlike that for myristoylation (N-terminal glycine in the sequence MGXXXS/T (M, Met; G, Gly; S, Ser; T, Thr; X, a variety of amino acids)) or prenylation (cysteine in the C-terminal CaaX motif; a, aliphatic amino acid). Therefore, it long remained unclear how many proteins are palmitoylated in cells and whether the reactions of palmitoylation and depalmitoylation are actually meditated by the enzymes.

The recent development of new proteomic methods combined with chemical biology approaches and the discovery of enzymes catalyzing palmitoylation (DHHC (aspartate-histidine-histidine-cysteine) proteins) caused a paradigm shift in the field of palmitoylation. Here, we review the recent proteome-wide identification of palmitoylated proteins in various species and tissues/cells and the pathophysiological relevance of DHHC proteins. In addition, we discuss the role of local palmitoylation cycles that could be a clue to solve a fundamental question in cell biology, "how do cells create, maintain, and (re)organize the specialized membrane domains?"

2. PALMITOYLATED PROTEINS IN VARIOUS SPECIES AND IN SPECIFIC SUBCELLULAR ORGANELLES

How many proteins are actually palmitoylated in cells or tissues? The development of purification methods of palmitoylated proteins opens the avenue for addressing this long-lasting question. In 2004, Green and Drisdel established the acyl–biotinyl exchange (ABE) method for purification of palmitoylated proteins from complex protein extracts, even tissue samples (Drisdel & Green, 2004). In the ABE method, nonpalmitoylated free cysteine thiols of proteins are blocked by N-ethylmaleimide in advance and the palmitoyl–thioester bonds are cleaved (hydrolyzed) by neutral hydroxylamine. The newly generated free thiols are then specifically labeled with sulfhydryl-reactive biotin reagents (e.g., biotin-HPDP), allowing enrichment of biotinylated (formerly palmitoylated) proteins on the streptavidin-agarose. Combining the ABE method with quantitative mass spectrometry, Davis's group the first performed global analysis of protein palmitoylation in the yeast *Saccharomyces cerevisiae* and identified 35 new proteins that are palmitoylated, in addition to 12 known ones (Roth et al., 2006). This approach was then applied to rat brain synaptosomal fractions to find 68 known proteins and more than 200 new palmitoyl–protein candidates (Kang et al., 2008). These experiments identified a substantial number of unexpected palmitoyl–proteins, such as many SNARE proteins, amino acid permeases, NMDA receptor subunits (GluN2A and GluN2B), and a brain-specific Cdc42 splice variant (Kang et al., 2008; Roth et al., 2006). The ABE-based method has been modified by several groups to enrich palmitoylated proteins/peptides more specifically and sensitively (Zhou, An, Freeman, & Yang, 2014). For example, the acyl-resin-assisted capture method, in which samples treated with hydroxylamine are incubated directly with thiol-reactive resin (e.g., thiopropyl sepharose), was used for purifying and identifying palmitoylated proteins in HEK293 cells (Forrester et al., 2011) and mouse adipocytes (Ren, Jhala, & Du, 2013). These types of comprehensive profiling of palmitoylated proteins (called palmitoyl-proteomics or palmitoylome profiling) have been increasingly applied to various cells/tissues and various species (Table 1): human endothelial cells (Marin, Derakhshan, Lam, Davalos, & Sessa, 2012; Wei, Song, & Semenkovich, 2014), human B cells (Ivaldi et al., 2012), human DU145 prostate cancer cells (Yang, Di Vizio, Kirchner, Steen, & Freeman, 2010), human platelets (Dowal, Yang, Freeman, Steen, & Flaumenhaft, 2011), mouse macrophages (Merrick et al., 2011), *Arabidopsis thaliana* root cells (Hemsley, Weimar, Lilley, Dupree, & Grierson, 2013), *Trypanosoma brucei* parasites (Emmer et al., 2011), and *Plasmodium falciparum* parasites (Jones, Collins, Goulding, Choudhary, & Rayner, 2012).

Independently, several groups developed nonradioactive metabolic labeling methods of palmitoyl-proteins by taking advantage of bioorthogonal palmitate analogues, such as ω-azide- or ω-alkynyl-palmitate (Charron et al., 2009; Hang et al., 2007; Hannoush & Arenas-Ramirez, 2009; Hannoush & Sun, 2010; Kostiuk et al., 2008; Martin & Cravatt, 2009). Cravatt and Martin used this click chemistry-based metabolic labeling method to enrich palmitoylated proteins for proteomic analysis and comprehensively identified about 125 palmitoyl-protein candidates in human Jurkat T cells (Martin & Cravatt, 2009). Similar approaches were done in mouse dendritic cells (Yount et al., 2010), rat liver mitochondria fraction (Kostiuk et al., 2008), and *Saccharomyces pombe* (Zhang, Wu, Kelly, Nurse, & Hang, 2013) (Table 1). The two major methods, the ABE and bioorthogonal metabolic labeling, have individual advantages and disadvantages and are mutually complementary (Storck, Serwa, & Tate, 2013; Zhou et al., 2014). By combining the two methods, more comprehensive and confident palmitoylome is established for *Plasmodium* (Jones et al., 2012). These palmitoyl-proteomics methods tremendously expanded the knowledge about the number and types of palmitoyl-proteins covering various species and their tissues. Accumulated data sets show that the palmitoylome size reaches around 400 in mammalian cells and tissues and that protein palmitoylation is highly conserved posttranslational modification to regulate diverse cellular functions (Table 1).

Although it was speculated that most of palmitoylated proteins are residents at the plasma membrane or at the Golgi apparatus, recent evidence including the palmitoylome profiling unexpectedly tells us that a substantial number of proteins, known to be localized at other specific organelles or subcellular compartments, are also modified by palmitoylation as summarized below (Figure 1).

Endoplasmic Reticulum (ER): Calnexin, an ER chaperone involved in glycoprotein folding, was listed in palmitoylome data sets (Kang et al., 2008; Martin & Cravatt, 2009). Calnexin is indeed palmitoylated by zinc-finger DHHC6 (ZDHHC6), a member of the DHHC protein family, in HeLa cells (Lakkaraju et al., 2012). Interestingly, palmitoylated calnexin preferentially partitions into the perinuclear ER to associate with the ribosome—translocon complex and efficiently interacts with nascent glycoproteins. Another group reported that calnexin is targeted to the

Table 1 Protein S-palmitoylation studied in various species

Species	Tested organism/tissue/cell-type	Palmitoylome study	
		Number of candidate substrates	Predicted DHHC family member (gene name)
Human	Jurkat T cells,[1] DU145 prostate cancer cells,[2] platelets,[3] HEK293T cells,[4] EA.hy926 endothelial cells,[5] lymphoid B cells,[6] and endothelial cells.[7]	$95^6-393^{2,a}$	23 (ZDHHC1–9, 11–24; UniGene)[22,23]
Mouse	DC2.4 dendritic cells,[8] Raw2.4 macrophages,[9] BW5147 T cells,[10] neural stem cells,[11] adipose tissue plus 3T3-L1 adipocytes,[12] and brain.[13]	$101^9-338^{10,a}$	24 (ZDHHC1–9, 11–25; UniGene)[24]
Rat	Liver mitochondrial fraction[14] and embryonic cortical neurons and brain.[15]	$21^{14}-495^{15,a}$	24 (ZDHHC1–9, 11–25; UniGene)[25,26]
Fruit fly	Drosophila melanogaster	n.t.	22[25,26]
Nematode	Caenorhabditis elegans	n.t.	15 (DHHC1–14, spe-10)[26]
Plants	Arabidopsis thaliana root cell culture	259[16]	24 (AtPAT1–24)[27,28]
	Oryza sativa	n.t.	30 (OsPAT1–30)[29]
	Zea mays	n.t.	38 (ZmPAT1–38)[29]
	Glycine max	n.t.	42 (GmPAT1–42)[29]
Protozoa	Trypanosoma brucei	124[17]	12 (TbPAT1–12)[30]
	Plasmodium falciparum	409[18]	12 (PfDHHC1–12)[31]
	Plasmodium berghei	n.t.	11 (PbPAT1–11)[31]
	Toxoplasma gondii	n.t.	18 (TgDHHC1–18)[31]

Yeast	Saccharomyces cerevisiae	47[19]	7 (Akr1/2, Erf2, Swf1, Pfa3-5)[19,32]
	Saccharomyces pombe	238[20]	5 (Akr1, Erf2, Swf1, Pfa3/5)[20]
	Cryptococcus neoformans	427[21]	7 (including Pfa4 and Akr1)[21]

ZDHHC, zinc-finger DHHC; PAT, palmitoyl acyltransferase; n.t., not tested.

[a]The smallest and largest number of candidates determined by the quantification methods are shown with the individual reference.

[1]Martin and Cravatt (2009).
[2]W. Yang et al. (2010).
[3]Dowal et al. (2011).
[4]Forrester et al. (2011).
[5]Marin et al. (2012).
[6]Ivaldi et al. (2012).
[7]Wei et al. (2014).
[8]Yount et al. (2010).
[9]Merrick et al. (2011).
[10]Martin et al. (2012).
[11]Li et al. (2010).
[12]W. Ren et al. (2013).
[13]Wan et al. (2013).
[14]Kostiuk et al. (2008).
[15]Kang et al. (2008).
[16]Hemsley et al. (2013).
[17]Emmer et al. (2011).
[18]Jones et al. (2012).
[19]Roth et al. (2006).
[20]Zhang et al. (2013).
[21]Santiago–Tirado et al. (2015).
[22]Ohno et al. (2006).
[23]Korycka et al. (2012).
[24]Fukata et al. (2004).
[25]Bannan et al. (2008).
[26]Edmonds and Morgan (2014).
[27]Hemsley et al. (2005).
[28]Batistic (2012).
[29]Yuan et al. (2013).
[30]Emmer et al. (2009).
[31]Frenal et al. (2014).
[32]Lobo et al. (2002).

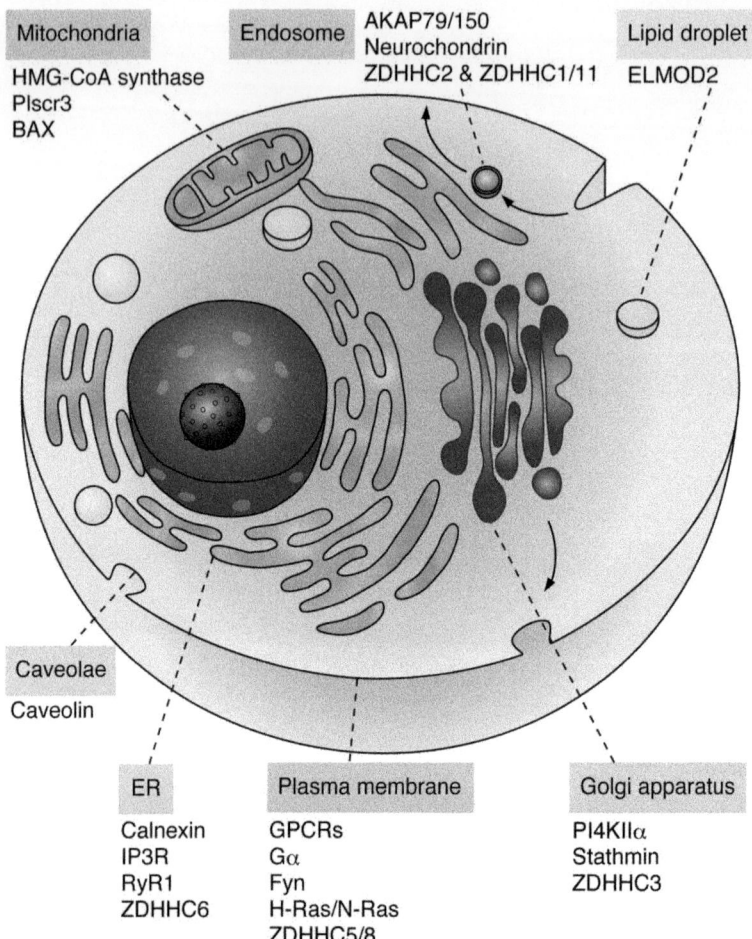

Figure 1 Palmitoylated proteins in various subcellular organelles in animal cells. Classically, most of palmitoylated proteins are thought to be localized at the Golgi apparatus and plasma membranes. Recent palmitoylome analyses show that numerous palmitoylated proteins and DHHC proteins are distributed at the various subcellular organelles including the ER, mitochondria, endosome, lipid droplet, and so on. Several representative substrates are shown. DHHC, aspartate-histidine-histidine-cysteine; ER, endoplasmic reticulum.

mitochondria-associated membrane (MAM), a specialized subdomain of the ER, and shuttles between the rough ER (i.e., perinuclear and nuclear membrane ER) and the MAM in its palmitoylation-dependent manner (Lynes et al., 2012). These reports suggest that palmitoylation controls calnexin localization within the complex ER network for the proper glycoprotein folding. In addition, a recent study suggests that inositol 1,4,5-trisphosphate receptor (IP3R), a calcium channel leading to calcium release from the ER, is a possible palmitoyl-substrate for ZDHHC6 (Fredericks et al., 2014). Related to IP3R, the ryanodine receptor/Ca^{2+}-release channel 1 (RyR1), which is localized in the sarcoplasmic reticulum of skeletal striated muscle, is also palmitoylated (Chaube et al., 2014). Inhibition of RyR1, palmitoylation significantly reduces the electrically evoked calcium release through RyR1.

Mitochondria: Taking advantage of bioorthogonal labeling with azido-palmitoyl-CoA, 21 mitochondrial proteins including 3-hydroxy-3-methylglutaryl-CoA synthase were identified as putative palmitoyl-proteins from rat liver mitochondria preparation (Kostiuk et al., 2008). In addition, phospholipid scramblase 3 (PLSCR3) (Merrick et al., 2011) and BCL-2-associated X (BAX) (Frohlich, Dejanovic, Kashkar, Schwarz, & Nussberger, 2014) involved in the regulation of apoptosis were reported as palmitoyl-mitochondrial proteins. Palmitoylation-deficient mutations in PLSCR3 cause its mislocalization from the mitochondrion to the nucleus and reduce macrophage apoptosis upon the etoposide treatment. Likewise, loss of palmitoylation of BAX reduces its mitochondrial targeting and apoptosis. Thus, palmitoylation of mitochondrial proteins emerges as a novel regulatory mechanism for apoptosis.

Lipid droplet (LD): LD is an organelle found in many eukaryotic cells that stores neutral lipids such as triacylglycerols and regulates intracellular lipid metabolism. LD is composed of a lipid ester core and a surface phospholipid monolayer. Recent studies have revealed many new proteins associated with LDs. However, their targeting mechanism to LDs remains incompletely understood. Very recently, Fujimoto's group the first found that an LD-localized noncanonical Arf-GTPase-activating protein, ELMOD2, is modified by palmitoylation (Suzuki et al., 2015). Importantly, ELMOD2 is necessary to recruit adipocyte triglyceride lipase (ATGL), a major LD-localized enzyme to hydrolyze triglyceride, to LDs, probably by modulating Arf1-COP1 activity. Palmitoylation-deficient ELMOD2 fails to be localized at LDs and fails to reconstitute the LD-targeting of ATGL after the ELMOD2 knockdown. Further analysis will reveal additional palmitoyl-proteins involved in lipid metabolism in LDs.

Endosomes: Endocytic processes involve dynamic transformation/maturation of endosomes: the early endosome, the recycling endosome, and the late endosome. Some palmitoylated proteins are enriched on the specific endosomes. Examples include neurochondrin/norbin in

Rab5-positive early endosomes (Oku, Takahashi, Fukata, & Fukata, 2013) and GRIP1 (Thomas, Hayashi, Chiu, Chen, & Huganir, 2012) and A-kinase anchoring protein 79/150 (AKAP79/150) (Woolfrey, Sanderson, & Dell'Acqua, 2015) in Rab11-positive recycling endosomes. Yeast vacuoles are the lysosomal-like organelles to degrade endocytic components and undergo dynamic fragmentation and fusion cycles. Vac8 is localized at vacuole membranes in its myristoylation and palmitoylation-dependent manner. Palmitoylation of Vac8 plays an essential role in homotypic vacuole fusion (Veit, Laage, Dietrich, Wang, & Ungermann, 2001; Wang, Kauffman, Duex, & Weisman, 2001).

Synaptic vesicle (SV): SVs are small neurotransmitter-filled secretory vesicles observed at the axon terminal of neurons. Each SV contains about 50 different integral membrane proteins and several soluble proteins at its surface (Takamori et al., 2006). Importantly, many SV proteins such as v-SNAREs (synaptobrevin/VAMP and synaptotagmin), cysteine-string protein (CSP), glutamic acid decarboxylase (GAD65), and SCAMP1 are modified by palmitoylation. Interestingly, t-SNARE syntaxin and SNAP25, which are localized at the target nerve-terminal membrane (i.e., presynaptic active zone), are also palmitoyl-proteins (Prescott, Gorleku, Greaves, & Chamberlain, 2009). It is tempting to know how much palmitoylation contributes to the very specific segregation of these proteins into SVs and active zones in neurons (see Section 5.2).

Cilium and Flagellum: Cilia are the hairlike extension of cell surfaces, serving as sensory organelles for chemo-, photo-, and mechanotransduction. Arl13b/ARL-13, the small GTPase required for cilium biogenesis, is enriched in the restricted region of a proximal ciliary membrane via its palmitoylation in *Caenorhabditis elegans* (Cevik et al., 2010). Flagellum is a longer version of cilium and palmitoylation of the flagellar calcium-binding protein of *Trypanosoma cruzi* is necessary for its flagellar membrane targeting (Maric et al., 2011).

In addition to these various membrane-bound organelles, palmitoylated proteins are often found in specialized membrane domains: caveolin in the caveolae, a flask-shaped small invagination at the cell surface to form endocytic vesicles (Dietzen, Hastings, & Lublin, 1995); ankyrin-G (He, Jenkins, & Bennett, 2012) and neurofascin (Ren & Bennett, 1998) at the axon initial segment of neurons; claudin at the tight junction (Van Itallie, Gambling, Carson, & Anderson, 2005); integrin (Yang et al., 2004) and zyxin (Oku et al., 2013) at the focal adhesion; and PSD-95 (Topinka & Bredt, 1998), AMPA receptor (Hayashi, Rumbaugh, & Huganir, 2005), NMDA receptor (Hayashi, Thomas, & Huganir, 2009), and gephyrin (Dejanovic et al., 2014) at the postsynaptic membrane.

The list of the palmitoylated proteins localized to various organelles and membrane domains is expected to grow further. Furthermore, a recently developed method for in situ imaging of the palmitoylome with clickable probes and proximity ligation will provide us with valuable information about substrate localization in a single cell (Gao & Hannoush, 2014a, 2014b). These new findings raise additional questions. How is the final destination of individual palmitoyl-proteins determined especially in highly polarized cells such as neurons? Where and when are substrate proteins palmitoylated? We will discuss these issues later using PSD-95 as a model substrate (see Section 5.1).

3. ROLES OF PROTEIN PALMITOYLATION AND REVERSIBLE PALMITOYL CYCLES

3.1 Protein trafficking and partitioning into specialized membrane compartments

In most cases of peripheral membrane proteins (originally soluble proteins), such as PSD-95 (Topinka & Bredt, 1998) and Gαq (Linder et al., 1993; Wedegaertner, Chu, Wilson, Levis, & Bourne, 1993), palmitoylation plays an essential role in their trafficking and anchoring to the plasma membrane by increasing the hydrophobicity of proteins. For example, PSD-95 is specifically enriched just beneath the plasma membrane of postsynapses in neurons, and palmitoylation-deficient mutant of PSD-95 is completely diffused into the cytoplasm (Topinka & Bredt, 1998). In contrast, most of transmembrane proteins are trafficked to the plasma membrane irrespective of their palmitoylation. Although there are controversies about the effects of palmitoylation on integral transmembrane proteins (also see Section 3.3), palmitoylation could control protein partitioning into the specialized domains in the membranes. In support of this idea, partitioning of the transmembrane linker for activation of T cells (LAT) to the liquid-ordered fluid phase domain (there referred to as "raft phase") in the unique giant plasma membrane vesicles (GPMVs) is dependent on its palmitoylation (Levental, Lingwood, Grzybek, Coskun, & Simons, 2010). Palmitoylation may play a key role in the segregation or clustering of proteins at distinct membrane domains for the efficient signal transduction. Such typical concentrating platforms are found in neuronal synapses and immunological synapses. Examples for palmitoyl-proteins at the excitatory synapses are presynaptic synaptotagmin, synaptobrevin, syntaxin, and SNAP25; and postsynaptic AMPA receptor, NMDA receptor, and PSD-95. These integral and peripheral membrane proteins have to cluster together in the same compartments to efficiently exert their physiological functions (i.e., synaptic transmission). Therefore, palmitoylation could function as the common tag to assemble the proper compartments. Indeed, palmitoylated PSD-95 is not laterally diffused along the dendritic plasma membrane of neurons, but is highly restricted into the specific compartment of the postsynaptic membrane, so-called postsynaptic densities (PSDs) (Fukata et al., 2013). By analogy, the GPCR signaling pathway may take advantage of this modification for efficient and specific signal transduction. In fact, many components of GPCR signaling pathway including GPCRs (Tobin & Wheatley, 2004), trimeric Gα subunits (Linder et al., 1993; Wedegaertner et al., 1993), regulator of G-protein signal (RGS) and its binding protein, R7BP (De Vries, Elenko, Hubler, Jones, & Farquhar, 1996; Druey et al., 1999; Rose et al., 2000; Srinivasa, Bernstein, Blumer, & Linder, 1998; Tu, Popov, Slaughter, & Ross, 1999), phosphodiesterase (Charych, Jiang, Lo, Sullivan, & Brandon, 2010), and a Gαq-regulated RhoGEF (p63RhoGEF) (Aittaleb, Nishimura, Linder, & Tesmer, 2011) are modified by palmitoylation.

Multipass transmembrane proteins are often modified by palmitoylation at more than one domain (i.e., the juxta- or intratransmembrane, intracellular loop, and cytoplasmic tail). Recently, it has become evident that the multisite palmitoylation in the ion channel protein controls distinct channel functions (channel kinetics and modulation by other types of modification as well as subcellular trafficking) (Jeffries, Tian, McClafferty, & Shipston, 2012; Lin et al., 2009), as previously reviewed in detail (Shipston, 2014; Thomas & Huganir, 2013). Palmitoylation should control more aspects of ion channel physiology than previously expected.

3.2 Palmitoylation—depalmitoylation cycles and subcellular protein shuttling

The unique reversibility of palmitoylation modification allows proteins to dynamically shuttle between intracellular compartments (Figure 2(A)). For example, H/N-Ras are localized at the plasma membrane as well as at the Golgi apparatus in mammalian cells. H/N-Ras constitutively shuttle between the plasma membrane and the Golgi driven by their own constitutive palmitoylation—depalmitoylation cycles (Rocks, Peyker, & Bastiaens, 2006; Rocks et al., 2005). Fluorescence recovery after photobleaching (FRAP) analysis showed that YFP-H-Ras fluorescence in the Golgi rapidly

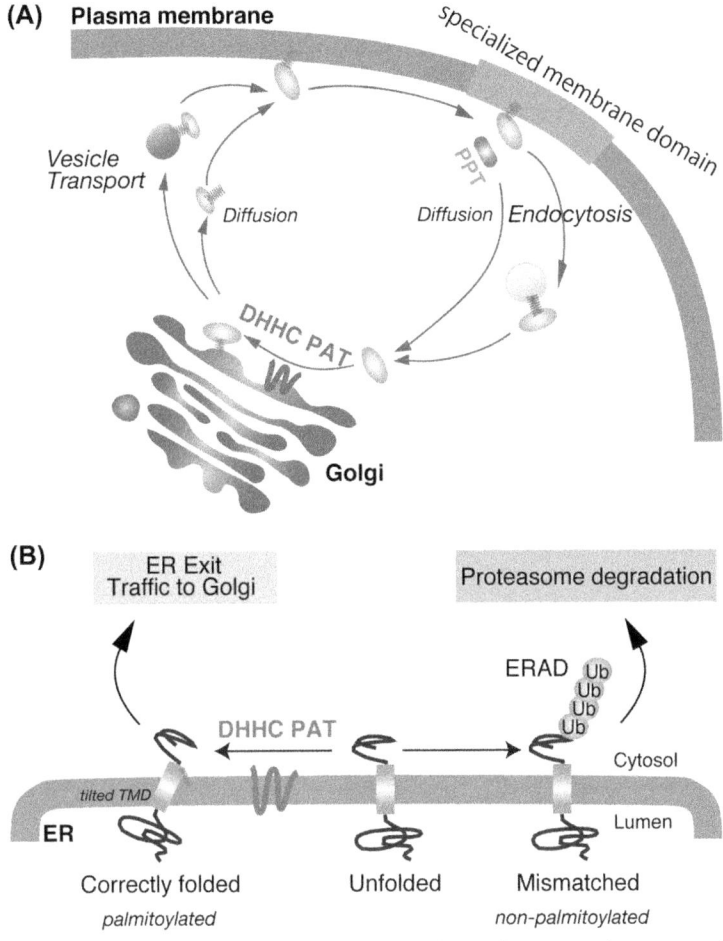

Figure 2 Roles of protein palmitoylation. (A) Palmitoylation mediates protein trafficking to the plasma membrane and protein partitioning into specialized membrane domains (e.g., detergent-insoluble lipid-enriched domains). In addition, palmitoylation—depalmitoylation cycles allow several proteins to shuttle between the intermembrane compartments. The shuttling route could be determined by loci of the responsible DHHC PAT and PPT (unidentified depalmitoylating enzyme). (B) Palmitoylation of integral membrane proteins regulates their conformation and stability. In this case, nonpalmitoylated proteins are recognized as misfolded proteins by the ER quality control system, do not exit ER, and finally enter the ERAD pathway to be degraded. DHHC, aspartate-histidine-histidine-cysteine; ER, endoplasmic reticulum; PAT, palmitoyl acyltransferase; PPT, palmitoyl-protein thioesterase; ERAD, ER-associated degradation.

recovers with a half-life of 6 min. Photoactivation of H-Ras at the plasma membrane reveals the retrograde plasma membrane-Golgi trafficking. Importantly, treatment of cells with an inhibitor of protein palmitoylation, 2-bromopalmitate, blocks this shuttling. Also, treatment of cells by an inhibitor of a putative depalmitoylating enzyme, acyl-protein thioesterase 1 (APT1), causes redistribution of H-Ras over endomembranes (Dekker et al., 2010). These studies propose a concept that the continuous palmitoyl-cycling on Ras protein prevents Ras from nonspecific residence on other compartments, thereby maintaining the specific plasma membrane-Golgi compartmentalization (Rocks et al., 2010). Gαs is also a representative palmitoyl-protein that undergoes rapid palmitoyl cycles and importantly the agonist stimulation of GPCRs increases Gαs depalmitoylation inducing shifts of Gαs to the cytosol (Wedegaertner & Bourne, 1994). Gαq (Tsutsumi et al., 2009) and Gαo (Chisari, Saini, Kalyanaraman, & Gautam, 2007) also shuttle between the plasma membrane and the Golgi by palmitoylation—depalmitoylation cycles (Fukata & Fukata, 2010). Another example is reported for R7BP. R7BP complexed with RGS7-Gβ5 dynamically moves between the plasma membrane and the nucleus in an R7BP palmitoylation-dependent manner (Drenan et al., 2005) and inhibition of palmitate turnover on R7BP redistributes R7BP-RGS complexes from the plasma membrane to endomembranes (Jia et al., 2014). In cultured neurons, palmitate on PSD-95 turns over with a half-life of about 2 h and this process is controlled in a synaptic-activity-dependent manner (El-Husseini Ael et al., 2002). By visualizing the palmitoylated state of endogenous PSD-95 at the postsynapse we found that PSD-95 undergoes the spatially restricted palmitoyl cycles and thereby PSD-95 shuttles between the postsynaptic membrane and the cytoplasm in a single dendritic spine (a small dendritic protrusion harboring the PSD) (Fukata et al., 2013) (see Section 5.1). GRIP1 is another neuronal palmitoyl-protein with rapid palmitoyl cycles, which may regulate its endosomal targeting from the cytoplasm (Thomas et al., 2012). In contrast, a recent global pulse-chase proteomic study suggests that the fast palmitate cycling might be exceptional under the basal conditions (<25% tested proteins showing palmitoyl turnover), but instead the context-dependent dynamic palmitate cycling as shown for Gα and PSD-95 may occur on other palmitoyl-proteins (Martin, Wang, Adibekian, Tully, & Cravatt, 2012). Because (1) most of evidence for the palmitoyl-cycling or protein-shuttling is obtained from overexpressed palmitoyl-proteins, (2) the real depalmitoylating enzyme is still unknown, and (3) it remains elusive where in the cell palmitoylation and depalmitoylation of individual proteins actually occur, it will take a little more to fully understand the roles and regulation of palmitoyl cycles in cells.

3.3 Protein conformation and stability

Accumulating evidence indicates another function of palmitoylation for the transmembrane proteins. Palmitoylation of integral membrane proteins is often required for their proper conformation, protects them from the ER-associated degradation (ERAD) pathway, and contributes to their stability and subcellular trafficking to the plasma membrane (Blaskovic, Blanc, & van der Goot, 2013; Linder & Deschenes, 2007) (Figure 2(B)). The essential role of palmitoylation in protein stability is demonstrated by the shortened half-life and reduced cell-surface expression of palmitoylation-deficient transmembrane proteins. Examples include: A_1 adenosine receptor (Gao, Ni, Szabo, & Linden, 1999), chemokine receptor CCR5 (Blanpain et al., 2001; Percherancier et al., 2001), the envelope glycoprotein of Rous sarcoma virus (Ochsenbauer-Jambor, Miller, Roberts, Rhee, & Hunter, 2001), LAT (Tanimura, Saitoh, Kawano, Kosugi, & Miyake, 2006), LRP6 (Abrami, Kunz, Iacovache, & van der Goot, 2008), dopamine transporter (Foster & Vaughan, 2011), APP (Bhattacharyya, Barren, & Kovacs, 2013), and IFITM1 (Abrami et al., 2008).

How does palmitoylation affect the protein conformation? Importantly, palmitoylation of transmembrane proteins often occurs on cysteines very adjacent to and even in the transmembrane domain (TMD) (Fukata & Fukata, 2010). Biophysical analysis with liposomes proposed that palmitoylation in the TMD may tilt the TMD to the bilayer surface from its normal perpendicular orientation (Joseph & Nagaraj, 1995). van der Goot and coworkers looked into this proposal with lipoprotein-receptor-related protein 6 (LRP6). LRP6 has a long 23-residue TMD (i.e., two residues longer than that of most single-pass transmembrane proteins) and is retained in the ER in the absence of palmitoylation at two juxtamembrane cysteines (Abrami et al., 2008). They hypothesized that a long TMD causes a hydrophobic mismatch in the ER membrane and subsequent recognition by the ER quality control machinery and that palmitoylation may change the TMD conformation (Figure 2(B)). Consistently, in the absence of palmitoylation shortening TMD of LRP6 relieves its ER retention but in the presence of palmitoylation shortening does not. Thus, it is speculated that palmitoylation plays a general role in fine-tuning the protein conformation. What mediates/regulates precise and timely palmitoylation of transmembrane proteins at the ER during their biosynthesis is another important issue to be solved.

Interestingly, stability and function of several viral and host transmembrane proteins are known to be regulated by palmitoylation (Veit, 2012). The Rous sarcoma virus transmembrane glycoprotein is palmitoylated within the TMD. The palmitoylation-deficient envelope glycoprotein shows the reduced steady-state surface expression and increased internalization, and viruses containing this palmitoylation-deficient glycoprotein appear to be poorly infectious or noninfectious (Ochsenbauer-Jambor et al., 2001). Given that this envelope glycoprotein is not partitioned into the detergent-resistant membrane domains, they proposed that palmitoylation could affect its TMD topology in the plasma membrane and help its membrane anchoring. Regarding the host protein involved in the viral infection pathway, palmitoylation regulates degradation and stability of the HIV receptor CCR5 and inhibition of its palmitoylation may render cells resistant against HIV-1 infection (Blanpain et al., 2001; Percherancier et al., 2001). Thus, pharmacological modifiers of their palmitoylation status might become a target for new antiviral drugs.

Levental's group recently reported that palmitoylation at the juxtamembrane cysteines and TMD length of LAT independently determine its specific partitioning within the GPMVs and control its destinations, plasma membrane targeting (with palmitoylation or long TMD), or lysosomal degradation (without palmitoylation or with short TMD) (Diaz-Rohrer, Levental, Simons, & Levental, 2014). Growing number of studies will reveal how palmitoylation, partitioning/segregation within the membrane, and conformation/stability of transmembrane proteins are related to each other.

4. THE DHHC/ZDHHC PROTEIN FAMILY OF PALMITOYLATING ENZYMES

Elegant genetic screens in yeast identified Erf2/Erf4 (Bartels, Mitchell, Dong, & Deschenes, 1999; Lobo, Greentree, Linder, & Deschenes, 2002) and Akr1 (Roth, Feng, Chen, & Davis, 2002) as the first PATs for Ras2 and Yck2, respectively. Erf2 and Akr1 have four- or six-pass TMDs and share a common domain referred to as a ZDHHC-CRD (cysteine rich domain), a ~50 aa sequence with a conserved DHHC signature motif. Subsequent studies have established that the evolutionarily conserved DHHC-CRD-containing protein (DHHC/ZDHHC protein) family members mediate protein S-palmitoylation. In addition to yeast, the large DHHC/ZDHHC protein family has been found in mammalian genomes (Fukata, Fukata, Adesnik, Nicoll, & Bredt, 2004; Korycka et al., 2012; Ohno, Kihara, Sano, & Igarashi, 2006), nematode (Edmonds & Morgan, 2014), fruit fly (Bannan et al., 2008), plant (Batistic, 2012; Hemsley & Grierson, 2008; Hemsley, Kemp, & Grierson, 2005; Yuan et al., 2013), and protozoa (Emmer et al., 2009; Frenal, Kemp, & Soldati-Favre, 2014) (Table 1).

4.1 Enzymology and substrate specificity of DHHC proteins

The mutation analysis and topology study of yeast DHHC proteins (Bartels et al., 1999; Lobo et al., 2002) showed that the DHHC-CRD facing the cytoplasm is responsible for the enzymatic activity of DHHC PATs. Although autopalmitoylation of DHHC proteins had been detected and its importance was suggested as a palmitoyl-DHHC intermediate (Fukata et al., 2004; Roth et al., 2002; Swarthout et al., 2005), the enzymatic mechanism for DHHC PATs had remained poorly characterized due to the difficulty in in vitro reconstitution assay. Mitchell and Deschenes recently demonstrated a two-step reaction mechanism for protein palmitoylation, which is composed of (1) autopalmitoylation of the DHHC motif sequence (of Erf2) to produce a palmitoyl-DHHC intermediate and (2) the transfer of the palmitoyl moiety to the substrate protein (Ras2) (Mitchell, Mitchell, Ling, Budde, & Deschenes, 2010) (Figure 3(A)). Isolation of Erf2 yeast mutants allowed to dissect the two steps: mutating Cys-203 of the core DHHC motif of Erf2 inhibits its initial autopalmitoylation, whereas mutating His-201 does not affect the autopalmitoylation of Erf2 but reduces the rate of palmitoyl-Erf2 thioester hydrolysis (release of palmitate from Erf2), causing uncoupling of Erf2 autopalmitoylation from Ras2 palmitoylation. Linder's group directly verified the two-step reaction. They mixed autopalmitoylated ZDHHC2 or ZDHHC3 labeled with [3H]-palmitate together with the substrates in the absence of [3H]-palmitoyl-CoA and found the coupled events, the loss of radiolabel from DHHC PATs, and the gain of radiolabel on the substrate. Thus, DHHC protein-mediated S-palmitoylation uses the ping-pong kinetic mechanism (Jennings & Linder, 2012).

Previous biochemical studies by the overexpressed multiple sets of DHHC proteins indicate that DHHC proteins show distinct but overlapping substrate specificities (Fukata et al., 2004; Huang et al., 2009). Genetic evidence in yeast also indicates overlapping DHHC PAT functionalities. Using the genetic deletion of DHHC proteins in S. cerevisiae, essential roles of individual DHHC PATs in palmitoylation of specific substrate proteins have been demonstrated: Erf2 DHHC protein toward Ras2 (Lobo et al., 2002),

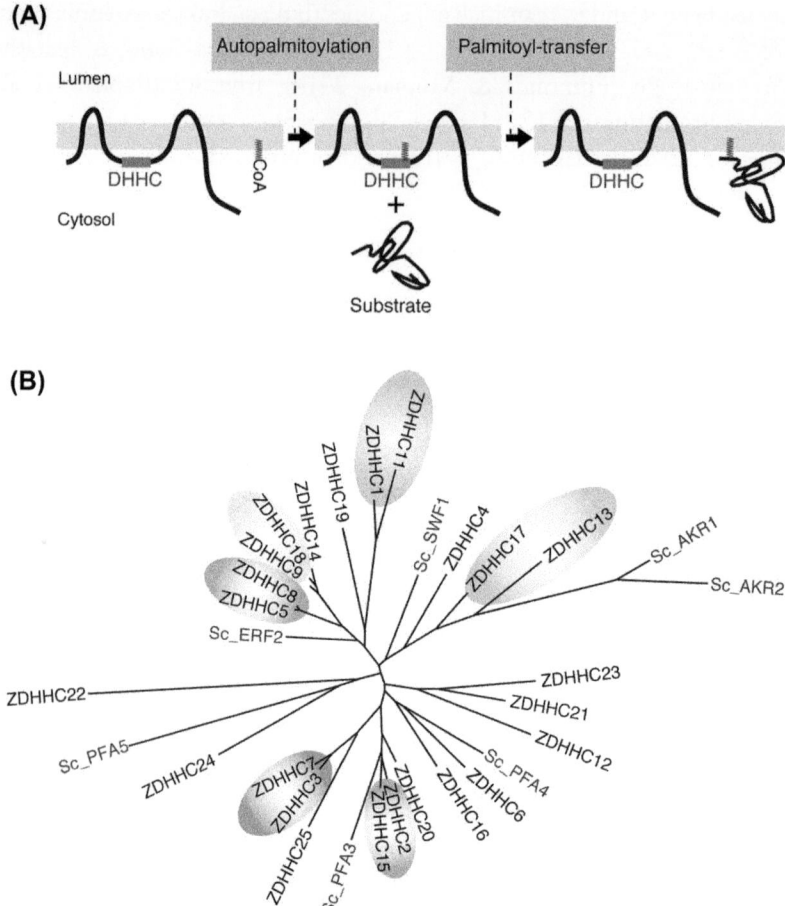

Figure 3 DHHC/ZDHHC palmitoylating enzyme family. (A) A two-step reaction mechanism for DHHC protein-mediated protein palmitoylation. DHHC protein has four (or six) transmembrane domains and a conserved CRD containing a DHHC motif in the cytoplasmic loop (left). The DHHC proteins palmitoylate themselves producing a palmitoyl enzyme intermediate (middle), and then transfer the palmitate to the substrate proteins. (B) Phylogenetic tree of the DHHC protein family in *Mus musculus* (black) and *Saccharomyces cerevisiae* (green (gray in print versions)). The tree is based on alignment of the DHHC-CRD core domains. The 24 family members of mouse DHHC proteins are classified into several subfamilies based on their sequence similarity and known substrate preference (colored ovals). DHHC, aspartate-histidine-histidine-cysteine; ZDHHC, zinc-finger DHCC; CRD, cysteine-rich domain.

Swf1 toward Tlg1 (Valdez-Taubas & Pelham, 2005), Akr1 toward Yck1/2 (Roth et al., 2002), Pfa4 toward Chs3 (Lam et al., 2006), and Pfa3 toward Vac8 (Smotrys, Schoenfish, Stutz, & Linder, 2005). However, residual level of palmitoylation was still detected upon a single DHHC gene depletion. Taking advantage of palmitoylome analysis with yeast strains in which the DHHC genes are deleted individually or in combination (comparative palmitoylome analysis), Davis's group systematically examined the effects of individual DHHC protein deletion on the substrate palmitoylation level (Roth et al., 2002). They found that palmitoylation of some substrates completely depends on a single-specific DHHC protein (e.g., Akr1 for Yck1/2 palmitoylation). In contrast, the palmitoylation level of most palmitoyl-proteins of *S. cerevisiae* yeast including Ras2 and Vac8 are only partially reduced when one DHHC protein is deleted and their palmitoylation levels are lost only when multiple DHHC proteins are deleted. In addition, rescue experiments of Erf2, Akr1, Pfa4, and Pfa5 enzymes in the Erf2 deletion background demonstrated that each restores the plasma membrane localization of Ras2 (Hou, John Peter, Meiringer, Subramanian, & Ungermann, 2009). A very recent study in pathogenic yeast *Cryptococcus neoformans* reported that deletion of Pfa4 DHHC protein causes robust defects in cryptococcal cell morphology and virulence and reduces palmitoylation levels of 21 proteins (Santiago-Tirado, Peng, Yang, Hang, & Doering, 2015). This substrate list includes the proteins known to be palmitoylated by other DHHC enzymes in *S. cerevisiae* yeast (i.e., Vac8, Yck1/2, and others). Given the strong phenotype of Pfa4 deletion, however, there should be one or more critical substrates palmitoylated only by cryptococcal Pfa4. Thus, some substrates are palmitoylated specifically by a single DHHC enzyme in (yeast) cells, but most of substrates are shared by multiple DHHC proteins, causing the functional redundancy in the DHHC protein family. To precisely determine the substrate specificity and palmitoyl-consensus sequence for individual DHHC PATs, we still need more examples of DHHC enzyme—substrate relationships and more information about the corresponding palmitoylation sites. In addition, simple, sensitive, and specific in vitro PAT assay with the purified DHHC protein is absolutely required.

4.2 Expanded DHHC enzyme—substrate pairs

The larger size of the mammalian DHHC protein family (Figure 3(B)) with several subfamilies shows more complicated functional redundancies. Therefore, careful approaches are necessary to know the authentic enzyme—substrate pairs. Currently, the following systematic/integrative approaches are often used (Fukata, Iwanaga, & Fukata, 2006): (1) screening of the mammalian DHHC protein library to find DHHC PAT clones that quantitatively enhance the palmitoylation level of the coexpressed substrate by metabolic labeling with [^{3}H]-palmitate or ABE methods (Fukata et al., 2004, 2006), (2) expression analysis of candidate DHHC proteins and the substrate by real-time PCR or in situ hybridization, and (3) loss-of-function analysis using siRNAs or knockout (KO) cells/animals for specific DHHC candidates. Instead of the overexpression screening with the DHHC protein library, siRNA libraries of DHHC proteins have been successfully used for some substrates (Lakkaraju et al., 2012; Tian, McClafferty, Jeffries, & Shipston, 2010).

Such approaches have identified many enzyme–substrate pairs in mammals (Fukata & Fukata, 2010; Greaves & Chamberlain, 2011; Ohno et al., 2012). For example, Gαq is palmitoylated by ZDHHC3 and ZDHHC7 in HeLa cells, which belong to the same subfamily and show broader substrate specificity than other DHHC proteins (Tsutsumi et al., 2009). PSD-95 is quantitatively palmitoylated at least by two different subfamily members of DHHC proteins, ZDHHC3/7 and ZDHHC2/15 subfamilies when they are overexpressed in HEK293T cells (Fukata et al., 2004). In the hippocampal neurons, ZDHHC3 and ZDHHC2 are predominantly expressed over ZDHH7 and ZDHHC15, and PSD-95 palmitoylation is differently regulated by Golgi-resident ZDHHC3 and dendritic mobile ZDHHC2 (Noritake et al., 2009). In addition, some other efforts such as (1) bioinformatic prediction of palmitoyl proteins (i.e., CSS-Palm prediction program (Ren et al., 2008)), (2) detection of physical interactions with DHHC proteins, and (3) assessment of subcellular colocalization with DHHC proteins have been combined with the above approaches and contributed to finding new palmitoyl proteins and their responsible DHHC enzymes (Frenal et al., 2013; George, Soares, Montersino, Beique, & Thomas, 2015; Howie et al., 2014; Keller et al., 2004; Oku et al., 2013; Thomas et al., 2012; Thomas, Hayashi, Huganir, & Linden, 2013).

As an alternative approach, the comparative palmitoylome analysis has been used to identify the critical substrates for an individual DHHC protein of interest in various species as done in yeast (Roth et al., 2006). In this approach, the palmitoylome of cells/animals with a certain DHHC gene manipulated (KO or knockdown) is quantitatively compared with that of wild-type ones. Representative examples include reports for mammalian cultured cells with a specific DHHC gene knocked-down (Zhang et al., 2008) and for the specific DHHC gene-disrupted plant (Hemsley et al.,

2013) and mice (Li, Martin, Cravatt, & Hofmann, 2012; Wan et al., 2013) (see Section 4.4). Identified enzyme–substrate pairs include: CKAP4 for ZDHHC2 (Zhang et al., 2008), PECAM for ZDHHC21 (Marin et al., 2012), flotillin-2 for ZDHHC5 (Li et al., 2012), and 11 proteins including flotillin-1/2 for ZDHHC17 (Wan et al., 2013). Then, it appears to be hard to identify substrates for DHHC enzymes by proteomics approaches in mammalian samples, again due to the functional redundancy.

The identification and characterization of DHHC enzyme–substrate pairs have highlighted the physiological roles of protein palmitoylation. For example, knockdown of ZDHHC2 and ZDHHC3 accelerates the degradation of tetraspanins CD9/CD151 (Sharma, Yang, & Hemler, 2008) and integrin α6β4 (Sharma, Rabinovitz, & Hemler, 2012), respectively. Loss of yeast Swf1 and Pfa4 causes mislocalization and instability of palmitoyl-transmembrane proteins: v-SNARE Tlg1 (Valdez-Taubas & Pelham, 2005) and chitin synthase Chs3 (Lam et al., 2006), respectively. These results support the idea that palmitoylation of integral membrane proteins protects them from the ERAD pathway and promotes their efficient ER exit, contributing protein stability (see Section 3.3).

4.3 Subcellular localization of DHHC proteins

The exact subcellular location where protein palmitoylation occurs had been controversial before a large family of DHHC palmitoylating enzymes was discovered. In 2006, Igarashi's group systematically characterized the subcellular localization of tagged DHHC proteins in S. cerevisiae and HEK293T cells (Linder & Deschenes, 2007; Ohno et al., 2006). In S. cerevisiae, seven DHHC proteins are localized differently: Erf2, Swf1, and Pfa4 at the ER; Akr1 and Akr2 at the Golgi apparatus; Pfa3 at the vacuole; and Pfa5 at the plasma membrane. In HEK293T cells, they found that most of the DHHC proteins are localized in the ER and/or Golgi compartments and a few DHHC proteins are localized at the plasma membrane. Importantly, the cellular locus of DHHC proteins may depend on the cellular contexts. For example, ZDHHC8 is localized at the plasma membrane in MDCK epithelial cells (He, Abdi, & Bennett, 2014) although ZDHHC8 is distributed at the Golgi in HEK293T cells. Furthermore, our group found that two PSD-95 palmitoylating enzymes, ZDHHC3 and ZDHHC2, are differently distributed and regulated in differentiated neurons (Fukata & Fukata, 2010; Noritake et al., 2009). ZDHHC3 is localized essentially only at the Golgi apparatus in the cell body and may mediate initial palmitoylation of various newly synthesized substrates, including

PSD-95 and Gαq (Tsutsumi et al., 2009). In contrast, ZDHHC2 is localized on the Rab11-positive recycling endosomes in dendrites far from the cell body and even inserted at dendritic spine membranes (Fukata et al., 2013; Greaves, Carmichael, & Chamberlain, 2011). Importantly, a dendritically localized ZDHHC2, but not a Golgi-resident ZDHHC3, mediates synaptic activity-sensitive palmitoylation of PSD-95. When synaptic activity is inhibited by treatment with tetrodotoxin, ZDHHC2 on the recycling endosomes translocates to the postsynapse to palmitoylate PSD-95 (Noritake et al., 2009) (see Section 5.1). Thus, palmitoylation occurs not only at the ER and Golgi apparatus that are spatially close to the protein synthesis machinery but also at the plasma membrane. The subcellular localization of DHHC proteins and its dynamic relocation provide another regulatory mechanism for protein palmitoylation, contributing to diverse cellular events that should occur at the right location and at the right moment.

4.4 Pathophysiological significance of DHHC proteins

Irrespective of the molecular redundancy, accumulating genetic evidence in mice and humans has begun to unveil the pathophysiological significance of DHHC proteins in vivo. So far, KO mice of ZDHHC5, ZDHHC8, ZDHHC13/HIP14L, and ZDHHC17/HIP14 have been characterized (Table 2). In addition, potential links between mutations of DHHC proteins and human disorders have been reported.

ZDHHC5: Mice homozygous for the ZDHHC5 hypomorphic allele (<7% protein expression compared with wild-type mouse) are born at half the expected rate and display a significant deficit in contextual fear conditioning and hippocampus-dependent learning (Li et al., 2010). Palmitoyl-proteomics with this KO mouse revealed that flotillin-2 is a substrate for ZDHHC5 (Li et al., 2012). Flotillin is a marker protein of the detergent-insoluble lipid-enriched membrane domains and its plasma membrane association is palmitoylation-dependent (Morrow et al., 2002). Flotillin-1 is shown to be essential for dopamine- and glutamate-transporter internalization (Cremona et al., 2011). A next step will be to examine how ZDHHC5-mediated palmitoylation of flotillin-2 is involved in the learning phenotype. Given the previously known substrates of ZDHHC5, GRIP1 (Thomas et al., 2012) and δ-catenin (Brigidi et al., 2014) may be responsible for the learning phenotype of the ZDHHC5 KO mouse as GRIP1 is genetically associated with neuropsychiatric conditions and autism (Mejias et al., 2011) and δ-catenin KO mice have learning deficits in contextual fear

Table 2 Genetic studies of DHHC proteins in mammals

ZDHHC gene	Gene manipulation	Mouse		References
		Phenotype and altered palmitoylation	Related human disease	
ZDHHC2	—	—	Colorectal cancer, hepatocellular carcinoma, nonsmall cell lung cancer (mapped to 8p21.3–22 microdeletion, with reduced expression; somatic mutations)	Oyama et al. (2000)
ZDHHC5	Gene-targeting (gt, hypomorphic with <7% in the brain)	Embryonic lethal (50% the expected ratio at birth), deficit in contextual fear conditioning and defective hippocampal–dependent learning; palmitoylation of flotillin-2 is abolished in neuronal stem cells	—	Li et al. (2010, 2012)
ZDHHC8	Gene-targeting (knockout (KO))	Sexual dimorphic defects (female-specific) in prepulse inhibition and exploratory behavior; defective spatial working memory (heterozygous KO); reduction in mushroom spines, glutamatergic synapses, and dendritic complexity; altered axonal growth and brain connectivity; most of postsynaptic and axonal defects are similar to those in the mouse model of 22q11 microdeletion;	Schizophrenia (mapped to 22q11 microdeletion; SNP variants)	Mukai et al. (2008, 2004, 2015)

(Continued)

Table 2 Genetic studies of DHHC proteins in mammals—cont'd

ZDHHC gene	Gene manipulation	Mouse Phenotype and altered palmitoylation	Related human disease	References
ZDHHC9	—	reduced palmitoylation of PSD-95, GRIP1, SCG10, Cdc42, Rac1, and GAP43		
	—	X-linked intellectual disability (mapped to Xq26.1; frameshift, splice-site, and missense mutations affecting the DHHC domain; a nonsense mutation downstream of the DHHC domain)	Masurel-Paulet et al. (2014) and Raymond et al. (2007)
ZDHHC13/HIP14L	Gene-targeting (gt, completely abolished)	Reduced body weight, skin and hair abnormalities; adult-onset and progressive neurological defects; reduced brain weight, medium spiny neuron (MSN) loss in the striatum, motor impairment, motor learning deficits; reduced palmitoylation of SNAP-25	Huntington disease (the ZDHHC13-gt mouse phenotypes are more similar to those in the mouse model of Huntington disease with the mutated huntingtin (YAC128) than the ZDHHC17-gt mouse ones)	Sutton et al. (2013)
	ENU-induced mutagenesis (a nonsense mutation upstream of DHHC domain, RNA decay	Shortened life-span, reduced body weight, skin and hair abnormalities, osteoporosis, kyphosis, and systemic amyloidosis		Saleem et al. (2010) and Song et al. (2014)

ZDHHC15	— due to premature stop codon)	—	Nonsyndromic severe intellectual disability (a balanced translocation, the breakpoint at the first exon, loss of expression)	Mansouri et al. (2005)
ZDHHC17/ HIP14	Gene-targeting (gt, hypomorphic with <10% in the brain)	Early-onset defects, striatal volume loss, increased cell death and MSN loss in the striatum, motor and cognitive dysfunction; reduced excitability of MSNs, reduced spine density and excitatory synaptic transmission, and impaired LTP in the hippocampus; reduced expression and palmitoylation of flotillin-1/2; reduced palmitoylation of SNAP-25, CSP, and synaptotagmin I (<10% reduction) in the brain	Huntington disease (identified as one of huntingtin-interacting proteins; most of the ZDHHC17-gt mouse phenotypes are similar to those in the mouse model of Huntington disease with the mutated huntingtin (YAC128))	Milnerwood et al. (2013), Singaraja et al. (2011) and Wan et al. (2013)
ZDHHC21	A spontaneous mutation, depilated (dep) (a single amino acid deletion, inactivated enzyme activity)	Recessive phenotypes of hair loss with thinner and shorted hairs; hyperplasia of interfollicular epidermis and sebaceous glands and delayed differentiation of hair shafts; mislocalization of Fyn in the hair follicle	—	Mill et al. (2009)

gt, gene-trap; CSP, cysteine–string protein.

conditioning and severe impairments in cognitive function (Israely et al., 2004). Consistently with this genetic evidence, δ-catenin palmitoylation is required for activity-induced stabilization of N-cadherin in the dendritic spine heads, spine remodeling, and AMPA receptor insertion into the synaptic membrane (Brigidi et al., 2014). Ankyrin-G that is palmitoylated by ZDHHC5 and ZDHHC8 (He et al., 2014) is another candidate as a subset of ankyrin-G forms nanodomain structures in the dendritic spine and regulates AMPA receptor-mediated synaptic transmission (Smith et al., 2014). Genetic studies using knock-in mice carrying the palmitoylation site mutation in individual substrates will clarify the roles of their palmitoylation.

ZDHHC8: ZDHHC8 KO mice have reduced mushroom spine density and dendritic complexity in its enzymatic activity-dependent manner (Mukai et al., 2008). Behavioral studies modeling schizophrenia phenotypes show that ZDHHC8-KO female mice have a deficit in prepulse inhibition, a decrease in exploratory activity in a new environment, and a decreased sensitivity to the locomotor stimulatory effect of the psychomimetic drug dizocilpine (MK801) (Mukai et al., 2004). The same group subsequently reported that ZDHHC8 KO mice also show reduced axonal growth and brain connectivity in a gene-dose-dependent manner and that heterozygous ZDHHC8 KO mice have impaired working memory (Mukai et al., 2015). Importantly, ZDHHC8 rs175174 polymorphism is associated with the risk of schizophrenia and putatively produces a truncated inactive form of ZDHHC8. In addition, the ZDHHC8 gene is located in 22q11.2 region whose microdeletion is often detected in emotional problems, a spectrum of cognitive deficits, and schizophrenia. Consistently, a mouse model of 22q11 microdeletion has similar neurodevelopmental deficits observed in ZDHHC8 KO mouse, and the defects are rescued by the expression of wild-type ZDHHC8 but not by the enzymatically inactive one. Thus, ZDHHC8-mediated palmitoylation in neurons is strongly associated with schizophrenia although some reports do not support this hypothesis (Glaser et al., 2006; Saito et al., 2005). The known neuronal substrates for ZDHHC8 involve PSD-95 (Mukai et al., 2008), GRIP1 (Thomas et al., 2012), PICK1 (Thomas et al., 2013), ankyrin-G (He et al., 2014), and Cdc42 (Mukai et al., 2015). To identify the critical substrates for ZDHHC8 in the brain and clarify the pathophysiological roles of ZDHHC8 in neurological disorders, palmitoylome analysis using ZDHHC8 KO mice is awaited. Given that ZDHHC5 and ZDHHC8 belong to the same subfamily and are localized at the plasma membrane, it is speculated that ZDHHC5 and ZDHHC8 redundantly function under certain situations. It will be worthwhile to examine the palmitoylome analysis using the double KO mice of ZDHHC5 and ZDHHC8.

ZDHHC17/HIP14 and **ZDHHC13/HIP14L**: ZDHHC17/HIP14 (huntingtin-interacting protein 14) was originally reported as an interacting protein with huntingtin, which is mutated in Huntington disease (HD). ZDHHC17/HIP14 and its paralog ZDHHC13/HIP14L (HIP14-like) uniquely have an ankyrin repeat domain that mediates the interaction with huntingtin and palmitoylate huntingtin (Huang et al., 2009). Mice homozygous for the ZDHHC17 gene-trap allele (<10% protein expression compared with wild-type mouse) have been intensively characterized as this KO mouse displays neuropathological defects that are reminiscent of HD: decreased striatal volume, loss of striatal medium spiny neurons, reduced excitatory synapses, impairment of hippocampal LTP and hippocampal-dependent memory, and deficits in motor-coordination (Milnerwood et al., 2013; Singaraja et al., 2011). These neurological phenotypes are similar to those of the mouse model of HD (YAC128) with the mutated huntingtin protein (Singaraja et al., 2011). Although it has been expected that features of HD in ZDHHC17 KO mice are due to the impaired palmitoylation of huntingtin, palmitoylation of huntingtin is not affected but SNAP-25 and PSD-95 palmitoylation is reduced in the KO mouse brain (Singaraja et al., 2011). Subsequent global palmitoylome analysis using ZDHHC17 KO brains shows that most of known neuronal palmitoyl-proteins do not show substantial changes in ZDHHC17 KO versus wild-type brain (small but significant 5—8% reduction in SNAP-25, CSP, and syntaxin 1 palmitoylation) (Wan et al., 2013). Instead, this study identified more promising substrate candidates for ZDHHC17, including flotillins and a glial protein, carbonic anhydrase II. Importantly, palmitoylome patterns (i.e., palmitoylation levels of proteome) are not well correlated between ZDHHC17 KO and YAC128 mice. In contrast, genetic studies in the fly revealed an essential role of ZDHHC17/HIP14 in CSP palmitoylation and presynaptic functions (Ohyama et al., 2007; Stowers & Isacoff, 2007). Thus, the DHHC protein—substrate relationship in vivo is not simple. The spatiotemporal expression levels of DHHC proteins and the substrates within individual cells must be considered. For example, it may be worthwhile to examine the palmitoylation levels of huntingtin, SNAP-25, and CSP in the striatum (the most affected brain region) of ZDHHC17 KO mice, but not in the whole brain.

ZDHHC13 KO mice (the ZDHHC13 gene-trap allele) display more similar neuropathological features to those in the HD mouse model than

ZDHHC17 KO mice (Sutton et al., 2013) (Table 2). SNAP-25 is commonly palmitoylated by ZDHHC17 and 13 in vitro (Huang et al., 2009) and its palmitoylation level is partially decreased in both KO mouse brains (about 30% reduction) (Sutton et al., 2013; Singaraja et al., 2011). Direct comparison of palmitoylome of the two mice will reveal how much the contribution of each DHHC protein to HD pathogenesis is. In contrast, using N-ethyl-N-nitrosourea-mutagenesis another group identified mice that have a nonsense mutation in the ZDHHC13 gene and produce a truncated protein (Saleem et al., 2010; Song et al., 2014). The mutant mice display severe phenotypes: failure to thrive, alopecia, osteoporosis, systemic amyloidosis, and early death. The group confirmed these phenotypes by generating a gene-trap allele of ZDHHC13. Further studies are required to reconcile the different phenotypes of ZDHHC13 KO mice reported by two groups.

ZDHHC21: A spontaneous mutant mouse with a recessive mutation in ZDHHC21, the depilated (*dep*) mutant mouse, shows delayed hair shaft differentiation and hyperplasia of the interfollicular epidermis and sebaceous glands (Mill et al., 2009). The mutation causes a single amino acid deletion in ZDHHC21 protein and completely impairs its palmitoylating activity toward Fyn. Considering that ZDHHC21 favors ubiquitously expressed substrates with cysteine residues located near the N-terminal myristoyl glycine including Lck, Gαi2, Fyn, and eNOS (Fukata & Fukata, 2010), the limited phenotypes in the epidermal and hair follicule of *dep* mutant mice suggest that the functional redundancy in DHHC proteins may mask other physiological functions of ZDHHC21.

Other genetic studies in protein palmitoylation: ZDHHC2 was originally named as reduced expression associated with metastasis protein (REAM), and the gene is located in chromosome 8p21.3−22, where frequent loss of heterozygosity has been detected for many types of metastatic cancers such as colorectal cancer, hepatocellular carcinoma, and nonsmall cell lung cancer (Oyama et al., 2000). Recent clinical studies have supported the idea that ZDHHC2 is a tumor suppressor (Yan et al., 2013). Consistently, knockdown of ZDHHC2 in A431 epithelial cells causes dispersed, single cell morphology (like epithelial−mesenchymal transition in tumor metastasis) (Sharma et al., 2008). In this type of epithelial cells, ZDHHC2 is essential for CD9 and CD151 palmitoylation and their protein stability. Because ZDHHC2 mediates palmitoylation of various substrates including PSD-95, eNOS, SNAP25, and CKAP4 as well as CD9 and CD151 (Fukata & Fukata, 2010), additional phenotypes might be expected in the ZDHHC2 KO mouse.

Mutations in ZDHHC9 at Xq26.1 were identified as the cause of X-linked intellectual disability (XLID; previously, X-linked mental retardation) associated with a Marfanoid habitus in 4 out of 250 families examined (Raymond et al., 2007). Mutations include one frameshift and one splice-site mutations, both deleting the DHHC catalytic domain. Importantly, the rest of two missense mutations are mapped in the catalytic DHHC domain (R148W and P150S), each of which reduces the steady-state level of the ZDHHC9-palmitate intermediate (autopalmitoylation) through different mechanisms (Mitchell et al., 2014). By sequencing of all X-chromosome exons of additional family with XLID, another nonsense mutation of ZDHHC9 (R298X) is identified (Masurel-Paulet et al., 2014). Given that ZDHHC9 mediates palmitoylation of H/N-Ras and the Ras small GTPase plays an important role in synaptic plasticity (Zhu, Qin, Zhao, Van Aelst, & Malinow, 2002), reduced Ras palmitoylation may contribute to the mechanism by which ZDHHC9 mutations cause intellectual disability. Furthermore, loss of ZDHHC15 expression was reported in a patient with XLID (Mansouri et al., 2005).

Thus, emerging evidence suggests that DHHC proteins may be associated with specific human disorders. The generation of conventional and conditional KO mouse models of DHHC proteins should provide the valuable information about their pathophysiological functions. Because DHHC proteins display apparent overlapping substrate specificity, the knock-in approach targeting the palmitoylation-site of substrates is also essential to demonstrate the importance of palmitoylation. Such knock-in approaches may also unveil unexpected significant roles of palmitoylation. For example, it was recently shown that membrane-localized palmitoylated estrogen receptor α is required for specific actions of estrogen in ovarian function and vascular physiology although it had been believed that nuclear localized estrogen receptor α plays a central role in the estrogen actions (Adlanmerini et al., 2014; Pedram, Razandi, Lewis, Hammes, & Levin, 2014). A powerful genome-editing tool such as CRISPR/Cas9 system should be useful to address these issues.

5. SPECIALIZED MEMBRANE DOMAINS ORGANIZED BY LOCAL PALMITOYLATION CYCLES

Recently, our group has proposed a pivotal role of local palmitoylation machinery in specialized membrane domain (re)organization, that is, each organelle or membrane compartment may contain its own characteristic sets of DHHC PAT and unidentified PPT to recruit and keep specific substrates. In this section, we will summarize recent studies on local palmitoylation cycles in organizing specialized membrane domains (Figure 4).

5.1 Postsynaptic nanodomain organization of palmitoylated PSD-95

The PSD is a highly specialized membrane region that is small (~ 0.4 μm in average diameter) but abundant ($\sim 10,000-100,000$/neuron) in dendrites of differentiated neurons (Sheng & Hoogenraad, 2007). PSD-95 is a major constituent of the PSD and plays a central role in the PSD organization. However, it has still not been completely understood how palmitoylated PSD-95 gets restricted in the PSD. Although the PSDs have been thought to be homogenous and static architecture where PSD-95 is randomly accumulated, recent studies using super-resolution imaging methods have revealed more heterogeneous and dynamic property of the PSD organization (Fukata et al., 2013; MacGillavry, Song, Raghavachari, & Blanpied, 2013; Nair et al., 2013). We combined an intrabody against palmitoylated PSD-95 (PF11–GFP, a recombinant antibody tagged with GFP) with live-cell super-resolution imaging and discovered subsynaptic nanodomains composed of palmitoylated PSD-95 (Fukata et al., 2013) (Figure 4(A and B)). That is, an individual PSD was shown to be a collective of multiple nanodomains with a diameter of ~ 200 nm (Fukata et al., 2013). Interestingly, FRAP analysis using PF11–GFP demonstrates that PSD-95 in individual nanodomains undergoes rapid continuous cycles of depalmitoylation and repalmitoylation. Both two reactions occur within a single dendritic spine without significant supply of new PSD-95 molecule from the outside of the spine. This strongly suggests that an individual nanodomain (or at least individual dendritic spine) should contain two enzymes: palmitoylating (PAT) and depalmitoylating (PPT) enzymes for PSD-95 (Figure 4(B)). Among PSD-95 palmitoylating enzymes, ZDHHC2, but not Golgi-resident ZDHHC3, is distributed on the Rab11–positive recycling endosomes in the dendrites (Fukata et al., 2013; Greaves et al., 2011), and some pool of ZDHHC2 is directly inserted into the plasma membrane as clusters, overlapping with PSD-95 clusters (Fukata et al., 2013). Importantly, acutely induced plasma membrane-insertion of ZDHHC2 triggers specific accumulation of PSD-95 at the same plasma membrane in a palmitoylating activity-dependent manner. In addition, molecular replacement of ZDHHC2 with the ER-retained ZDHHC2 mutant reduces the number

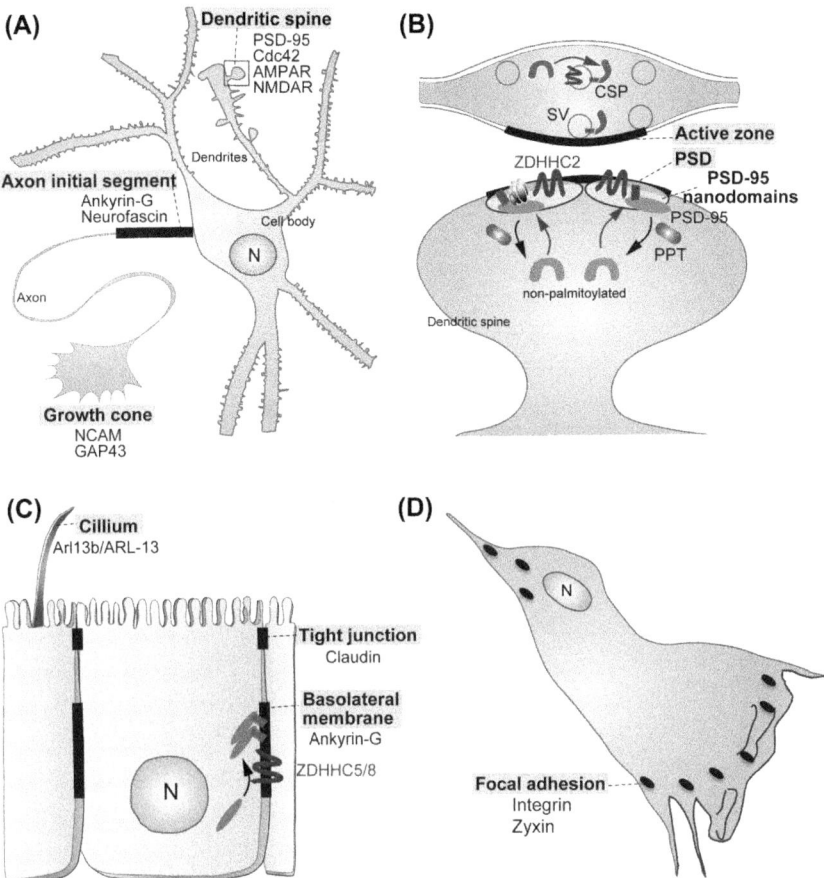

Figure 4 Organization of specialized membrane domains by local palmitoylation cycles. (A) Neurons develop highly polarized morphology with two distinct processes, an axon and dendrites. Palmitoylation is often used to target functional proteins to specialized membrane compartments such as an axon initial segment (e.g., ankyrin-G and neurofascin), growth cones (e.g., NCAM and GAP43), and dendritic spines (e.g., Cdc42 and PSD-95). (B) Local palmitoylation cycles of PSD-95 at excitatory postsynapses. Palmitoylated PSD-95 is partitioned into discrete nanodomains (two ovals) to constitute a PSD. Spine-membrane inserted ZDHHC2 and an unidentified depalmitoylating enzyme (PPT) locally generate PSD-95 palmitoylation cycles and maintain/reorganize the individual nanodomains. At the presynapse, CSP may be palmitoylated by SV-resident DHHC protein and trapped on the SVs. (C and D) Locally expressed DHHC proteins may organize membrane specializations in epithelial cells (C) and migrating fibroblasts (D). ZDHHC5/8 locally palmitoylates ankyrin-G to determine ankyrin-G distribution to the lateral membrane in the epithelial cells (C). PSD, postsynaptic scaffolding proteins; PPT, palmitoyl-protein thioesterase; CSP, cysteine-string protein; SV, synaptic vesicle; DHHC, aspartate-histidine-histidine-cysteine; ZDHHC, zinc-finger DHCC.

of PSD-95 nanodomains (Fukata et al., 2013). These results strongly suggest that ZDHHC2 would palmitoylate PSD-95 locally at the spine membrane to directly nucleate and maintain PSD-95 nanodomains. It is conceivable that such a local (intraspine) palmitoyl-cycling may prevent lateral diffusion or endocytosis of membrane-bound PSD-95 and contribute to keeping discrete PSD-95 clusters. In addition, synaptic activity-dependent reorganization of PSD-95 nanodomains through local ZDHHC2 could allow the individual PSD to reorganize its structure and molecular composition in a synapse-specific manner. Still unknown is what is the depalmitoylating enzyme that mediates local depalmitoylation of PSD-95.

5.2 Micro-/nanodomain organization potentially regulated by DHHC proteins

Given that palmitoylation often occurs on the proteins enriched in specialized membrane domains, such as cell-cell or cell-extracellular matrix contact sites, basolateral membrane, SVs, and specific endosomes as well as PSDs (Figure 4), the presence and activity of local palmitoylation machinery may explain a generalized molecular mechanism for membrane specialization and compartmentalization as in the case of the PSD.

One of possible examples in which a local palmitoylating activity may determine the substrate localization was shown for ankyrin-G at the basolateral membrane domains (Figure 4(C)). Ankyrin-G is a representative membrane skeleton protein, contributes to promoting epithelial lateral membrane assembly, and plays an essential role in building neuronal axon initial segments (Bennett & Healy, 2008; Bennett & Lorenzo, 2013). In 1987, ankyrin-R (ANK1) expressed in erythrocyte was shown to be reversibly palmitoylated with a half-life, ~50 min (Staufenbiel, 1987). Recently, Bennett's group showed that cysteine 70 of ankyrin-G (ANK3), conserved among all three ankyrin family members, is palmitoylated and this lipid modification is required for its membrane targeting and its function in membrane domain biogenesis at epithelial lateral membranes and axon initial segments (He et al., 2012). Then, they identified ZDHHC5 and ZDHHC8 as ankyrin-G palmitoylating enzymes in MDCK epithelial cells (He et al., 2014). Double knockdown of ZDHHC5 and ZDHHC8, but not the single knockdown, reduces ankyrin-G palmitoylation, causes disassociation of ankyrin-G from the basolateral membrane domain, and reduces lateral membrane height in MDCK cells as in the case of loss of ankyrin-G. Interestingly, 3D deconvolution microscopy unveiled the highly organized micrometer-scale subdomains in the ranges of $0.5-2 \ \mu m^2$ where

ZDHHC5/8, ankyrin-G, and its binding protein βII-spectrin are together concentrated, strongly suggesting that ZDHHC5/8 locally palmitoylates ankyrin-G to determine ankyrin-G distribution to the lateral membrane.

As described in Section 2, several SV proteins such as synaptobrevin/ VAMP, synaptotagmin, CSP, and SCAMP1 are palmitoylated. CSP is a peripheral membrane protein and is required for maintaining the releasable vesicular pool and SV recycling. In *Drosophila melanogaster*, palmitoylation of CSP mediated by dHIP14 is essential for its targeting to SVs (Ohyama et al., 2007). Given that most SVs are regenerated by local recycling from the presynaptic plasma membrane in the presynaptic terminals, it is conceivable that certain DHHC proteins (such as dHIP14) are specifically localized on the SV and recruit their substrate proteins on the same SVs (Stowers & Isacoff, 2007) (Figure 4(B)).

Examples supporting the organelle-specific palmitoylation are: (1) neurochondrin/norbin, a neuron-specific cytoplasmic protein, is palmitoylated and targeted to the Rab5-positive, dendritic early endosomes in its palmitoylation-dependent manner (Oku et al., 2013). Interestingly, palmitoylating enzymes responsible for neurochondrin/norbin, ZDHHC1/11 subfamily members, are colocalized with the Rab5-positive early endosomes, strongly suggesting that ZDHHC1/11 may locally palmitoylate neurochondrin and actively trap it on the early endosomes; (2) Calnexin-PAT, ZDHHC6, is localized mainly in the ER (Gorleku, Barns, Prescott, Greaves, & Chamberlain, 2011; Ohno et al., 2006), raising a possibility that ZDHHC6 represents an ER-PAT for the ER resident proteins and/or nascent glycoproteins (Lakkaraju et al., 2012); and (3) it was reported that AKAP79/150, localized at the recycling endosomes, is palmitoylated by an endosomal PAT, ZDHHC2 (Woolfrey et al., 2015). Taken together, organelles/vesicles or membrane compartments where palmitoylated proteins are specifically targeted may have a specific set of DHHC proteins and still unidentified depalmitoylating enzymes for organization of specialized membrane domains.

6. CONCLUSIONS

In this review, we reviewed recent advances in research on protein palmitoylation. The most remarkable advances in the last decade are: (1) development of purification method for palmitoylated proteins and subsequent establishment of proteomic approaches (palmitoyl-proteomics;

Table 1) and (2) identification of DHHC-type palmitoylating enzymes in various species. The proteomic analysis has revealed the importance of protein palmitoylation beyond species from virus, yeast, protozoa, plants, nematode, and fruit fly to mammals. Combining with the cells/animals with a certain DHHC gene manipulated (knockdown or KO), it is becoming possible to identify the enzyme—substrate pairs in vivo. However, molecular redundancy often prevents us to identify substrates for DHHC enzymes by proteomics approaches in mammalian samples. To overcome this issue, it might be helpful to take advantage of the recent genome-editing methods, which allows us to generate multiple KO animals in a short period.

Recent studies have begun to show that polarized cells, such as neurons, have more than one cellular locus of protein palmitoylation. Given that mammalian cells harbor 23—24 kinds of DHHC palmitoylating enzymes with divergent substrate specificities, the hypothesis that locally expressed DHHC proteins directly nucleate membrane specializations may explain compartmentalized protein assembly processes in presynapses, tight junctions, and focal adhesions, where many palmitoylated proteins are enriched. Furthermore, to maintain and reorganize specialized membrane domains, polarized cells should establish local palmitoyl cycles on individual palmitoyl proteins. The next decade of the research will be to understand the roles and regulation of local palmitoyl cycles on different proteins in various cellular contexts (i.e., cell types, extracellular signals, cell differentiation and polarization, tissue development, and diseases).

The field, however, still has many questions to be solved. (1) What is the exact subcellular localization of endogenous DHHC proteins in (polarized) cells and how are their specific localizations regulated? Good antibodies to specifically detect (or distinguish between) individual DHHC family members and super-resolution microscopy analysis are definitively required for their submicron-scale localizations in the membrane. (2) What are depalmitoylating enzymes? Are there any other depalmitoylating enzymes than APT? Judging from our study that PSD-95 depalmitoylation occurs in the local synapses (postsynaptic membrane) (Fukata et al., 2013), there should exist certain depalmitoylating enzymes in or near the postsynaptic membrane. Identification of the enzyme(s) that can govern depalmitoylation of many types of palmitoylated proteins in various cellular locations is long awaited. (3) What is the palmitoylation stoichiometry on proteins in cells or tissues? It is worthwhile to develop general methods to quantify palmitoylation stoichiometry of endogenous proteins, like the phosphorylation profiling with Phos-tag SDS-PAGE (Kinoshita, Kinoshita-Kikuta, & Koike,

2015). (4) What is the lipid composition of the specialized membrane domains and how is the lipid composition related to the enrichment of palmitoylated proteins in such domains? Lipid profiling of these domains, for example, the PSD or PSD-95 nanodomains, is challenging, but should give us a clue to understand how lipids and (palmitoylated) proteins interact and orchestrate to establish the complex membrane domains. (5) In spite of intensive characterization of DHHC enzymes, no structural analysis by NMR and X-ray crystallography has been done probably due to the complex topology of DHHC proteins (e.g., tetra- or penta-transmembrane spanning and multimer formation). Recent advances in cryo-electron microscopy analysis may provide the insight into their structure and function. (6) Development of pharmacological modifiers specific for the DHHC protein members is awaited as a present available inhibitor, 2-bromopalmitate, has off-targets other than DHHC proteins (Davda et al., 2013). The structural information of DHHC proteins, in vitro enzyme assays, and chemical biology techniques will contribute to developing pharmacological modifiers. Given that numerous synaptic proteins associated with neuropsychiatric disorders are palmitoylated (e.g., serotonin receptors and glutamate receptors), it will be important to develop pharmacological modifiers of the DHHC protein family and depalmitoylating enzyme family to determine the physiological and pathological roles for these enzymes. As is clear from all the questions raised above, we have just started to understand the significance of this expanding field.

ACKNOWLEDGMENTS

We thank Mie Watanabe (NIPS) for helping with a part of illustrations. We apologize to researchers whose work could not be cited owing to space constraints. Y.F. is supported by grants from the Ministry of Education, Culture, Sports, Science, and Technology (MEXT, 15H04279; 15H01299). T.M. is a research fellow of Japan Society for the Promotion of Science. M.F. is supported by grants from MEXT (26291045; 15H01570), Daiichi Sankyo Foundation of Life Science, Ono Medical Research Foundation, The Uehara Memorial Foundation, and The Naito Foundation.

REFERENCES

Abrami, L., Kunz, B., Iacovache, I., & van der Goot, F. G. (2008). Palmitoylation and ubiquitination regulate exit of the Wnt signaling protein LRP6 from the endoplasmic reticulum. *Proceedings of the National Academy of Sciences of the United States of America, 105*, 5384–5389.

Adlanmerini, M., Solinhac, R., Abot, A., Fabre, A., Raymond-Letron, I., Guihot, A. L., et al. (2014). Mutation of the palmitoylation site of estrogen receptor alpha in vivo reveals tissue-specific roles for membrane versus nuclear actions. *Proceedings of the National Academy of Sciences of the United States of America, 111*, E283–E290.

Aittaleb, M., Nishimura, A., Linder, M. E., & Tesmer, J. J. (2011). Plasma membrane association of p63 Rho guanine nucleotide exchange factor (p63RhoGEF) is mediated by
palmitoylation and is required for basal activity in cells. The Journal of Biological Chemistry,
286, 34448–34456.

Bannan, B. A., Van Etten, J., Kohler, J. A., Tsoi, Y., Hansen, N. M., Sigmon, S., et al. (2008).
The Drosophila protein palmitoylome: characterizing palmitoyl-thioesterases and DHHC
palmitoyl-transferases. Fly (Austin), 2, 198–214.

Bartels, D. J., Mitchell, D. A., Dong, X., & Deschenes, R. J. (1999). Erf2, a novel gene product that affects the localization and palmitoylation of Ras2 in Saccharomyces cerevisiae. Molecular and Cellular Biology, 19, 6775–6787.

Batistic, O. (2012). Genomics and localization of the Arabidopsis DHHC-cysteine-rich
domain S-acyltransferase protein family. Plant Physiology, 160, 1597–1612.

Bennett, V., & Healy, J. (2008). Organizing the fluid membrane bilayer: diseases linked to
spectrin and ankyrin. Trends in Molecular Medicine, 14, 28–36.

Bennett, V., & Lorenzo, D. N. (2013). Spectrin- and ankyrin-based membrane domains and
the evolution of vertebrates. Current Topics in Membranes, 72, 1–37.

Bhattacharyya, R., Barren, C., & Kovacs, D. M. (2013). Palmitoylation of amyloid precursor
protein regulates amyloidogenic processing in lipid rafts. The Journal of Neuroscience, 33,
11169–11183.

Blanpain, C., Wittamer, V., Vanderwinden, J. M., Boom, A., Renneboog, B., Lee, B.,
et al. (2001). Palmitoylation of CCR5 is critical for receptor trafficking and efficient
activation of intracellular signaling pathways. The Journal of Biological Chemistry, 276,
23795–23804.

Blaskovic, S., Blanc, M., & van der Goot, F. G. (2013). What does S-palmitoylation do to
membrane proteins? FEBS Journal, 280, 2766–2774.

Brigidi, G. S., Sun, Y., Beccano-Kelly, D., Pitman, K., Mobasser, M., Borgland, S. L., et al.
(2014). Palmitoylation of delta-catenin by DHHC5 mediates activity-induced synapse
plasticity. Nature Neuroscience, 17, 522–532.

Cevik, S., Hori, Y., Kaplan, O. I., Kida, K., Toivenon, T., Foley-Fisher, C., et al. (2010).
Joubert syndrome Arl13b functions at ciliary membranes and stabilizes protein transport
in Caenorhabditis elegans. The Journal of Cell Biology, 188, 953–969.

Chamberlain, L. H., & Shipston, M. J. (2015). The physiology of protein S-acylation. Physiological Reviews, 95, 341–376.

Charron, G., Zhang, M. M., Yount, J. S., Wilson, J., Raghavan, A. S., Shamir, E., et al.
(2009). Robust fluorescent detection of protein fatty-acylation with chemical
reporters. Journal of the American Chemical Society, 131, 4967–4975.

Charych, E. I., Jiang, L. X., Lo, F., Sullivan, K., & Brandon, N. J. (2010). Interplay of palmitoylation and phosphorylation in the trafficking and localization of phosphodiesterase
10A: implications for the treatment of schizophrenia. The Journal of Neuroscience, 30,
9027–9037.

Chaube, R., Hess, D. T., Wang, Y. J., Plummer, B., Sun, Q. A., Laurita, K., et al. (2014).
Regulation of the skeletal muscle ryanodine receptor/Ca^{2+}-release channel RyR1 by
S-palmitoylation. The Journal of Biological Chemistry, 289, 8612–8619.

Chisari, M., Saini, D. K., Kalyanaraman, V., & Gautam, N. (2007). Shuttling of G protein
subunits between the plasma membrane and intracellular membranes. The Journal of Biological Chemistry, 282, 24092–24098.

Cremona, M. L., Matthies, H. J., Pau, K., Bowton, E., Speed, N., Lute, B. J., et al. (2011).
Flotillin-1 is essential for PKC-triggered endocytosis and membrane microdomain localization of DAT. Nature Neuroscience, 14, 469–477.

Davda, D., El Azzouny, M. A., Tom, C. T., Hernandez, J. L., Majmudar, J. D.,
Kennedy, R. T., et al. (2013). Profiling targets of the irreversible palmitoylation inhibitor
2-bromopalmitate. ACS Chemical Biology, 8, 1912–1917.

De Vries, L., Elenko, E., Hubler, L., Jones, T. L., & Farquhar, M. G. (1996). GAIP is membrane-anchored by palmitoylation and interacts with the activated (GTP-bound) form of G alpha i subunits. *Proceedings of the National Academy of Sciences of the United States of America, 93*, 15203—15208.

Dejanovic, B., Semtner, M., Ebert, S., Lamkemeyer, T., Neuser, F., Luscher, B., et al. (2014). Palmitoylation of gephyrin controls receptor clustering and plasticity of GABAergic synapses. *PLoS Biology, 12*, e1001908.

Dekker, F. J., Rocks, O., Vartak, N., Menninger, S., Hedberg, C., Balamurugan, R., et al. (2010). Small-molecule inhibition of APT1 affects Ras localization and signaling. *Nature Chemical Biology, 6*, 449—456.

Diaz-Rohrer, B. B., Leventhal, K. R., Simons, K., & Leventhal, I. (2014). Membrane raft association is a determinant of plasma membrane localization. *Proceedings of the National Academy of Sciences of the United States of America, 111*, 8500—8505.

Dietzen, D. J., Hastings, W. R., & Lublin, D. M. (1995). Caveolin is palmitoylated on multiple cysteine residues. Palmitoylation is not necessary for localization of caveolin to caveolae. *The Journal of Biological Chemistry, 270*, 6838—6842.

Dowal, L., Yang, W., Freeman, M. R., Steen, H., & Flaumenhaft, R. (2011). Proteomic analysis of palmitoylated platelet proteins. *Blood, 118*, e62—73.

Drenan, R. M., Doupnik, C. A., Boyle, M. P., Muglia, L. J., Huettner, J. E., Linder, M. E., et al. (2005). Palmitoylation regulates plasma membrane-nuclear shuttling of R7BP, a novel membrane anchor for the RGS7 family. *The Journal of Cell Biology, 169*, 623—633.

Drisdel, R. C., & Green, W. N. (2004). Labeling and quantifying sites of protein palmitoylation. *Biotechniques, 36*, 276—285.

Druey, K. M., Ugur, O., Caron, J. M., Chen, C. K., Backlund, P. S., & Jones, T. L. (1999). Amino-terminal cysteine residues of RGS16 are required for palmitoylation and modulation of Gi- and Gq-mediated signaling. *The Journal of Biological Chemistry, 274*, 18836—18842.

Edmonds, M. J., & Morgan, A. (2014). A systematic analysis of protein palmitoylation in *Caenorhabditis elegans*. *BMC Genomics, 15*, 841.

El-Husseini Ael, D., & Bredt, D. S. (2002). Protein palmitoylation: a regulator of neuronal development and function. *Nature Reviews Neuroscience, 3*, 791—802.

El-Husseini Ael, D., Schnell, E., Dakoji, S., Sweeney, N., Zhou, Q., Prange, O., et al. (2002). Synaptic strength regulated by palmitate cycling on PSD-95. *Cell, 108*, 849—863.

Emmer, B. T., Nakayasu, E. S., Souther, C., Choi, H., Sobreira, T. J., Epting, C. L., et al. (2011). Global analysis of protein palmitoylation in African trypanosomes. *Eukaryotic Cell, 10*, 455—463.

Emmer, B. T., Souther, C., Toriello, K. M., Olson, C. L., Epting, C. L., & Engman, D. M. (2009). Identification of a palmitoyl acyltransferase required for protein sorting to the flagellar membrane. *Journal of Cell Science, 122*, 867—874.

Forrester, M. T., Hess, D. T., Thompson, J. W., Hultman, R., Moseley, M. A., Stamler, J. S., et al. (2011). Site-specific analysis of protein S-acylation by resin-assisted capture. *The Journal of Lipid Research, 52*, 393—398.

Foster, J. D., & Vaughan, R. A. (2011). Palmitoylation controls dopamine transporter kinetics, degradation, and protein kinase C-dependent regulation. *The Journal of Biological Chemistry, 286*, 5175—5186.

Fredericks, G. J., Hoffmann, F. W., Rose, A. H., Osterheld, H. J., Hess, F. M., Mercier, F., et al. (2014). Stable expression and function of the inositol 1,4,5-triphosphate receptor requires palmitoylation by a DHHC6/selenoprotein K complex. *Proceedings of the National Academy of Sciences of the United States of America, 111*, 16478—16483.

Frenal, K., Kemp, L. E., & Soldati-Favre, D. (2014). Emerging roles for protein S-palmitoylation in *Toxoplasma* biology. *International Journal for Parasitology, 44*, 121—131.

Frenal, K., Tay, C. L., Mueller, C., Bushell, E. S., Jia, Y., Graindorge, A., et al. (2013). Global analysis of apicomplexan protein S-acyl transferases reveals an enzyme essential for invasion. *Traffic, 14,* 895–911.

Frohlich, M., Dejanovic, B., Kashkar, H., Schwarz, G., & Nussberger, S. (2014). S-palmitoylation represents a novel mechanism regulating the mitochondrial targeting of BAX and initiation of apoptosis. *Cell Death and Disease, 5,* e1057.

Fukata, Y., Dimitrov, A., Boncompain, G., Vielemeyer, O., Perez, F., & Fukata, M. (2013). Local palmitoylation cycles define activity-regulated postsynaptic subdomains. *The Journal of Cell Biology, 202,* 145–161.

Fukata, Y., & Fukata, M. (2010). Protein palmitoylation in neuronal development and synaptic plasticity. *Nature Reviews Neuroscience, 11,* 161–175.

Fukata, M., Fukata, Y., Adesnik, H., Nicoll, R. A., & Bredt, D. S. (2004). Identification of PSD-95 palmitoylating enzymes. *Neuron, 44,* 987–996.

Fukata, Y., Iwanaga, T., & Fukata, M. (2006). Systematic screening for palmitoyl transferase activity of the DHHC protein family in mammalian cells. *Methods, 40,* 177–182.

Gao, X., & Hannoush, R. N. (2014a). Method for cellular imaging of palmitoylated proteins with clickable probes and proximity ligation applied to Hedgehog, tubulin, and Ras. *Journal of the American Chemical Society, 136,* 4544–4550.

Gao, X., & Hannoush, R. N. (2014b). Single-cell imaging of Wnt palmitoylation by the acyltransferase porcupine. *Nature Chemical Biology, 10,* 61–68.

Gao, Z., Ni, Y., Szabo, G., & Linden, J. (1999). Palmitoylation of the recombinant human A_1 adenosine receptor: enhanced proteolysis of palmitoylation-deficient mutant receptors. *Biochemical Journal, 342,* 387–395.

George, J., Soares, C., Montersino, A., Beique, J. C., & Thomas, G. M. (2015). Palmitoylation of LIM Kinase-1 ensures spine-specific actin polymerization and morphological plasticity. *eLife, 4,* e06327.

Glaser, B., Moskvina, V., Kirov, G., Murphy, K. C., Williams, H., Williams, N., et al. (2006). Analysis of ProDH, COMT and ZDHHC8 risk variants does not support individual or interactive effects on schizophrenia susceptibility. *Schizophrenia Research, 87,* 21–27.

Gorleku, O. A., Barns, A. M., Prescott, G. R., Greaves, J., & Chamberlain, L. H. (2011). Endoplasmic reticulum localization of DHHC palmitoyltransferases mediated by lysine-based sorting signals. *The Journal of Biological Chemistry, 286,* 39573–39584.

Greaves, J., Carmichael, J. A., & Chamberlain, L. H. (2011). The palmitoyl transferase DHHC2 targets a dynamic membrane cycling pathway: regulation by a C-terminal domain. *Molecular Biology of the Cell, 22,* 1887–1895.

Greaves, J., & Chamberlain, L. H. (2011). DHHC palmitoyl transferases: substrate interactions and (patho)physiology. *Trends in Biochemical Sciences, 36,* 245–253.

Hang, H. C., Geutjes, E. J., Grotenbreg, G., Pollington, A. M., Bijlmakers, M. J., & Ploegh, H. L. (2007). Chemical probes for the rapid detection of fatty-acylated proteins in mammalian cells. *Journal of the American Chemical Society, 129,* 2744–2745.

Hannoush, R. N., & Arenas-Ramirez, N. (2009). Imaging the lipidome: omega-alkynyl fatty acids for detection and cellular visualization of lipid-modified proteins. *ACS Chemical Biology, 4,* 581–587.

Hannoush, R. N., & Sun, J. (2010). The chemical toolbox for monitoring protein fatty acylation and prenylation. *Nature Chemical Biology, 6,* 498–506.

Hayashi, T., Rumbaugh, G., & Huganir, R. L. (2005). Differential regulation of AMPA receptor subunit trafficking by palmitoylation of two distinct sites. *Neuron, 47,* 709–723.

Hayashi, T., Thomas, G. M., & Huganir, R. L. (2009). Dual palmitoylation of NR2 subunits regulates NMDA receptor trafficking. *Neuron, 64,* 213–226.

He, M., Abdi, K. M., & Bennett, V. (2014). Ankyrin-G palmitoylation and betaII-spectrin binding to phosphoinositide lipids drive lateral membrane assembly. *The Journal of Cell Biology, 206,* 273–288.

He, M., Jenkins, P., & Bennett, V. (2012). Cysteine 70 of ankyrin-G is S-palmitoylated and is required for function of ankyrin-G in membrane domain assembly. *The Journal of Biological Chemistry, 287*, 43995–44005.

Hemsley, P. A., & Grierson, C. S. (2008). Multiple roles for protein palmitoylation in plants. *Trends in Plant Science, 13*, 295–302.

Hemsley, P. A., Kemp, A. C., & Grierson, C. S. (2005). The TIP GROWTH DEFEC-TIVE1 S-acyl transferase regulates plant cell growth in *Arabidopsis. Plant Cell, 17*, 2554–2563.

Hemsley, P. A., Weimar, T., Lilley, K. S., Dupree, P., & Grierson, C. S. (2013). A proteomic approach identifies many novel palmitoylated proteins in *Arabidopsis. New Phytologist, 197*, 805–814.

Hou, H., John Peter, A. T., Meiringer, C., Subramanian, K., & Ungermann, C. (2009). Analysis of DHHC acyltransferases implies overlapping substrate specificity and a two-step reaction mechanism. *Traffic, 10*, 1061–1073.

Howie, J., Reilly, L., Fraser, N. J., Vlachaki Walker, J. M., Wypijewski, K. J., Ashford, M. L., et al. (2014). Substrate recognition by the cell surface palmitoyl transferase DHHC5. *Proceedings of the National Academy of Sciences of the United States of America, 111*, 17534–17539.

Huang, K., Sanders, S., Singaraja, R., Orban, P., Cijsouw, T., Arstikaitis, P., et al. (2009). Neuronal palmitoyl acyl transferases exhibit distinct substrate specificity. *FASEB Journal, 23*, 2605–2615.

Israely, I., Costa, R. M., Xie, C. W., Silva, A. J., Kosik, K. S., & Liu, X. (2004). Deletion of the neuron-specific protein delta-catenin leads to severe cognitive and synaptic dysfunction. *Current Biology, 14*, 1657–1663.

Ivaldi, C., Martin, B. R., Kieffer-Jaquinod, S., Chapel, A., Levade, T., Garin, J., et al. (2012). Proteomic analysis of S-acylated proteins in human B cells reveals palmitoylation of the immune regulators CD20 and CD23. *PLoS One, 7*, e37187.

Jeffries, O., Tian, L., McClafferty, H., & Shipston, M. J. (2012). An electrostatic switch controls palmitoylation of the large conductance voltage- and calcium-activated potassium (BK) channel. *The Journal of Biological Chemistry, 287*, 1468–1477.

Jennings, B. C., & Linder, M. E. (2012). DHHC protein S-acyltransferases use similar ping-pong kinetic mechanisms but display different acyl-CoA specificities. *The Journal of Biological Chemistry, 287*, 7236–7245.

Jia, L., Chisari, M., Maktabi, M. H., Sobieski, C., Zhou, H., Konopko, A. M., et al. (2014). A mechanism regulating G protein-coupled receptor signaling that requires cycles of protein palmitoylation and depalmitoylation. *The Journal of Biological Chemistry, 289*, 6249–6257.

Jones, M. L., Collins, M. O., Goulding, D., Choudhary, J. S., & Rayner, J. C. (2012). Analysis of protein palmitoylation reveals a pervasive role in *Plasmodium* development and pathogenesis. *Cell Host and Microbe, 12*, 246–258.

Joseph, M., & Nagaraj, R. (1995). Conformations of peptides corresponding to fatty acylation sites in proteins. A circular dichroism study. *The Journal of Biological Chemistry, 270*, 19439–19445.

Kang, R., Wan, J., Arstikaitis, P., Takahashi, H., Huang, K., Bailey, A. O., et al. (2008). Neural palmitoyl-proteomics reveals dynamic synaptic palmitoylation. *Nature, 456*, 904–909.

Keller, C. A., Yuan, X., Panzanelli, P., Martin, M. L., Alldred, M., Sassoe-Pognetto, M., et al. (2004). The gamma2 subunit of GABA$_A$ receptors is a substrate for palmitoylation by GODZ. *The Journal of Neuroscience, 24*, 5881–5891.

Kinoshita, E., Kinoshita-Kikuta, E., & Koike, T. (2015). Advances in Phos-tag-based methodologies for separation and detection of the phosphoproteome. *Biochimica et Biophysica Acta, 1854*, 601–608.

Korycka, J., Lach, A., Heger, E., Boguslawska, D. M., Wolny, M., Toporkiewicz, M., et al. (2012). Human DHHC proteins: a spotlight on the hidden player of palmitoylation. *European Journal of Cell Biology, 91*, 107—117.

Kostiuk, M. A., Corvi, M. M., Keller, B. O., Plummer, G., Prescher, J. A., Hangauer, M. J., et al. (2008). Identification of palmitoylated mitochondrial proteins using a bio-orthogonal azido-palmitate analogue. *FASEB Journal, 22*, 721—732.

Lakkaraju, A. K., Abrami, L., Lemmin, T., Blaskovic, S., Kunz, B., Kihara, A., et al. (2012). Palmitoylated calnexin is a key component of the ribosome-translocon complex. *EMBO Journal, 31*, 1823—1835.

Lam, K. K., Davey, M., Sun, B., Roth, A. F., Davis, N. G., & Conibear, E. (2006). Palmitoylation by the DHHC protein Pfa4 regulates the ER exit of Chs3. *The Journal of Cell Biology, 174*, 19—25.

Levental, I., Lingwood, D., Grzybek, M., Coskun, U., & Simons, K. (2010). Palmitoylation regulates raft affinity for the majority of integral raft proteins. *Proceedings of the National Academy of Sciences of the United States of America, 107*, 22050—22054.

Li, Y., Hu, J., Hofer, K., Wong, A. M., Cooper, J. D., Birnbaum, S. G., et al. (2010). DHHC5 interacts with PDZ domain 3 of post-synaptic density-95 (PSD-95) protein and plays a role in learning and memory. *The Journal of Biological Chemistry, 285*, 13022—13031.

Li, Y., Martin, B. R., Cravatt, B. F., & Hofmann, S. L. (2012). DHHC5 protein palmitoylates flotillin-2 and is rapidly degraded on induction of neuronal differentiation in cultured cells. *The Journal of Biological Chemistry, 287*, 523—530.

Lin, D. T., Makino, Y., Sharma, K., Hayashi, T., Neve, R., Takamiya, K., et al. (2009). Regulation of AMPA receptor extrasynaptic insertion by 4.1N, phosphorylation and palmitoylation. *Nature Neuroscience, 12*, 879—887.

Linder, M. E., & Deschenes, R. J. (2007). Palmitoylation: policing protein stability and traffic. *Nature Reviews Molecular Cell Biology, 8*, 74—84.

Linder, M. E., Middleton, P., Hepler, J. R., Taussig, R., Gilman, A. G., & Mumby, S. M. (1993). Lipid modifications of G proteins: alpha subunits are palmitoylated. *Proceedings of the National Academy of Sciences of the United States of America, 90*, 3675—3679.

Lobo, S., Greentree, W. K., Linder, M. E., & Deschenes, R. J. (2002). Identification of a Ras palmitoyltransferase in *Saccharomyces cerevisiae*. *The Journal of Biological Chemistry, 277*, 41268—41273.

Lynes, E. M., Bui, M., Yap, M. C., Benson, M. D., Schneider, B., Ellgaard, L., et al. (2012). Palmitoylated TMX and calnexin target to the mitochondria-associated membrane. *EMBO Journal, 31*, 457—470.

MacGillavry, H. D., Song, Y., Raghavachari, S., & Blanpied, T. A. (2013). Nanoscale scaffolding domains within the postsynaptic density concentrate synaptic AMPA receptors. *Neuron, 78*, 615—622.

Mansouri, M. R., Marklund, L., Gustavsson, P., Davey, E., Carlsson, B., Larsson, C., et al. (2005). Loss of ZDHHC15 expression in a woman with a balanced translocation t(X; 15)(q13.3;cen) and severe mental retardation. *European Journal of Human Genetics, 13*, 970—977.

Maric, D., McGwire, B. S., Buchanan, K. T., Olson, C. L., Emmer, B. T., Epting, C. L., et al. (2011). Molecular determinants of ciliary membrane localization of *Trypanosoma cruzi* flagellar calcium-binding protein. *The Journal of Biological Chemistry, 286*, 33109—33117.

Marin, E. P., Derakhshan, B., Lam, T. T., Davalos, A., & Sessa, W. C. (2012). Endothelial cell palmitoylproteomic identifies novel lipid-modified targets and potential substrates for protein acyl transferases. *Circulation Research, 110*, 1336—1344.

Martin, B. R., & Cravatt, B. F. (2009). Large-scale profiling of protein palmitoylation in mammalian cells. *Nature Methods, 6*, 135—138.

Martin, B. R., Wang, C., Adibekian, A., Tully, S. E., & Cravatt, B. F. (2012). Global profiling of dynamic protein palmitoylation. *Nature Methods, 9*, 84−89.

Masurel-Paulet, A., Kalscheuer, V. M., Lebrun, N., Hu, H., Levy, F., Thauvin-Robinet, C., et al. (2014). Expanding the clinical phenotype of patients with a ZDHHC9 mutation. *American Journal of Medical Genetics. Part A, 164A*, 789−795.

Mejias, R., Adamczyk, A., Anggono, V., Niranjan, T., Thomas, G. M., Sharma, K., et al. (2011). Gain-of-function glutamate receptor interacting protein 1 variants alter GluA2 recycling and surface distribution in patients with autism. *Proceedings of the National Academy of Sciences of the United States of America, 108*, 4920−4925.

Merrick, B. A., Dhungana, S., Williams, J. G., Aloor, J. J., Peddada, S., Tomer, K. B., et al. (2011). Proteomic profiling of S-acylated macrophage proteins identifies a role for palmitoylation in mitochondrial targeting of phospholipid scramblase 3. *Molecular and Cellular Proteomics, 10*. M110 006007.

Mill, P., Lee, A. W., Fukata, Y., Tsutsumi, R., Fukata, M., Keighren, M., et al. (2009). Palmitoylation regulates epidermal homeostasis and hair follicle differentiation. *PLoS Genetics, 5*, e1000748.

Milnerwood, A. J., Parsons, M. P., Young, F. B., Singaraja, R. R., Franciosi, S., Volta, M., et al. (2013). Memory and synaptic deficits in Hip14/DHHC17 knockout mice. *Proceedings of the National Academy of Sciences of the United States of America, 110*, 20296−20301.

Mitchell, D. A., Hamel, L. D., Reddy, K. D., Farh, L., Rettew, L. M., Sanchez, P. R., et al. (2014). Mutations in the X-linked intellectual disability gene, zDHHC9, alter autopalmitoylation activity by distinct mechanisms. *The Journal of Biological Chemistry, 289*, 18582−18592.

Mitchell, D. A., Mitchell, G., Ling, Y., Budde, C., & Deschenes, R. J. (2010). Mutational analysis of *Saccharomyces cerevisiae* Erf2 reveals a two-step reaction mechanism for protein palmitoylation by DHHC enzymes. *The Journal of Biological Chemistry, 285*, 38104−38114.

Morrow, I. C., Rea, S., Martin, S., Prior, I. A., Prohaska, R., Hancock, J. F., et al. (2002). Flotillin-1/reggie-2 traffics to surface raft domains via a novel golgi-independent pathway. Identification of a novel membrane targeting domain and a role for palmitoylation. *The Journal of Biological Chemistry, 277*, 48834−48841.

Mukai, J., Dhilla, A., Drew, L. J., Stark, K. L., Cao, L., MacDermott, A. B., et al. (2008). Palmitoylation-dependent neurodevelopmental deficits in a mouse model of 22q11 microdeletion. *Nature Neuroscience, 11*, 1302−1310.

Mukai, J., Liu, H., Burt, R. A., Swor, D. E., Lai, W. S., Karayiorgou, M., et al. (2004). Evidence that the gene encoding ZDHHC8 contributes to the risk of schizophrenia. *Nature Genetics, 36*, 725−731.

Mukai, J., Tamura, M., Fenelon, K., Rosen, A. M., Spellman, T. J., Kang, R., et al. (2015). Molecular substrates of altered axonal growth and brain connectivity in a mouse model of schizophrenia. *Neuron, 86*, 680−695.

Nair, D., Hosy, E., Petersen, J. D., Constals, A., Giannone, G., Choquet, D., et al. (2013). Super-resolution imaging reveals that AMPA receptors inside synapses are dynamically organized in nanodomains regulated by PSD95. *The Journal of Neuroscience, 33*, 13204−13224.

Noritake, J., Fukata, Y., Iwanaga, T., Hosomi, N., Tsutsumi, R., Matsuda, N., et al. (2009). Mobile DHHC palmitoylating enzyme mediates activity-sensitive synaptic targeting of PSD-95. *The Journal of Cell Biology, 186*, 147−160.

Ochsenbauer-Jambor, C., Miller, D. C., Roberts, C. R., Rhee, S. S., & Hunter, E. (2001). Palmitoylation of the *Rous sarcoma* virus transmembrane glycoprotein is required for protein stability and virus infectivity. *Journal of Virology, 75*, 11544−11554.

Ohno, Y., Kashio, A., Ogata, R., Ishitomi, A., Yamazaki, Y., & Kihara, A. (2012). Analysis of substrate specificity of human DHHC protein acyltransferases using a yeast expression system. *Molecular Biology of the Cell, 23*, 4543−4551.

Ohno, Y., Kihara, A., Sano, T., & Igarashi, Y. (2006). Intracellular localization and tissue-specific distribution of human and yeast DHHC cysteine-rich domain-containing proteins. *Biochimica et Biophysica Acta, 1761,* 474—483.

Ohyama, T., Verstreken, P., Ly, C. V., Rosenmund, T., Rajan, A., Tien, A. C., et al. (2007). Huntingtin-interacting protein 14, a palmitoyl transferase required for exocytosis and targeting of CSP to synaptic vesicles. *The Journal of Cell Biology, 179,* 1481—1496.

Oku, S., Takahashi, N., Fukata, Y., & Fukata, M. (2013). In silico screening for palmitoyl substrates reveals a role for DHHC1/3/10 (zDHHC1/3/11)-mediated neurochondrin palmitoylation in its targeting to Rab5-positive endosomes. *The Journal of Biological Chemistry, 288,* 19816—19829.

Oyama, T., Miyoshi, Y., Koyama, K., Nakagawa, H., Yamori, T., Ito, T., et al. (2000). Isolation of a novel gene on 8p21.3-22 whose expression is reduced significantly in human colorectal cancers with liver metastasis. *Genes, Chromosomes and Cancer, 29,* 9—15.

Pedram, A., Razandi, M., Lewis, M., Hammes, S., & Levin, E. R. (2014). Membrane-localized estrogen receptor alpha is required for normal organ development and function. *Developmental Cell, 29,* 482—490.

Percherancier, Y., Planchenault, T., Valenzuela-Fernandez, A., Virelizier, J. L., Arenzana-Seisdedos, F., & Bachelerie, F. (2001). Palmitoylation-dependent control of degradation, life span, and membrane expression of the CCR5 receptor. *The Journal of Biological Chemistry, 276,* 31936—31944.

Prescott, G. R., Gorleku, O. A., Greaves, J., & Chamberlain, L. H. (2009). Palmitoylation of the synaptic vesicle fusion machinery. *Journal of Neurochemistry, 110,* 1135—1149.

Raymond, F. L., Tarpey, P. S., Edkins, S., Tofts, C., O'Meara, S., Teague, J., et al. (2007). Mutations in ZDHHC9, which encodes a palmitoyltransferase of NRAS and HRAS, cause X-linked mental retardation associated with a Marfanoid habitus. *American Journal of Human Genetics, 80,* 982—987.

Ren, Q., & Bennett, V. (1998). Palmitoylation of neurofascin at a site in the membrane-spanning domain highly conserved among the L1 family of cell adhesion molecules. *Journal of Neurochemistry, 70,* 1839—1849.

Ren, W., Jhala, U. S., & Du, K. (2013). Proteomic analysis of protein palmitoylation in adipocytes. *Adipocyte, 2,* 17—28.

Ren, J., Wen, L., Gao, X., Jin, C., Xue, Y., & Yao, X. (2008). CSS-Palm 2.0: an updated software for palmitoylation sites prediction. *Protein Engineering, Design and Selection, 21,* 639—644.

Rocks, O., Gerauer, M., Vartak, N., Koch, S., Huang, Z. P., Pechlivanis, M., et al. (2010). The palmitoylation machinery is a spatially organizing system for peripheral membrane proteins. *Cell, 141,* 458—471.

Rocks, O., Peyker, A., & Bastiaens, P. I. (2006). Spatio-temporal segregation of Ras signals: one ship, three anchors, many harbors. *Current Opinion in Cell Biology, 18,* 351—357.

Rocks, O., Peyker, A., Kahms, M., Verveer, P. J., Koerner, C., Lumbierres, M., et al. (2005). An acylation cycle regulates localization and activity of palmitoylated Ras isoforms. *Science, 307,* 1746—1752.

Rose, J. J., Taylor, J. B., Shi, J., Cockett, M. I., Jones, P. G., & Hepler, J. R. (2000). RGS7 is palmitoylated and exists as biochemically distinct forms. *Journal of Neurochemistry, 75,* 2103—2112.

Roth, A. F., Feng, Y., Chen, L., & Davis, N. G. (2002). The yeast DHHC cysteine-rich domain protein Akr1p is a palmitoyl transferase. *The Journal of Cell Biology, 159,* 23—28.

Roth, A. F., Wan, J., Bailey, A. O., Sun, B., Kuchar, J. A., Green, W. N., et al. (2006). Global analysis of protein palmitoylation in yeast. *Cell, 125,* 1003—1013.

Saito, S., Ikeda, M., Iwata, N., Suzuki, T., Kitajima, T., Yamanouchi, Y., et al. (2005). No association was found between a functional SNP in ZDHHC8 and schizophrenia in a Japanese case-control population. *Neuroscience Letters, 374,* 21—24.

Saleem, A. N., Chen, Y. H., Baek, H. J., Hsiao, Y. W., Huang, H. W., Kao, H. J., et al. (2010). Mice with alopecia, osteoporosis, and systemic amyloidosis due to mutation in Zdhhc13, a gene coding for palmitoyl acyltransferase. *PLoS Genetics, 6,* e1000985.

Santiago-Tirado, F. H., Peng, T., Yang, M., Hang, H. C., & Doering, T. L. (2015). A single protein s-acyl transferase acts through diverse substrates to determine cryptococcal morphology, stress tolerance, and pathogenic outcome. *PLoS Pathogens, 11,* e1004908.

Schmidt, M. F., Bracha, M., & Schlesinger, M. J. (1979). Evidence for covalent attachment of fatty acids to Sindbis virus glycoproteins. *Proceedings of the National Academy of Sciences of the United States of America, 76,* 1687—1691.

Schmidt, M. F., & Schlesinger, M. J. (1979). Fatty acid binding to vesicular stomatitis virus glycoprotein: a new type of post-translational modification of the viral glycoprotein. *Cell, 17,* 813—819.

Sharma, C., Rabinovitz, I., & Hemler, M. E. (2012). Palmitoylation by DHHC3 is critical for the function, expression, and stability of integrin alpha6beta4. *Cellular and Molecular Life Sciences, 69,* 2233—2244.

Sharma, C., Yang, X. H., & Hemler, M. E. (2008). DHHC2 affects palmitoylation, stability, and functions of tetraspanins CD9 and CD151. *Molecular Biology of the Cell, 19,* 3415—3425.

Sheng, M., & Hoogenraad, C. C. (2007). The postsynaptic architecture of excitatory synapses: a more quantitative view. *Annual Review of Biochemistry, 76,* 823—847.

Shipston, M. J. (2014). Ion channel regulation by protein S-acylation. *Journal of General Physiology, 143,* 659—678.

Singaraja, R. R., Huang, K., Sanders, S. S., Milnerwood, A. J., Hines, R., Lerch, J. P., et al. (2011). Altered palmitoylation and neuropathological deficits in mice lacking HIP14. *Human Molecular Genetics, 20,* 3899—3909.

Smith, K. R., Kopeikina, K. J., Fawcett-Patel, J. M., Leaderbrand, K., Gao, R., Schurmann, B., et al. (2014). Psychiatric risk factor ANK3/ankyrin-G nanodomains regulate the structure and function of glutamatergic synapses. *Neuron, 84,* 399—415.

Smotrys, J. E., Schoenfish, M. J., Stutz, M. A., & Linder, M. E. (2005). The vacuolar DHHC-CRD protein Pfa3p is a protein acyltransferase for Vac8p. *The Journal of Cell Biology, 170,* 1091—1099.

Song, I. W., Li, W. R., Chen, L. Y., Shen, L. F., Liu, K. M., Yen, J. J., et al. (2014). Palmitoyl acyltransferase, Zdhhc13, facilitates bone mass acquisition by regulating postnatal epiphyseal development and endochondral ossification: a mouse model. *PLoS One, 9,* e92194.

Srinivasa, S. P., Bernstein, L. S., Blumer, K. J., & Linder, M. E. (1998). Plasma membrane localization is required for RGS4 function in *Saccharomyces cerevisiae. Proceedings of the National Academy of Sciences of the United States of America, 95,* 5584—5589.

Staufenbiel, M. (1987). Ankyrin-bound fatty acid turns over rapidly at the erythrocyte plasma membrane. *Molecular and Cellular Biology, 7,* 2981—2984.

Storck, E. M., Serwa, R. A., & Tate, E. W. (2013). Chemical proteomics: a powerful tool for exploring protein lipidation. *Biochemical Society Transactions, 41,* 56—61.

Stowers, R. S., & Isacoff, E. Y. (2007). *Drosophila* huntingtin-interacting protein 14 is a presynaptic protein required for photoreceptor synaptic transmission and expression of the palmitoylated proteins synaptosome-associated protein 25 and cysteine string protein. *The Journal of Neuroscience, 27,* 12874—12883.

Sutton, L. M., Sanders, S. S., Butland, S. L., Singaraja, R. R., Franciosi, S., Southwell, A. L., et al. (2013). Hip14l-deficient mice develop neuropathological and behavioural features of Huntington disease. *Human Molecular Genetics, 22,* 452—465.

Suzuki, M., Murakami, T., Cheng, J., Kano, H., Fukata, M., & Fujimoto, T. (2015). ELMOD2 is anchored to lipid droplets by palmitoylation and regulates ATGL recruitment. *Molecular Biology of the Cell, 26,* 2333—2342.

Swarthout, J. T., Lobo, S., Farh, L., Croke, M. R., Greentree, W. K., Deschenes, R. J., et al. (2005). DHHC9 and GCP16 constitute a human protein fatty acyltransferase with specificity for H- and N-Ras. *The Journal of Biological Chemistry, 280*, 31141–31148.

Takamori, S., Holt, M., Stenius, K., Lemke, E. A., Gronborg, M., Riedel, D., et al. (2006). Molecular anatomy of a trafficking organelle. *Cell, 127*, 831–846.

Tanimura, N., Saitoh, S., Kawano, S., Kosugi, A., & Miyake, K. (2006). Palmitoylation of LAT contributes to its subcellular localization and stability. *Biochemical and Biophysical Research Communications, 341*, 1177–1183.

Thomas, G. M., Hayashi, T., Chiu, S. L., Chen, C. M., & Huganir, R. L. (2012). Palmitoylation by DHHC5/8 targets GRIP1 to dendritic endosomes to regulate AMPA-R trafficking. *Neuron, 73*, 482–496.

Thomas, G. M., Hayashi, T., Huganir, R. L., & Linden, D. J. (2013). DHHC8-dependent PICK1 palmitoylation is required for induction of cerebellar long-term synaptic depression. *The Journal of Neuroscience, 33*, 15401–15407.

Thomas, G. M., & Huganir, R. L. (2013). Palmitoylation-dependent regulation of glutamate receptors and their PDZ domain-containing partners. *Biochemical Society Transactions, 41*, 72–78.

Tian, L., McClafferty, H., Jeffries, O., & Shipston, M. J. (2010). Multiple palmitoyltransferases are required for palmitoylation-dependent regulation of large conductance calcium- and voltage-activated potassium channels. *The Journal of Biological Chemistry, 285*, 23954–23962.

Tobin, A. B., & Wheatley, M. (2004). G-protein-coupled receptor phosphorylation and palmitoylation. *Methods in Molecular Biology, 259*, 275–281.

Topinka, J. R., & Bredt, D. S. (1998). N-terminal palmitoylation of PSD-95 regulates association with cell membranes and interaction with K^+ channel, Kv1.4. *Neuron, 20*, 125–134.

Tsutsumi, R., Fukata, Y., Noritake, J., Iwanaga, T., Perez, F., & Fukata, M. (2009). Identification of G protein alpha subunit-palmitoylating enzyme. *Molecular and Cellular Biology, 29*, 435–447.

Tu, Y., Popov, S., Slaughter, C., & Ross, E. M. (1999). Palmitoylation of a conserved cysteine in the regulator of G protein signaling (RGS) domain modulates the GTPase-activating activity of RGS4 and RGS10. *The Journal of Biological Chemistry, 274*, 38260–38267.

Valdez-Taubas, J., & Pelham, H. (2005). Swf1-dependent palmitoylation of the SNARE Tlg1 prevents its ubiquitination and degradation. *EMBO Journal, 24*, 2524–2532.

Van Itallie, C. M., Gambling, T. M., Carson, J. L., & Anderson, J. M. (2005). Palmitoylation of claudins is required for efficient tight-junction localization. *Journal of Cell Science, 118*, 1427–1436.

Veit, M. (2012). Palmitoylation of virus proteins. *Biology of the Cell, 104*, 493–515.

Veit, M., Laage, R., Dietrich, L., Wang, L., & Ungermann, C. (2001). Vac8p release from the SNARE complex and its palmitoylation are coupled and essential for vacuole fusion. *EMBO Journal, 20*, 3145–3155.

Wan, J., Savas, J. N., Roth, A. F., Sanders, S. S., Singaraja, R. R., Hayden, M. R., et al. (2013). Tracking brain palmitoylation change: predominance of glial change in a mouse model of Huntington's disease. *Chemistry and Biology, 20*, 1421–1434.

Wang, Y. X., Kauffman, E. J., Duex, J. E., & Weisman, L. S. (2001). Fusion of docked membranes requires the armadillo repeat protein Vac8p. *The Journal of Biological Chemistry, 276*, 35133–35140.

Wedegaertner, P. B., & Bourne, H. R. (1994). Activation and depalmitoylation of Gs alpha. *Cell, 77*, 1063–1070.

Wedegaertner, P. B., Chu, D. H., Wilson, P. T., Levis, M. J., & Bourne, H. R. (1993). Palmitoylation is required for signaling functions and membrane attachment of Gq alpha and Gs alpha. *The Journal of Biological Chemistry, 268*, 25001–25008.

Wei, X., Song, H., & Semenkovich, C. F. (2014). Insulin-regulated protein palmitoylation impacts endothelial cell function. *Arteriosclerosis, Thrombosis, and Vascular Biology, 34*, 346−354.

Woolfrey, K. M., Sanderson, J. L., & Dell'Acqua, M. L. (2015). The palmitoyl acyltransferase DHHC2 regulates recycling endosome exocytosis and synaptic potentiation through palmitoylation of AKAP79/150. *The Journal of Neuroscience, 35*, 442−456.

Yan, S. M., Tang, J. J., Huang, C. Y., Xi, S. Y., Huang, M. Y., Liang, J. Z., et al. (2013). Reduced expression of ZDHHC2 is associated with lymph node metastasis and poor prognosis in gastric adenocarcinoma. *PLoS One, 8*, e56366.

Yang, W., Di Vizio, D., Kirchner, M., Steen, H., & Freeman, M. R. (2010). Proteome scale characterization of human S-acylated proteins in lipid raft-enriched and non-raft membranes. *Molecular and Cellular Proteomics, 9*, 54−70.

Yang, X., Kovalenko, O. V., Tang, W., Claas, C., Stipp, C. S., & Hemler, M. E. (2004). Palmitoylation supports assembly and function of integrin-tetraspanin complexes. *The Journal of Cell Biology, 167*, 1231−1240.

Yount, J. S., Moltedo, B., Yang, Y. Y., Charron, G., Moran, T. M., Lopez, C. B., et al. (2010). Palmitoylome profiling reveals S-palmitoylation-dependent antiviral activity of IFITM3. *Nature Chemical Biology, 6*, 610−614.

Yuan, X., Zhang, S., Sun, M., Liu, S., Qi, B., & Li, X. (2013). Putative DHHC-cysteine-rich domain S-acyltransferase in plants. *PLoS One, 8*, e75985.

Zhang, J., Planey, S. L., Ceballos, C., Stevens, S. M., Jr., Keay, S. K., & Zacharias, D. A. (2008). Identification of CKAP4/p63 as a major substrate of the palmitoyl acyltransferase DHHC2, a putative tumor suppressor, using a novel proteomics method. *Molecular and Cellular Proteomics, 7*, 1378−1388.

Zhang, M. M., Wu, P. Y., Kelly, F. D., Nurse, P., & Hang, H. C. (2013). Quantitative control of protein S-palmitoylation regulates meiotic entry in fission yeast. *PLoS Biology, 11*, e1001597.

Zhou, B., An, M., Freeman, M. R., & Yang, W. (2014). Technologies and challenges in proteomic analysis of protein S-acylation. *Journal of Proteomics and Bioinformatics, 7*, 256−263.

Zhu, J. J., Qin, Y., Zhao, M., Van Aelst, L., & Malinow, R. (2002). Ras and Rap control AMPA receptor trafficking during synaptic plasticity. *Cell, 110*, 443−455.

CHAPTER FIVE

An Adaptable Spectrin/Ankyrin-Based Mechanism for Long-Range Organization of Plasma Membranes in Vertebrate Tissues

Vann Bennett* and Damaris N. Lorenzo

HHMI and Department of Biochemistry, Duke University Medical Center, Durham, NC, USA
*Corresponding author: E-mail: benne012@mc.duke.edu

Contents

Abstract

Ankyrins are membrane-associated proteins that together with their spectrin partners are responsible for micron-scale organization of vertebrate plasma membranes, including those of erythrocytes, excitable membranes of neurons and heart, lateral membrane domains of columnar epithelial cells, and striated muscle. Ankyrins coordinate functionally related membrane transporters and cell adhesion proteins (15 protein families identified so far) within plasma membrane compartments through independently evolved interactions of intrinsically disordered sequences with a highly conserved peptide-binding groove formed by the ANK repeat solenoid. Ankyrins are coupled to spectrins, which are elongated organelle-sized proteins that form mechanically resilient arrays through cross-linking by specialized actin filaments. In addition to protein interactions, cellular targeting and assembly of spectrin/ankyrin domains also

Current Topics in Membranes, Volume 77
ISSN 1063-5823
http://dx.doi.org/10.1016/bs.ctm.2015.10.001

143

critically depend on palmitoylation of ankyrin-G by aspartate—histidine—histidine—cysteine 5/8 palmitoyltransferases, as well as interaction of beta-2 spectrin with phosphoinositide lipids. These lipid-dependent spectrin/ankyrin domains are not static but are locally dynamic and determine membrane identity through opposing endocytosis of bulk lipids as well as specific proteins. A partnership between spectrin, ankyrin, and cell adhesion molecules first emerged in bilaterians over 500 million years ago. Ankyrin and spectrin may have been recruited to plasma membranes from more ancient roles in organelle transport. The basic bilaterian spectrin—ankyrin toolkit markedly expanded in vertebrates through gene duplications combined with variation in unstructured intramolecular regulatory sequences as well as independent evolution of ankyrin-binding activity by ion transporters involved in action potentials and calcium homeostasis. In addition, giant vertebrate ankyrins with specialized roles in axons acquired new coding sequences by exon shuffling. We speculate that early axon initial segments and epithelial lateral membranes initially were based on spectrin—ankyrin-cell adhesion molecule assemblies and subsequently served as "incubators," where ion transporters independently acquired ankyrin-binding activity through positive selection.

1. INTRODUCTION

Plasma membranes are composite materials comprised of a thin and fragile phospholipid bilayer coordinated with both bilayer-spanning and bilayer-associated proteins. Most current physical models for plasma membrane structure emphasize dynamic behavior of membrane-spanning proteins due to random diffusion and transient associations that are confined within 10—300 nm compartments by cortical actin filaments and lipid phase separations (Garcia-Parajo, Cambi, Torreno-Pina, Thompson, & Jacobson, 2014; Kusumi et al., 2012; Saka et al., 2014). These models primarily are based on imaging of two-dimensional cultures of thin cells such as fibroblasts, which are motile, do not form extensive cell—cell contacts and are not subject to disruptive mechanical forces.

Metazoan cells in their native environments must solve two problems requiring long-range organization of their plasma membranes that are not faced by cells in culture or those with cell walls such as fungi and plants. First, they require mechanical support for their intrinsically fragile phospholipid bilayers. For example, axons are very thin (0.2—5 μm in diameter), yet frequently extend for hundreds of microns, and are vulnerable to shearing. Mammalian enucleated erythrocytes, similarly, must survive for periods of months in a high-shear environment while circulating thousands of times through the vascular system. In addition to physical reinforcement, surfaces of most animal cells are also patterned into regions with distinct molecular composition. For example, many cells coordinate membrane transporters and other bilayer-spanning proteins within micron-scale domains to optimize their physiological functions. Vertebrate neurons, for instance, must cluster voltage-gated sodium channels at high density at axon initial segments and nodes of Ranvier in order to fire properly timed self-propagating action potentials. Similarly, transporting epithelia must concentrate membrane transporters in their apical and lateral membranes in order to achieve vectorial transport of ions and small nutrients.

Plasma membranes of enucleated mammalian erythrocytes have provided an experimentally accessible model that has provided a foundation for understanding how vertebrate cells pattern and stabilize their plasma membranes (Bennett & Lorenzo, 2013; Fowler, 2013). Erythrocyte plasma membranes are coated with a spectrin-based membrane skeleton which forms a polygonal network coupled by ankyrin to the cytoplasmic domain of the anion exchanger (Bennett & Stenbuck, 1979a, 1979b; Bennett and Stenbuck, 1980a; Byers & Branton, 1985). The spectrin skeleton provides mechanical support required to prevent hemolysis and anemia (Agre et al., 1985; Agre, Orringer, & Bennett, 1982; Bodine, Birkenmeier, & Barker, 1984) and also restricts lateral mobility of the major bilayer-spanning proteins (Fowler & Bennett, 1978; Golan & Veatch, 1980; Sheetz, Schindler, & Koppel, 1980). Despite their expression in nonerythroid tissues, spectrin and ankyrin are not major components of plasma membranes of many cells grown in two-dimensional cultures, and have largely escaped the attention of the biophysics community (Garcia-Parjo et al., 2014; Kusumi et al., 2012).

Spectrins and their ankyrin partners are central elements in assembly and maintenance of diverse plasma membrane domains that coordinate functionally related membrane-spanning proteins in multiple vertebrate tissues (Table 1; Figures 1–3) (Bennett & Healy, 2009; Bennett & Lorenzo, 2013). We will summarize these findings and present advances in understanding mechanisms underlying the function and evolution of spectrin and ankyrin. A recurring theme will be the dynamic nature of this system on timescales ranging from millions of years, with adaptive evolution of new partners and physiological applications, to minutes with new evidence that spectrin and ankyrin, in a major departure from the erythrocyte, are organized into locally dynamic microdomains that determine membrane identity through regulating membrane recycling.

Table 1 Ankyrin/spectrin-dependent plasma membrane domains

Membrane domain/Cell	Ankyrin	αβ-Spectrin	Ankyrin partners
Mammalian erythrocyte[1]	200 kDa Ankyrin-R	α1/β1 spectrin	Anion exchanger 1 RhB/G
Axon initial segment/Neuron[2]	480 kDa Ankyrin-G	α2/β4 spectrin	Voltage-gated sodium channels KCNO2/3, NRCAM186 kDa Neurofascin
Paranode/Schwann cells[3]	Ankyrin-B	α2/β2 spectrin	155 kDa Neurofascin
Paranode/Oligodendrocyte[3]	Ankyrin-G	α2/β2 spectrin	155 kDa Neurofascin
Node of Ranvier/Neuron[4]	480 kDa Ankyrin-G	α2/β4 spectrin	Voltage-gated sodium channels KCNQ2/3, Kv3.1b 186 kDa Neurofascin
Unmyelinated axon/Neuron[5]	440 kDa Ankyrin-B	α2/β2 spectrin	L1CAM
Dendritic spine neck/Neuron[6]	190 kDa Ankyrin-G	α2/β2 spectrin	?
Neuromuscular junction/Skeletal muscle[7,8]	190 kDa Ankyrin-G	α2/β2 spectrin	Voltage-gated sodium channels 186 kDa Neurofascin, Dystroglycan
Neuromuscular junction/Skeletal muscle[8]	220 kDa Ankyrin-B	α2/β2 spectrin	?
Costamere/Skeletal muscle[8]	190 kDa Ankyrin-G 220 kDa Ankyrin-B	α2/β2 spectrin α2/β1 spectrin	Dystroglycan
Intercalated disc/Cardiomyocyte[9]	190 kDa Ankyrin-G 220 kDa Ankyrin-B	α2/β4 spectrin α2/β2 spectrin	Voltage-gated sodium channels N-cadherin,Cav1.3

SR/T-tubule junction/Cardiomyocyte[10]	220 kDa Ankyrin-B	α2/β2 spectrin	Na/K ATPase, NaCaX IP3R, Kir6.2
Outer segment/Rod photoreceptor[11]	190 kDa Ankyrin-G	α2/β4 spectrin	CNGI31
Inner segment/Rod photoreceptor[12]	220 kDa Ankyrin-B	α2/β2 spectrin	Na/K ATPase NaCaX
Lateral membrane/MDCK cell/Collecting Duct Cells/Bronchial Epithelial cell[13]	190 kDa Ankyrin-G	α2/β2 spectrin	Na/K ATPase E-cadherin

[1]Bennett and Stenbuck (1979a) and Lopez et al. (2005).
[2]Jenkins, Kim, et al. (2015), Pan et al. (2006) and Davis et al. (1996).
[3]Chang et al. (2014).
[4]Jenkins, Kim, et al. (2015) and Davis et al., 1996.
[5]Chan et al. (1993), Kunimoto (1995) and Scotland et al. (1998).
[6]Smith et al. (2014),
[7]Flucher and Daniels, 1989, Kordeli, Ludosky, Deprette, Frappier, and Cartaud (1998) and Wood and Slater (1998).
[8]Ayalon, Davis, Scotland, and Bennett (2008) and Ayalon et al. (2011).
[9]Mohler, Rivolta, et al. (2004), Hund et al. (2010) and Makara et al. (2014).
[10]Mohler (2005) and Li, Kline, Hund, Anderson, and Mohler (2010).
[11]Kizhatil, Baker, Arshavsky, and Bennett (2009).
[12]Kizhatil, Sandhu, Peachy, and Bennett (2009).
[13]Nelson and Hammerton (1989), Nelson, Shore, Wang, and Hammerton (1990), Kizhatil, Yoon, et al. (2007), Kizhatil, Davis, et al. (2007) and Jenkins et al. (2013).

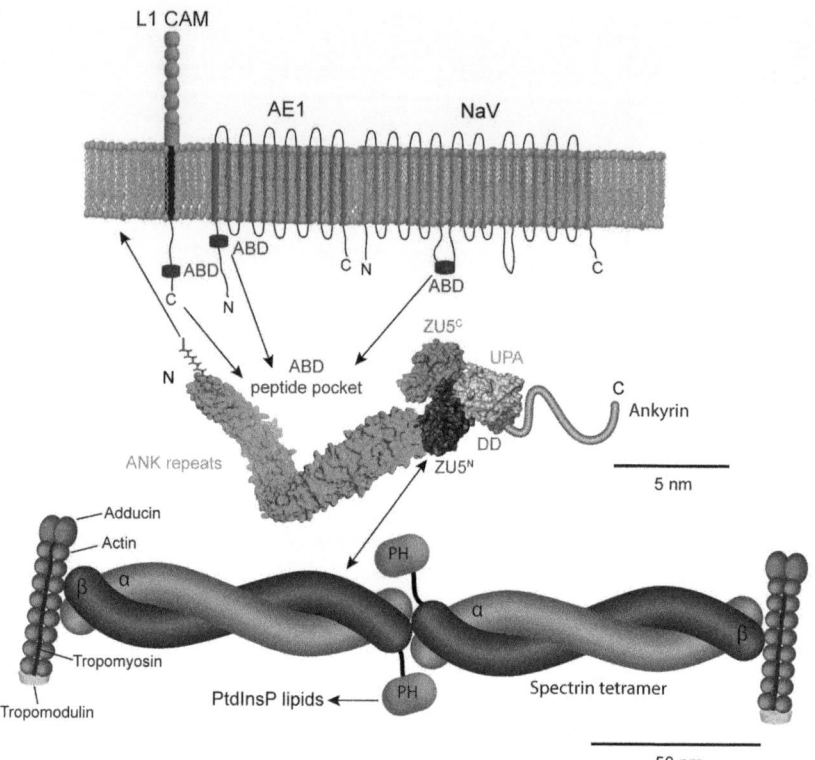

Figure 1 *Membrane-spanning proteins are coupled to spectrin—actin networks through association of independently evolved unstructured short peptide sequences to a conserved peptide-binding groove formed by ANK repeats.* Integral membrane proteins, including cell adhesion proteins, ion channels, and membrane transporters independently evolved intrinsically disordered ankyrin-binding sites (ABDs) within their cytoplasmic domains, which associate with a conserved peptide-binding pocket formed by the ANK repeat solenoid. Ankyrin with its membrane partners is coupled to extended spectrin—actin networks tightly associated with the plasma membrane. Spectrin polymerizes into membrane-associated networks through association with specialized actin filaments, which is promoted by accessory proteins.

2. A CORE SPECTRIN/ACTIN BUILDING SET FOR ASSEMBLY OF PLASMA MEMBRANE DOMAINS

Ankyrin and/or spectrin have been implicated in organization and/or stabilization of plasma membrane domains in multiple cell types including those in the nervous system, striated muscle, and epithelial tissues (Table 1; Figures 1—3). Although details of membrane partners, as well as genes encoding ankyrin and spectrin differ, a central logic underlies all of these

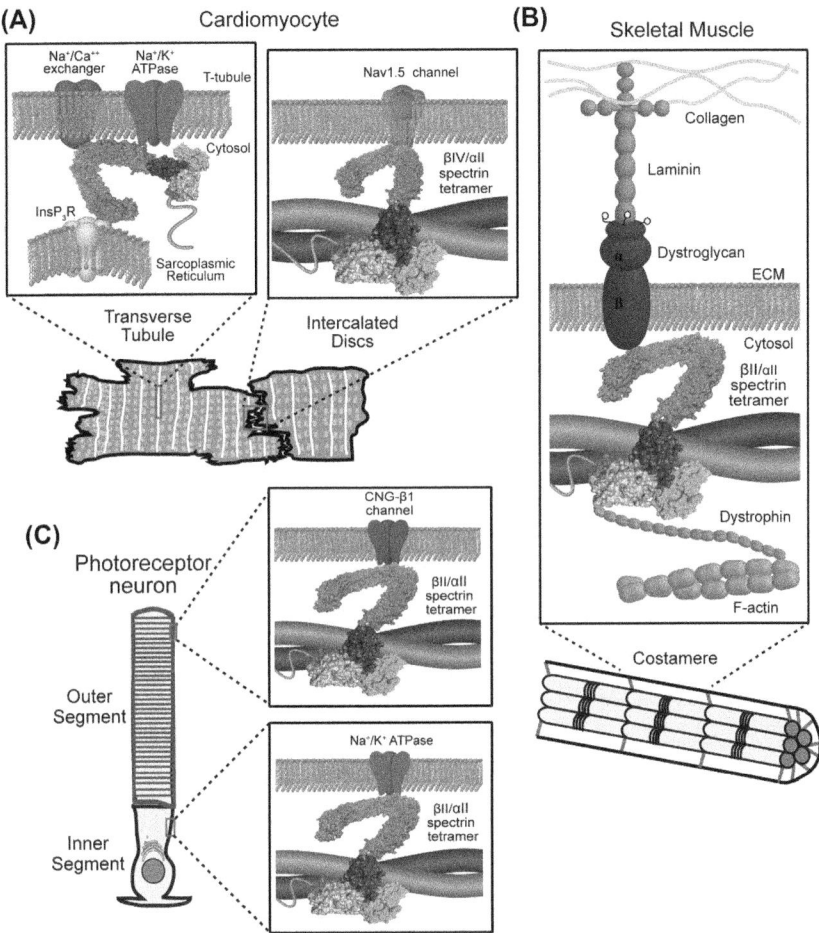

Figure 2 *Ankyrin/spectrin-based coordination of functionally related membrane-spanning proteins within plasma membrane domains.* Ankyrin—spectrin-based domains coordinate functionally related but structurally distinct membrane transporters, cell adhesion molecules to specialized, micron-scale membrane domains, including transverse tubules and intercalated discs of cardiomyocyte (A), costameres of skeletal muscle (B), and the outer and inner segments of photoreceptor neurons (C).

examples: functionally related membrane-spanning proteins, including cell adhesion proteins as well as membrane transporters, associate through easily and independently evolved interactions of intrinsically disordered regions of their cytoplasmic domains with a highly conserved peptide-binding pocket formed by the ANK repeat solenoid (Bennett & Healy, 2009; Wang et al., 2014) (Figure 1). Ankyrins, in turn, are coupled to spectrins, which

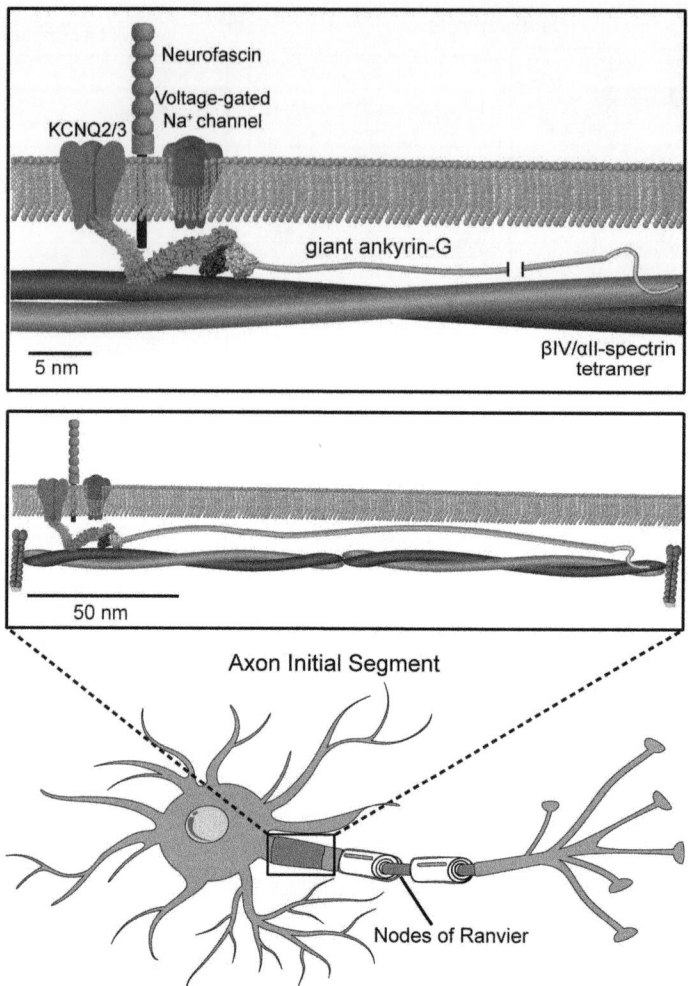

Figure 3 *Giant ankyrin-G coordinates cell adhesion molecules and membrane transporters at the axon initial segment.* Schematic representation drawn at scale illustrates the molecular composition of the axon initial segment. 480 kDa ankyrin-G extends for about 150 nm along the axis of the axon and functions as a master organizer of the axon initial segment required for the localization of all of the known axon initial segment components including voltage-gated sodium channels, KCNQ2/3 channels, neurofascin, β4-spectrin, as well as bundled microtubules.

are elongated organelle–scale proteins 190 nm in length (close to the length of mitochondria) (Bennett, Davis, & Fowler, 1982; Bennett & Lorenzo, 2013; Shotton, Burke, & Branton, 1979) (Figure 1). Spectrins assemble into extended networks through interaction at their ends with specialized configurations of actin, distinct from cortical actin filaments, that require accessory proteins such as adducin and protein 4.1 (Baines, Lu, & Bennett, 2014; Fowler, 2013; Matsuoka, Li, & Bennett, 2000).

It is challenging to image membrane-associated spectrin—actin structures, but this has been achieved at high resolution in both erythrocytes and axons. Electron microscopy of spreads of erythrocyte ghosts has resolved a polygonal network formed by association of 5—7 spectrins at 50—70 degree angles with short 40 nm actin protofilaments (Byers & Branton, 1985). Superresolution light microscopy of axons of cultured hippocampal neurons has imaged actin—adducin assemblies organized into periodic membrane-associated rings separated by 190 nm, which is precisely the length determined for brain spectrin tetramers by electron microscopy (Bennett et al., 1982; Xu, Zhong, & Zhuang, 2013; Zhong et al., 2014). Spectrin antibody labels sites located between the actin—adducin rings suggesting that these structures are interconnected by spectrin tetramers attached at a 90-degree angle (Xu et al., 2013). Spectrin—actin—adducin networks covering extensive areas of the plasma membrane so far have been resolved only in erythrocytes and axons.

A common misconception in the literature is that spectrins recruit extended actin filaments to the plasma membrane. Spectrin alone interacts weakly with actin filaments and requires accessory proteins to establish stable complexes (Gardner and Bennett, 1987; Kuhlman, Hughes, Bennett, & Fowler, 1996; Li, Matsuoka, & Bennett, 1998). Spectrin-associated accessory proteins in turn confer special properties on actin. For example, adducin, which is present in erythrocytes, axons, and epithelial lateral membranes, caps the fast-growing ends of actin filaments in addition to recruiting spectrin (Abdi & Bennett, 2008; Kuhlman et al., 1996; Li et al., 1998). Adducin-capped actin filaments are also capped on their slow-growing ends by tropomodulin working in concert with tropomyosin (Fowler, 2013). Although spectrin can associate with actin filaments with low affinity in the absence of adducin, adducin—spectrin—actin complexes predominate both in vitro and in vivo (Bennett, Gardner, & Steiner, 1988; Gardner & Bennett, 1987; Li et al., 1998).

We will next summarize the biochemical properties of spectrin and ankyrin that provide the basis for their plasma membrane functions. Spectrins are flexible elongated tetramers 190 nm in length comprised of alpha and beta subunits assembled side-to-side in an antiparallel orientation and head-to-head through association of alpha-spectrin with beta-spectrin (Figure 1) (Shotton et al., 1979). Alpha- and beta-spectrin subunits are both related to alpha-actinin (Djinovic-Carugo, Young, Gautel, & Saraste, 1999) but are extended to nearly 100 nm in length by multiple copies of a triple helical repeat (Yan et al., 1993). Head-to-head association between alpha—beta-spectrin dimers to form tetramers results from noncovalent assembly of partial triple helical repeats of each spectrin subunit (Ipsaro et al., 2010; Mehboob et al., 2010).

The beta subunit contributes the principal interactions of spectrins with other proteins. Beta-spectrin associates with F-actin through N-terminal tandem calponin homology domains (Banuelos, Saraste, & Carugo, 1998; Carugo, Banuelos, & Saraste, 1997), with ankyrin through the 14th and 15th triple helical repeats (Davis et al., 2009; Ipsaro, Huang, & Mondragón, 2009; Ipsaro & Mondragón, 2010; Stabach et al., 2009). Beta-spectrins also associate with phosphoinositide lipids through a C-terminal pleckstrin homology (PH) domain (Das et al., 2008; He, Abdi, & Bennett, 2014; Hyvonen et al., 1995; Macias et al., 1994) (Figure 1). Membrane interactions through the PH domain combined with ankyrin play important roles in polarized assembly of spectrin in epithelial cells (see below).

Ankyrin proteins are monomers with an N-terminal membrane-binding domain containing 24 tandem ANK repeats folded as a solenoid that mediate binding to membrane-spanning proteins and are S-palmitoylated at a single conserved cysteine (C70 in ankyrin-G) (He et al., 2014; He, Jenkins, & Bennett, 2012; Wang et al., 2014). ANK repeats are one of the most widely expressed protein motifs found in all phyla and have diverse roles in macromolecular recognition (Li et al., 2006). ANK repeats due to their stable folding and capacity for protein interaction have applications as therapeutic and diagnostic agents (Plückthun, 2015). ANK repeats of ankyrins are followed by two ZU5 domains and a UPA domain, a death domain, and an unstructured regulatory domain that modulates activities of other domains through autoinhibition (Figure 1). The ZU5 and UPA domains are folded into a tightly packed supermodule resembling a three-lobed clover leaf (Wang, Yu, Ye, Wei, & Zhang, 2012). The functional importance of the supermodule was established by an unbiased screen for loss-of-function mutations for ankyrin-B and ankyrin-G, which identified key residues located at subdomain interfaces in the supermodule structure (Kizhatil, Yoon, et al., 2007; Wang et al., 2012).

Ankyrins bind to beta-spectrins through their N-terminal ZU5 domain (Ipsaro & Mondragón, 2010; Mohler, Yoon, & Bennett, 2004). The C-terminal ZU5 domain of ankyrin-B contains basic pocket that is required for association with PtdIns(3)P lipids (Lorenzo et al., 2014; Wang et al.,

2012). Modeling suggests that the ankyrin-B ZU5-ZU5-UPA supermodule can bind simultaneously with spectrin and PtdIns(3)P lipids (Wang et al., 2012). Functions of the basic pocket are known so far only for ankyrin-B, which requires binding to PtdIns(3)P derived from the class 3 phosphoinositide kinase to promote organelle transport in axons (see below) (Lorenzo et al., 2014). However, the ZU5 domain basic sequences are likely highly conserved among both vertebrate and invertebrate ankyrins, suggesting that binding to phosphoinositide lipids is a core ankyrin function.

The same set of two ZU5 domains and UPA domain combined with a death domain occurs in the protein PIDD, which assembles into a large complex the promotes caspase-2-dependent apoptotic signaling (Nematolahi et al., 2015). Ankyrin-G has been implicated in regulation of anoikis (cell death due to detachment of epithelial cells from basement membranes), but roles of its death domain or death domains of other ankyrins are not known (Kumar et al., 2011).

Ankyrin and spectrin both experience stretching in erythrocytes under physiological levels of shear stress which is sufficient to expose buried cysteine residues (Krieger et al., 2011). A rationale for how these proteins can reversibly respond to shear is provided by atomic force microscopy measurements that reveal elastic behavior of ankyrin through its ANK repeats (Lee et al., 2006) and spectrin through its triple helical repeats (Rief, Pascual, Saraste, & Gaub, 1999). Moreover, deficiency of ankyrin and spectrin leads to fragile erythrocyte membranes (Eber & Lux, 2004) and as well as axons (Hammarlund, Jorgensen, & Bastiani, 2007). Ankyrin–spectrin assemblies thus can provide mechanical stability to lipid bilayers in addition to lateral segregation of membrane proteins.

3. VERTEBRATE EXPANSION OF THE BILATERIAN SPECTRIN–ANKYRIN GENETIC TOOLKIT

Ankyrins with all modern domains including a beta-spectrin-binding site as well as beta-spectrins with predicted ankyrin-binding sites (ABDs) are present throughout bilaterians (animals with bilateral symmetry including insects, nematodes, mollusks, etc.) (Bennett & Lorenzo, 2013). ABDs of the L1CAM family also are conserved among bilaterians and have been demonstrated to bind to ankyrin in the case of the *Caenorhabditis elegans* LAD-1 (Chen, Ong, & Bennett, 2001). Moreover, beta-dystroglycans bearing a recognizable ankyrin-binding sequence are present in bilaterians as well as Cnidarians (Bennett & Lorenzo, 2013). Thus, a spectrin/ankyrin system with potential for lateral organization of cell adhesion proteins into micron-scale mechanically resilient domains likely was in place in early bilaterians such as Kimberella, which is found in fossils dated to 555 million years ago in the pre-Cambrian period (Martin et al., 2000).

The basic bilaterian repertoire of single copies of alpha-spectrin, beta-spectrin with ankyrin-binding activity, and ankyrin, has been markedly expanded in vertebrates due to duplication and subsequent divergence of genes (Bennett & Lorenzo, 2013). Jawed fish and their vertebrate descendants have three beta-spectrin genes resulting from whole genome duplications: *SPTB*, encoding beta-1 spectrin, first characterized in red blood cells (Winkelmann, Chang, et al., 1990); *SPTBN1*, encoding beta-2 spectrin, first characterized in brain (Bennett et al., 1982; Hu, Watanabe, & Bennett, 1992); and *SPTBN4,* encoding beta-4 spectrin, which is localized with ankyrin-G at nodes of Ranvier and axon initial segments (Berghs et al., 2000; Jenkins & Bennett, 2001), as well as intercalated discs of cardiomyocytes (Hund et al., 2010). Vertebrates except for mammals have a single alpha-spectrin gene, *SPTAN1*, encoding alpha-2 spectrin, which is the generally expressed partner of beta-spectrins (Bennett et al., 1982; Davis & Bennett, 1983; Wasenius et al., 1989). In addition, mammals express a second alpha-spectrin gene, *SPTA1*, encoding alpha-1 spectrin, which is expressed primarily in erythrocytes and has reduced ability to assemble into tetramers (Mehboob et al., 2010; Salomeo et al., 2006).

Mammals have an additional beta-spectrin gene, *SPTBN2*, encoding beta-3 spectrin, which is most highly expressed in the cerebellum (Ohara, Ohara, Yamakawa, Nakajima, & Nakayama, 1998; Stankewich et al., 1998) and is mutated in autosomal dominant human spinocerebellar ataxia type 5 (Ikeda et al., 2006). A human family has been identified that is homozygous for beta-3 spectrin mutation and exhibits cognitive as well as motor deficits (Lise et al., 2012). Knockout of beta-3 spectrin in mice recapitulates human spinocerebellar ataxia symptoms and results in abnormal development of Purkinje cerebellar neuron dendrites as well as subtle changes in prefrontal neuron dendrites combined with cognitive deficits (Gao et al., 2011; Lise et al., 2012; Perkins et al., 2010). In contrast, knockout of the generally expressed beta-2 spectrin markedly impairs development and is embryonic lethal (Tang et al., 2003). Beta-3 spectrin is most closely related to beta-2 spectrin, and similar to alpha-1 spectrin, likely did not arise in the early whole genome duplication events.

Alternative splicing further increases the diversity of both alpha (Zhang et al., 2010) and beta-spectrins (Berghs et al., 2000; Hayes et al., 2000;

Winkelmann, Costa, Linzie, & Forget, 1990). Variations include deletion of PH domains in beta-1 and beta-2 spectrins (Hayes et al., 2000; Winkelmann, Costa, et al., 1990) and deletion of either N-terminal portions of beta-4 spectrin (Berghs et al., 2000; Uemoto et al., 2007).

Beta 1—3 spectrins are overall similar in sequence and domain organization, although beta-4 spectrin has additional sequence between the final spectrin repeat and the PH domain (Berghs et al., 2000). Beta-4 spectrin also associates with calmodulin-dependent protein kinase 2 (CAM kinase 2) through this sequence and recruits CAM kinase 2 to cardiac intercalated discs and axon initial segments (Hund et al., 2010, 2014).

The phenotypes of beta-spectrin knockout mice indicate that beta-2 spectrin is the generally expressed isoform with nonredundant functions, while beta-1 spectrin is essential for erythrocyte survival and beta-4 spectrin has specialized roles in neurons and cardiomyocytes. Mice lacking beta-2 spectrin die at embryonic day 12 (Tang et al., 2003), while mice lacking beta-1 spectrin are severely anemic and exhibit ataxia but are viable with a bone marrow transplant to restore normal erythrocyte survival (Barker, Deveau, & Wandersee, 2000). Mice deficient in beta-4 spectrin have neurological deficits and abnormal regulation of hearth rhythm, but are viable (Hund et al., 2010; Komada & Soriano, 2002; Parkinson et al., 2001). As noted above, beta-3 spectrin knockout mice and humans with beta-3 spectrin mutations have cerebellar deficits but are not otherwise markedly impaired (Lise et al., 2012).

Vertebrates starting with jawed fish have three ankyrin genes: *ANK1*, encoding ankyrin-R (Bennett & Stenbuck, 1979a, 1980b; Lux, John, & Bennett, 1990); *ANK2*, encoding ankyrin-B (Davis & Bennett, 1984; Otto, Kunimoto, McLaughlin, & Bennett, 1991); and *ANK3*, encoding ankyrin-G (Kordeli, Lambert, & Bennett, 1995; Peters et al., 1995). The first whole genome duplication event in early vertebrates, now represented by lampreys, resulted in *ANK1* and the *ANK2/ANK3* precursor, while the second event resulted in *ANK2* and *ANK3*, but loss of the duplicate of *ANK1* (Cai & Zhang, 2006).

Ankyrins-G, -B, and -R retain extensive sequence similarity in their core-folded domains, but are divergent in intrinsically unstructured regulatory sequences, and now have acquired distinct functions. A major site of variation is in their unstructured C-terminal domains that modulate interactions with membrane proteins and spectrin through direct interactions with ANK repeats and the spectrin-binding domain (Abdi, Mohler, Davis, & Bennett, 2006; Davis, Davis, & Bennett, 1992; Hall & Bennett, 1987;

Mohler, Gramolini, & Bennett, 2002; Wang et al., 2014). Another site of regulation and sequence divergence between ankyrin-G and ankyrin-B is located in the linker peptide connecting ANK repeats with the first ZU5 domain (He, Tseng, & Bennett, 2013). This peptide associates with ANK repeats and prevents binding of ankyrin-B with neurofascin and E-cadherin as well as association of ankyrin-B with the plasma membrane (He et al., 2013).

Alternative splicing adds functional diversity between members of the same ankyrin family (Cunha, Le Scouarnec, Schott, & Mohler, 2008; Cunha & Mohler, 2008; Hall & Bennett, 1987; Hoock et al., 2007; Hopitzan, Baines, & Kordeli, 2006; Hopitzan, Baines, Ludosky, Recouvreur, & Kordeli, 2005; Lux et al., 1990; Otto et al., 1991; Peters et al., 1995). For example, an in-frame splice in the regulatory domain of ankyrin-R (band 2.2 in erythrocyte membranes) results in elimination of an acidic 186-residue segment and increased affinity for the anion exchanger as well as for spectrin (Davis et al., 1992; Hall & Bennett, 1987; Lux et al., 1990). Splicing within the regulatory domain of ankyrin-B regulates association with obscurin and the co-chaperone Hsp40 (Cunha & Mohler, 2008). Ankyrin-G (and likely ankyrin-B) polypeptides include spliced variants lacking ANK repeats altogether (Hoock et al., 2007; Hopitzan et al., 2005; Peters et al., 1995). These truncated polypeptides retain spectrin-binding domains and associate with intracellular organelles (Hoock, Peters, & Lux, 1997), although their functions are not known.

4. GIANT ANKYRINS: GENETIC NOVELTY THROUGH EXON SHUFFLING

In addition to variations on an established bilaterian theme by gene duplication and divergence, vertebrate ankyrin genes also have benefited from introduction of completely new coding sequence. The early vertebrate ANK2/ANK3 gene resulting from the first round of whole genome duplication acquired a giant exon inserted between its UPA and death domains (Bennett & Lorenzo, 2013). Following the second whole genome duplication event, the ANK2/3 precursor gave rise to separate ANK2 and ANK3 genes that both retained the giant exon, which has been maintained as a single exon in most vertebrate genomes.

480 kDa ankyrin-G (ANK3) and 440 kDa ankyrin-B (ANK2) polypeptides contain this giant exon-encoded sequence and are conserved from jawed fish to mammals (Chan, Kordeli, & Bennett, 1993; Kordeli et al.,

1995; Kunimoto, 1995; Kunimoto, Otto, & Bennett, 1991). Both giant ankyrin polypeptides are selectively expressed in the nervous system with specialized roles in axons. 480 kDa ankyrin-G requires its giant exon to function as the master organizer of axon initial segments as well as stabilize somatodendritic GABA-A synapses (Jenkins, Kim et al., 2015; Tseng, Jenkins, Tanaka, Mooney, & Bennett, 2015). The ankyrin-G giant exon is required to recruit β4 spectrin to the axon initial segment and also interacts with GABARAP, which contributes to stabilization of GABA-A synapses (Jenkins, Kim et al., 2015; Tseng et al., 2015). 440 kDa ankyrin-B is expressed in premyelinated axons and together with its smaller 220 kDa-spliced variant lacking the giant exon is required for maintenance of long axon tracts in the mouse central nervous system (Lorenzo et al., 2014; Scotland, Zhou, Benveniste, & Bennett, 1998).

Acquisition of exons through exon swapping is a well-established mechanism for formation of modular proteins and has contributed to genetic novelty in metazoans (Patthy, 1999). Giant exons of ANK2 and ANK3 share sequence similarity with I-connectin, an extended titin-related protein expressed in arthropod muscle (Jenkins, Kim et al., 2015). The region of I-connectin present in ankyrins is outside of the FNIII/Ig-like repeats and predominantly within a 2700-amino acid stretch containing a series of 68-residue SEK repeats (Fukuzawa et al., 2001). Consistent with the sequence similarity to I-connectin, 480 kDa ankyrin-G is 150 nm in length based on images obtained from axon initial segments by platinum replica electron microscopy combined with immunolabeling (Jenkins, Kim et al., 2015; Jones, Korobova, & Svitkina, 2014).

Although *ANK2* and *ANK3* giant exons share extensive sequence similarity, the AnkG exon encodes an N-terminal 40 kDa region absent from ankyrin-B that is enriched in serine and threonine residues and is modified by O-GlucNac monosaccharide residues (Vosseller et al., 2006; Zhang & Bennett, 1996) as well as regions with sequence quite divergent from ankyrin-B. The ankyrin giant exons, while sharing overall shape and folded domains, likely have evolved distinct molecular partners and functions.

Mutations of giant vertebrate ankyrins can be anticipated to be relatively common due to their large size and result in survivable nervous system-specific phenotypes due to their restricted patterns of expression. Interestingly, a frameshift mutation within the giant exon of ankyrin-G results in major cognitive and behavioral deficits in humans (Iqbal et al., 2013). Moreover, mutation of the large exon of ankyrin-B has been linked to autism spectrum disorder (De Rubeis et al., 2014; Iossifov et al., 2014).

Giant axonal ankyrins are not unique to vertebrates. They also have been reported in *C. elegans* (the single worm ankyrin is termed Unc-44 for its role in axon guidance), as well as in *Drosophila* (Koch et al., 2008; Otsuka et al., 1995; Pielage et al., 2008; Stephan et al., 2015). However, while core ankyrin domains are highly conserved, there is little sequence similarity between the extended regions found in *Drosophila*, *C. elegans*, and vertebrate giant ankyrin-spliced variants. This lack of conservation is most likely not due to accelerated divergence, since the large exons of vertebrate ankyrin-G and ankyrin-B have retained substantial similarity over 450 million years. Moreover, giant exons of vertebrate, *Drosophila*, and *C. elegans* ankyrins are inserted at different sites within their genes (Bennett & Walder, 2015).

Axonal ankyrins in vertebrates, flies, and nematodes thus have independently acquired extended sequences encoded by giant exons. An implication is that the physical characteristic(s) shared by myelinated vertebrate and unmyelinated invertebrate axons created a molecular niche where ankyrins with extended domains would be subject to positive selection. Interestingly, *Drosophila* ank2-XL (the isoform containing a giant exon) and mammalian 480 kDa ankyrin-G both are required for normal spacing of axonal microtubules (Jenkins, Kim et al., 2015; Stephan et al., 2015). Microtubule organization provides an example of an activity that could depend on the intrinsic membrane association of ankyrin that was combined with extended length by acquisition of giant exons, thus allowing penetration into the axoplasm. Current functions of the giant exon of 480 kDa ankyrin-G also include recruitment of beta-4 spectrin (Jenkins, Kim et al., 2015) and GABARAP (Tseng et al., 2015) but these functions are specific to vertebrate ankyrin-G and likely evolved later than interactions with microtubules.

5. EVOLUTION OF ANKYRIN-BINDING ACTIVITY BY MEMBRANE PROTEINS

Physiological activity of ankyrin-based domains considered in this review depends on interactions of ankyrins with membrane partners that are functionally related although distinct in sequence. For example, axon initial segments are enriched in voltage-gated Na channels, which are modulated by KCNQ2/3 channels, and coupled to GABA-A synapses through 186 kDa neurofascin. Voltage-gated sodium channels, KCNQ2/3 channels, and neurofascin are all ankyrin-binding proteins, but are unrelated to each other (Figures 1 and 3) (Bennett & Lorenzo, 2013). Similarly ankyrin-B-based cardiomyocyte T-tubule/SR domains contain the Na/K ATPase together with the Na/Ca exchanger and IP3 receptors, which function cooperatively in calcium homeostasis and independently evolved ankyrin-binding activity (Table 1; Figure 2). Ankyrin-binding activity has independently evolved at least 15 times in unrelated families of membrane-spanning proteins, even though core-folded domains of ankyrins themselves are relatively invariant (Table 1). Together, these observations suggest the hypothesis that ankyrin-binding activity evolved multiple times due to positive selection for optimal physiological efficiency (Bennett & Chen, 2001).

Mapping of binding sites of ankyrin partners has provided insight into the structural basis for ankyrin recognition and how it might have evolved (Bennett & Healy, 2009). Ankyrin-binding motifs identified so far are relatively short peptide sequences (10–20 residues) that are distinct in their primary sequence but share a lack of secondary structure (Table 2). For example, the ankyrin-binding activity of the erythrocyte anion exchanger is due to two loops evident in its crystal structure (Chang & Low, 2003; Grey, Kodippili, Simon, & Low, 2012). Cytoplasmic domains of E-cadherin and L1CAMS are established to be natively unstructured by biophysical methods (Huber, Stewart, Laurents, Nelson, & Weis, 2001; Zhang, Davis, Carpenter, & Bennett, 1998), and sites of Nav channels, KCNQ2/3 channels, RhBG ammonium transporter, and beta-dystroglycan all are predicted to be unstructured (Bennett & Healy, 2009).

Natively unstructured motifs in general are widely utilized in protein recognition and are rapidly evolving in eukaryotic genomes (Dyson & Wright, 2005). Such a code offers multiple benefits, including ease of evolution of new protein interactions as well as capacity for integrating pathways through binding of unstructured motifs with multiple partners. It is of interest with respect to multitasking, that the ankyrin-binding motif of E-cadherin contains a dileucine motif that is required for clathrin-dependent endocytosis, but does not participate in ankyrin binding (Jenkins et al., 2013). The E-cadherin ankyrin-binding sequence thus is better described as a polarity motif that utilizes both ankyrin binding for retention and clathrin for editing to maintain E-cadherin apical–lateral polarity (Jenkins et al., 2013). Another advantage of unstructured peptides is that their affinity can vary depending on the physiological context. For example, the Kd for ankyrin is 10 nM for the erythrocyte anion exchanger where structural stability over a period of months is important, but only 500 nM for E-cadherin in epithelial cells which are much more dynamic (Bennett & Stenbuck, 1980a; Kizhatil, Davis, et al., 2007). Finally, intrinsically unstructured sequences are more tolerant of missense mutations than folded

Table 2 Independently evolved ankyrin-binding sites (See color plate)

Protein Function	Ankyrin Partner	Ankyrin binding motif[a]	Most recent organisms	~ mybp
Cell adhesion	Dystroglycan	GVPIIFADELDDSK[1]	Cnidarians	580
	L1CAM	QFNEDGSFIGQY[2]	Bilaterians	555
	E-cadherin	KEPLLPPEDDTRDNVYYYDEE[3]	Urochordates	520
Membrane transport	Nav-α	VPIAVGESDFE[4]	Cephalochordate	500
	KCNQ2	PYIAEGESDTDSD[5]	Jawed fish	440
	Kir6.2	VPIVAEED[6]	Jawed fish	440
	CNGβ	PEPGEQILSVKMPE[7]	Jawed fish	440
	Cav2.1	NLLASREALYGDAAERWPTT[8]	Jawed fish	440
	Cav2.2	NLRASCEALYSEMDPEERLR[8]	Jawed fish	440
	Nav-β-1a	KKIAATEAAAQENASEY[9]	Jawed fish	440
	AE1	ELVMDEKNQEL...PAVLTRSGDPS[10]	Mammals	200
	RhB/G	KLPFLDSPP[11]	Mammals	200

Ankyrin-binding sites in cell adhesion molecules and membrane transporters.
[a] Minimum ankyrin-binding sequences. Mutation of critical residues marked in red results in loss of ankyrin-binding activity.
[1] Ayalon et al. (2008).
[2] Zhang et al. (1998).
[3] Kizhatil, Davis, et al. (2007) and Jenkins et al. (2013) (leucine residues marked in blue are required for endocytosis but not for ankyrin-binding).
[4] Garrido et al. (2003), Lemaillet, Walker, and Lambert (2003) and Wang et al. (2014).
[5] Pan et al. (2006).
[6] Li et al. (2010).
[7] Kizhatil, Baker, et al. (2009).
[8] Kline, Scott, Curran, Hund, and Mohler (2014).
[9] Malhotra et al. (2002).
[10] Chang and Low (2003) and Grey et al. (2012).
[11] Lopez et al. (2005).

proteins and as a result evolve more rapidly (Brown, Johnson, Dunker, & Daughdrill, 2011).

Insight into how ankyrin associates with unstructured peptides has come from atomic structures of ANK repeats complexed with an internal linker peptide (Michaely, Tomchick, Machius, & Anderson, 2002) as well as with the binding peptide of Nav1.2 (Wang et al., 2014). The 24 ANK repeats are folded as a solenoid, which has a highly conserved 210 Å-long inner groove that can accommodate extended peptides (Michaely et al., 2002; Wang et al., 2014). The ANK repeat groove contains multiple quasi-independent binding sites that can associate with short regions of peptides in a combinatorial manner. For example, ankyrin-binding peptides of Nav1.2, KCNQ2, neurofascin, and Cav1.3 associate with distinct sets of sites along the solenoid groove (Wang et al., 2014). These as well as other ankyrin-binding peptides share features such as a critical glutamic acid, but it has not yet been possible to develop an algorithm for bioinformatic searches for ankyrin-binding candidates.

6. CELLULAR "INCUBATORS" IN EVOLUTION OF ANKYRIN-BINDING MEMBRANE PARTNERS

The hypothesis that ankyrin-binding activities arose through positive selection begs the question of the cellular and physiological contexts where such a process might have happened. One of the best examples of a potential "incubator" for evolution of ankyrin partners is axon initial segments that appeared in early vertebrates (Hill et al., 2008) (Figure 4). The phylogenetic record of appearance of ankyrin-binding motifs within protein families reveals that neurofascin as an L1CAM family member was the first initial segment component to acquire ankyrin-binding activity (Bennett & Lorenzo, 2013). L1CAM family members with a conserved ankyrin-binding

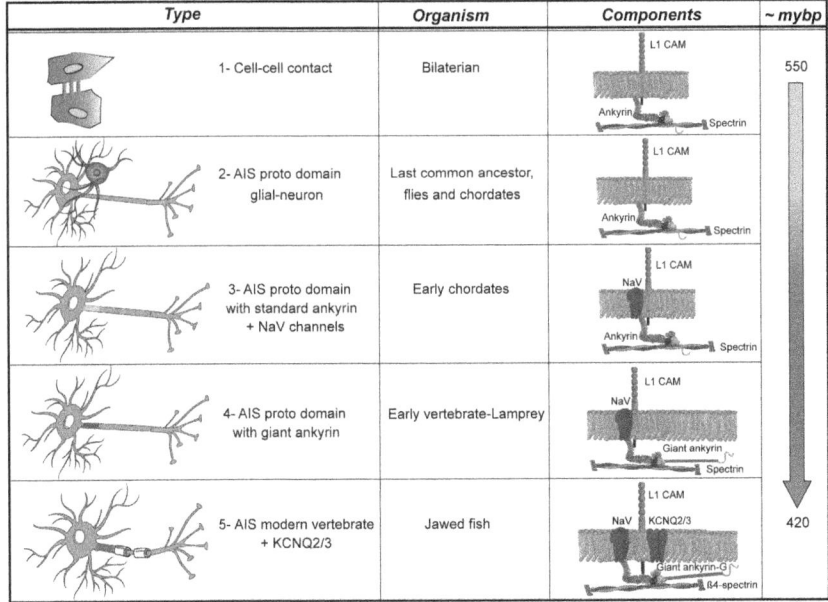

Figure 4 *Sequential evolution of the axon initial segment as a site for regulated fast signal integration.* Early axon initial segment proto-domains and epithelial lateral membranes were initially based on spectrin—ankyrin-cell adhesion molecule assemblies that subsequently became populated with clustered ion channels, specialized giant ankyrin-G, and β4-spectrin isoforms to facilitate faster signal integration and transduction.

motif are expressed throughout modern bilaterian organisms, and are represented in *C. elegans* by LAD-1, encoded by the *Sax-7* gene, and in *Drosophila*
by neuroglian (Chen et al., 2001; Hortsch, Nagaraj, & Godenschwege,
2009). The *C. elegans* L1CAM is localized with ankyrin (Unc44) at sites
of cell—cell contact in multiple cell types (Chen et al., 2001). Adhesive functions of the *C. elegans* L1CAM include correct positioning of neuronal cell
bodies and axons, and its ankyrin-binding motif is necessary for these
activities (Pocock, Bénard, Shapiro, & Hobert, 2008). The *Drosophila*
L1CAM, neuroglian, has roles in coordinating synaptic connections as
well as axon—axon interactions and again these functions require ankyrin-
binding activity (Enneking et al., 2013; Godenschwege, Kristiansen,
Uthaman, Hortsch, & Murphey, 2006; Hortsch et al., 2009; Siegenthaler,
Enneking, Moreno, & Pielage, 2015).

Cooper and colleagues have reported an extensive phylogenetic analysis
of ankyrin-binding motifs of both the voltage-gated sodium channels and
KCNQ2/3 channels (Hill et al., 2008). Voltage-gated sodium channels acquired an ankyrin-binding motif following evolution of the L1CAM site,
and this did not occur until early in chordate evolution in cephalochordates,
likely in the Cambrian period between 550 and 500 mybp (Hill et al., 2008).
KCNQ2/3 channels followed voltage-gated sodium channels in acquiring
an ankyrin-binding motif, and this only occurred in jawed fish at the end
of the Ordovician period around 450 mybp (Hill et al., 2008) (Figure 4).
Even though the ankyrin-binding motifs of KCNQ2/3 and voltage-gated
sodium channels are similar, these sequences evolved independently (Hill
et al., 2008).

The sequential and independent evolution of ankyrin-binding activities
beginning with L1 cell adhesion molecules suggests a scenario where a
proto-axon initial segment, containing ankyrin and an L1CAM evolved
first, and then was populated by ion channels, first by voltage-gated sodium
channels and later by KCNQ2/3 channels (Figure 4). In support of an initiating role of cell adhesion molecules and ankyrin, *Drosophila* mushroom
body neurons, although lacking an ankyrin-binding motif in their
voltage-gated sodium channels, contain a domain resembling the axon
initial segment that is located between the axon and dendrite and is enriched
in ankyrin-1 and the cell adhesion molecule fasciculin-1 (Trunova, Baek, &
Giniger, 2011).

The source of positive selection promoting evolution of vertebrate axon
initial segments presumably derived from the advantages of faster signaling

and its regulation. Higher concentration of sodium channels would have resulted in increased current density and greater amplitudes of depolarization that eventually became self-renewing action potentials. Evolutionary selection likely acted on ankyrin in parallel with its binding partners. For example, urochordates have sodium channels with an ankyrin-binding motif but only have a single ankyrin gene that lacks the axonal specific exons acquired later in vertebrate giant ankyrins. The giant exon of vertebrate 480 kDa ankyrin-G thus evolved in the context of axonal expression and clustering of sodium channels and now likely has acquired capabilities that are specifically related to the axon initial segment. Coevolution of ankyrin with the axon initial segment helps rationalize how 480 kDa ankyrin-G can be required for localization of all of the known axon initial segment components including voltage-gated sodium channels, KCNQ2/3 channels, 186 kDa neurofascin, beta-4 spectrin, GABA-A synapses, as well as bundled microtubules (Hedstrom, Ogawa, & Rasband, 2008; Jenkins, Kim et al., 2015; Sobotznik et al., 2009).

Primitive axon initial segments with clustered sodium channels are present in jawless fish, now represented by lampreys, and thus predated myelination and development of nodes of Ranvier, which appeared only later in jawed fish (Hill et al., 2008). Interestingly, even though nodes of Ranvier and initial segments are comprised of similar proteins (Davis, Lambert, & Bennett, 1996), these domains exhibit distinct mechanisms of assembly (Dzhashiashvili et al., 2007; Susuki et al., 2013). Thus, the basic ankyrin interactome of the axon initial segment was co-opted and modified during the process of myelination, resulting in closely related excitatory domains at nodes of Ranvier. A similar process of co-option likely resulted in utilization of ankyrin voltage-gated sodium channel interaction in mammalian cardiomyocytes, where Nav1.5 requires ankyrin-binding activity for targeting to intercalated discs (Lowe et al., 2008; Makara et al., 2014; Mohler, Rivolta, et al., 2004).

More generally, cell adhesion molecules are the earliest examples of membrane-spanning proteins with ankyrin-binding motifs (Table 2). Dystroglycan has an ABD that retains critical residues in Cnidarians, while the cadherin ABD is present in the urochordate *Ciona intestinalis* (Jenkins et al., 2013). In contrast, ion channel-binding sites have continued to evolve in terrestrial vertebrates, where the RhB/G ammonium transporter has an ankyrin-binding motif present in mammals but is not conserved at critical residues in chickens (Lopez et al., 2005) (Table 2).

These considerations suggest a scenario for ankyrin-based membrane domains where the earliest event was formation of a proto-domain through interaction of membrane-associated ankyrin and spectrin with membrane-spanning cell adhesion molecules such as L1CAMs, dystroglycan, and cadherins (Figure 4). These proto-domains then secondarily became populated with ion channels and likely other molecules that developed ankyrin-binding activity at sites determined by extracellular signals with selective advantage provided by optimization of physiological function (Figure 4). In the case of axon initial segments the physiological adaptation would be efficient generation of action potentials through clustering of voltage-gated sodium channels. Ankyrin, along with L1 cell adhesion molecules and cadherins, also is localized at sites of cell—cell contact in other cell types including epithelial tissues (Chen et al., 2001). Thus, it is conceivable that epithelial lateral membranes served as an "incubator" similar to the axon initial segment, where membrane transporters could have acquired ankyrin-binding activity with the advantage of increased efficiency through polarized fluxes of ions and nutrients.

Erythrocytes offer an apparent exception to the initiating role of cell adhesion molecules in evolution of ankyrin domains, as these cells do not engage in interactions with other cells. However, an ankyrin—anion exchanger—RhB/A ammonium transporter interaction has been reported in kidney epithelial cells (Genetet, Ripoche, Le Van Kim, Colin, & Lopez, 2015) and may have been secondarily co-opted by erythroid cells.

7. PARALLEL ROLES OF ANKYRIN AND SPECTRIN IN INTRACELLULAR ORGANELLE TRANSPORT

An implication of the proposed role of cell adhesion molecules in recruitment of ankyrin and spectrin to cell surface sites is that these proteins have other functions that predated their association with plasma membranes. Consistent with this idea, multiple lines of evidence support ancient roles of ankyrin and spectrin in polarized organelle transport. *Drosophila* spectrin, ankyrin, as well as the spectrin-associated protein adducin are all components of an intracellular microtubule-based structure termed the fusome, which delivers organelles and RNAs from nurse cells to oocytes (De Cuevas, Lee, & Spradling, 1996; Lighthouse, Buszczak, & Spradling, 2008; Lin, Yue, & Spradling, 1994). Ankyrin in *C. elegans* and *Drosophila* determines polarized organelle transport in dendrites and axons through organization of microtubules (Koch et al., 2008; Maniar et al., 2011; Pielage et al., 2008; Stephan et al., 2015). Spectrin also has been implicated in axonal transport in *Drosophila* (Lorenzo et al., 2010; Pielage, Fetter, & Davis, 2005). Ankyrin-G associates with kinesin-1 in neurons and contributes to axonal transport of voltage-gated sodium channels (Barry et al., 2014). Ankyrin-G also is required for directed delivery of voltage-gated sodium channels to axon initial segments in developing hippocampal neurons (Akin, Solé, Dib-Hajj, Waxman, & Tamkun, 2015).

At steady state, ankyrin-G is primarily localized to plasma membrane domains, while ankyrin-B associates with intracellular membranes despite a high level of sequence similarity to ankyrin-G (He et al., 2013). Dual roles of invertebrate ankyrins in both cell surface and intracellular functions thus appear to have largely been subdivided in vertebrates between ankyrin-G and ankyrin-B. Loss of plasma membrane association of ankyrin-B is due to an ankyrin-B-specific linker peptide connecting the ankyrin repeat domain to the ZU5-ZU5-UPA supermodule (He et al., 2013). This ankyrin-B-specific peptide inhibits binding of ankyrin-B to membrane protein partners E-cadherin and neurofascin and prevents association of ankyrin-B with epithelial lateral membranes as well as the axon initial segment (He et al., 2013). The residues of the ankyrin-B linker required for autoinhibition are encoded by a small exon that is highly divergent between ankyrin family members, but conserved in the ankyrin-B lineage. These considerations argue against neofunctionalization of ankyrin-B for intracellular functions and rather support partition of the dual functions of the ancestral single ankyrin following gene duplication in the vertebrate lineage (He et al., 2013).

Ankyrin-B interacts with Eps15 homology domain (EHD)-containing/receptor-mediated endocytosis (Rme) proteins, which are ATPases related to dynamin that are involved in endosomal recycling (Daumke et al., 2007; Gudmundsson et al., 2010). EHD/Rme 3 exhibits increased cardiac expression in ankyrin-B haploinsufficient mice and in acquired ankyrin-B deficiency, which occurs in heart failure (Gudmundsson et al., 2010; 2012). Moreover, ankyrin-B as well as NCX and L-type calcium channels are reduced in EHD3-deficient mice (Curran et al., 2014). These observations suggest that ankyrin-B, which has been characterized as a T-tubule-associated structural protein that stabilizes the Na/Ca exchanger and Na/K ATPase (Mohler, Davis, & Bennett, 2005; Mohler et al., 2003), also has a dynamic role in directing intracellular trafficking of these as well as other transporters.

Ankyrin–B promotes fast axonal transport of synaptic vesicles and other organelles in mammalian neurons and is required for normal axon elongation and formation of long axon tracts in mice (Lorenzo et al., 2014; Scotland et al., 1998). Ankyrin–B acts through recruiting the dynein/dynactin retrograde motor complex to diverse membrane cargos (Lorenzo et al., 2014). This function requires direct binding of ankyrin–B to both the p62 subunit of dynactin through the ZU5–ZU5–UPA supermodule and to membrane PtdIns(3)P lipids through a basic pocket located in the ZU5–C domain (Lorenzo et al., 2014; Wang et al., 2012) (Figure 5). The basic pocket of ankyrin–B resolved in the crystal structure of the ZU5–N–ZU5–C–UPA supermodule involves residues that are highly conserved among bilaterian ankyrins, suggesting association with PtdIns(3)P lipids is an ancient ankyrin function (Wang et al., 2012). In axons, the PtdIns(3)P lipids required to couple ankyrin–B and dynactin to organelles are generated by the class 3 phosphatidylinositol 3-kinase, PIK3C3 (Lorenzo et al., 2014). PIK3C3 also is referred to as Vps34 exists in a complex with beclin1 as well as its subunit Vps15 and has roles in autophagy as well as endosome trafficking (Funderburk, Wang, & Yue, 2010). Ankyrin–B, as a PtdIns(3)P effector, and dynactin adapter potentially could contribute many types of organelle trafficking in addition to axonal transport.

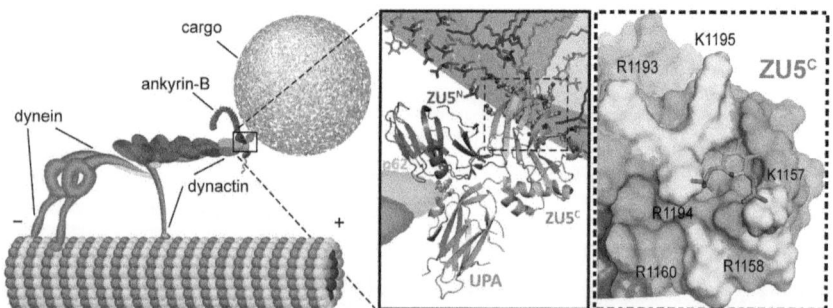

Figure 5 *Ankyrin-B's dual interactions with the dynactin complex and membrane PtdIns(3)P lipids promote multiorganelle transport in axons.* The schematic representation illustrates a proposed model for ankyrin-B coupling of the dynein—dynactin motor complex to cargoes. The center panel shows a homology model of the ZU5^N-ZU5^C-UPA-DD region of ankyrin-B highlighting binding sites (spheres) for the dynactin subunit p62 and for PtdIns(3)P lipids in membrane cargoes. A molecular docking simulation of PtdIns(3)P binding to ZU5^C domain highlights a positively charged surface (yellow) required for the interaction with PtdIns(3)P lipids (Lorenzo et al., 2014). (See color plate)

8. ROLES OF LIPIDS IN CELLULAR TARGETING OF SPECTRIN AND ANKYRIN

The discussion of plasma membrane roles of spectrin and ankyrin so far has been confined to genes and protein interactions centered around a common structural theme where diverse membrane-spanning proteins are "anchored" within the fluid bilayer by association with ankyrin, which in turn is coupled to an extended membrane-associated spectrin—actin network. However, ankyrin-G and beta-2 spectrin do not simply associate with protein partners in a preexisting plasma membrane domain since they are both required to form lateral membranes of columnar epithelial cells (Kizhatil & Bennett, 2004; Kizhatil, Davis, et al., 2007; Kizhatil, Yoon, et al., 2007), while giant ankyrin-G is required to form axon initial segments (Jenkins & Bennett, 2001; Jenkins, Kim et al., 2015). Moreover, ankyrin-G remains plasma membrane-associated following shift of epithelial cells to low calcium, even though this treatment results in internalization of all of its known protein partners including beta-2 spectrin and E-cadherin (He et al., 2014, 2012). Thus, protein-centric models, while descriptive of steady-state protein composition, do not explain how spectrin and ankyrin assemble in cells.

Recent findings have identified important roles for lipids in cellular targeting of both spectrin and ankyrin. Ankyrin-G requires palmitoylation at a conserved cysteine (C70) located in a loop at the N-terminus of the ANK repeat solenoid in order to assemble lateral epithelial membranes and for retention on plasma membranes of cells exposed to low calcium (He et al., 2014, 2012). An unbiased screen of the 23 mammalian aspartate—histidine—histidine—cysteine (DHHC) protein palmitoyltransferases revealed that DHHC 5 and 8 were the only enzymes confined to epithelial cell lateral membranes and the only ones that palmitoylated ankyrin-G (He et al., 2014). Interestingly, knockdown of both DHHC 5 and 8 in epithelial cells prevented palmitoylation of ankyrin-G as well as lateral membrane assembly and retention ankyrin-G on the plasma membrane in cells exposed to low calcium (He et al., 2014).

Beta-2 spectrin has a dual requirement for interaction with phosphoinositide lipids via a conserved PH domain and for association with ankyrin-G in order to localize to lateral membranes of epithelial cells (He et al., 2014). The phosphoinositide lipids interacting with beta-2 spectrin in vivo are PI(3,4)P2 and PI(3,4,5)P3, even though beta-2 spectrin associates with these lipids as well as PI(4,5)P2 in biochemical assays (He et al., 2014). Interestingly, PI(3,4,5)P3 confers basolateral membrane identity in some epithelial cell models (Martin-Belmonte et al., 2007; Shewan, Eastburn, & Mostov, 2011). βII-spectrin thus functions as a coincidence detector that requires recognition of both palmitoylated ankyrin-G and phosphoinositide lipids as a prerequisite for its localization to the lateral membrane (He et al., 2014) (Figure 6).

DHHC5/8 palmitoyltransferases acting through ankyrin-G and PI(3,4) P2 and PI(3,4,5)P3 phosphoinositides acting through beta-2 spectrin ensure that ankyrin-G and βII-spectrin localize only on lateral plasma membranes and not in the cytosol or on other membrane surfaces. Loss of either DHHC5/8, or interactions between ankyrin-G and βII-spectrin, or βII-spectrin recognition of phosphoinositide lipids all result in failure to build the lateral membrane, which underscores the functional importance of this pathway (He et al., 2014). The basis for targeting of DHHC 5 and 8 to epithelial lateral membranes is not currently known but likely is a key step in establishing the location of ankyrin-G.

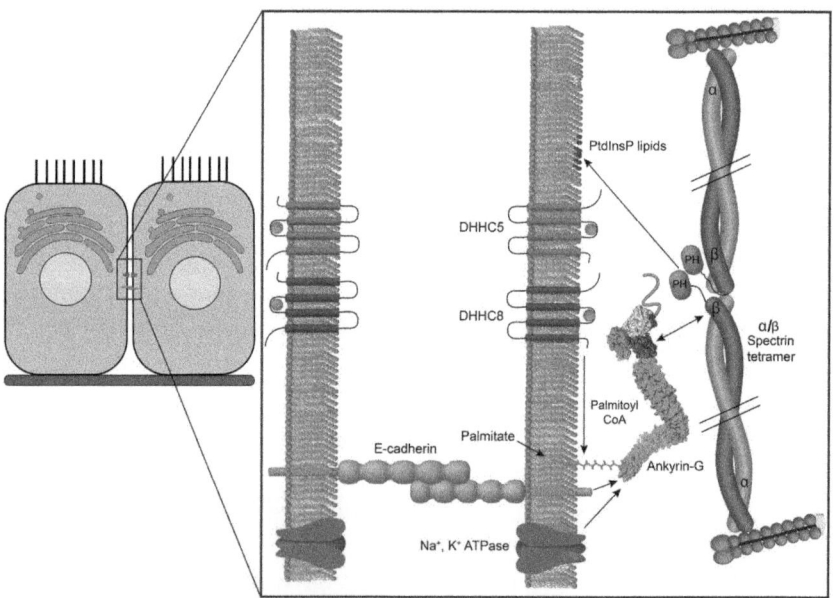

Figure 6 *Ankyrin-G palmitoylation and βII-spectrin binding to PI(4,5)P2 drive lateral membrane assembly.* A model summarizing roles of DHHC5/8 palmitoyltransferases, ankyrin-G, and βII-spectrin in assembling the lateral membrane of columnar epithelial cells. DHHC5/8 are localized to the lateral membrane, where they palmitoylate ankyrin-G. βII-Spectrin requires coincidence detection of ankyrin-G and phosphoinositide lipids to ensure its precise lateral membrane localization. Loss of any of the components or the functional relationship results in reduced lateral membrane height (He et al., 2014).

While functional requirements for palmitoylation and phosphoinositides have been demonstrated in columnar epithelial cells, several observations suggest that this is a feature conserved in other vertebrate ankyrin-based domains. Ankyrin-G requires C70 for assembly at axon initial segments (He et al., 2012). Moreover, ankyrin-R in erythrocytes incorporates palmitate (Staufenbiel, 1987), and the palmitoylated cysteine of ankyrin-G is conserved among all vertebrate ankyrins (He et al., 2012). Interestingly, although the cysteine is conserved in ankyrin-1 of *Drosophila*, this residue is not present in *C. elegans* ankyrin and *Drosophila* ankyrin-2. Thus, membrane association of ankyrins through palmitoylation likely occurs for all vertebrate ankyrins and in some contexts in *Drosophila*.

9. DYNAMIC SPECTRIN/ANKYRIN MICRODOMAINS

Recent high-resolution three-dimensional imaging has revealed micron-scale domains containing ankyrin-G and beta-2 spectrin as well as DHHC 5 and 8 palmitoyltransferases that cover 30—40% of the lateral membrane surface of Madin—Darby canine kidney (MDCK) cells (He et al., 2014; Jenkins, He, & Bennett, 2015) (Figure 7). Similar domains have been imaged in collecting duct cells in kidney tissue as well as neuronal cell bodies (Jenkins, He, et al., 2015; Tseng et al., 2015). Spectrin/ankyrin microdomains are distinct from lipid rafts since at 0.5—2 square microns they are much larger than rafts (Lingwood & Simons, 2010) and depend on interactions between ankyrin-G and beta-2 spectrin as well as palmitoylation of ankyrin-G and association of beta-2 spectrin with phosphoinositide lipids (He et al., 2014; Jenkins, He, & Bennett, 2015). Interestingly, microdomains containing 480 kDa ankyrin-G and beta-2 spectrin also are located on the somatodendritic surfaces of hippocampal neurons and are similar in size to those in epithelial cells (Tseng et al., 2015).

Live cell imaging has revealed that within a few minutes spectrin/ankyrin microdomains of MDCK cells change shape and move locally in a microtubule-dependent manner (Jenkins, He, et al., 2015). Within an hour, movement of spectrin—ankyrin microdomains is sufficient to contact most of the lateral membrane (Jenkins, He, et al., 2015). These new findings suggest that complete coverage of the inner surface of plasma membranes with a static spectrin skeleton observed in erythrocytes and neuronal axons may be the exception and that more generally spectrin and ankyrin are confined within locally motile microdomains.

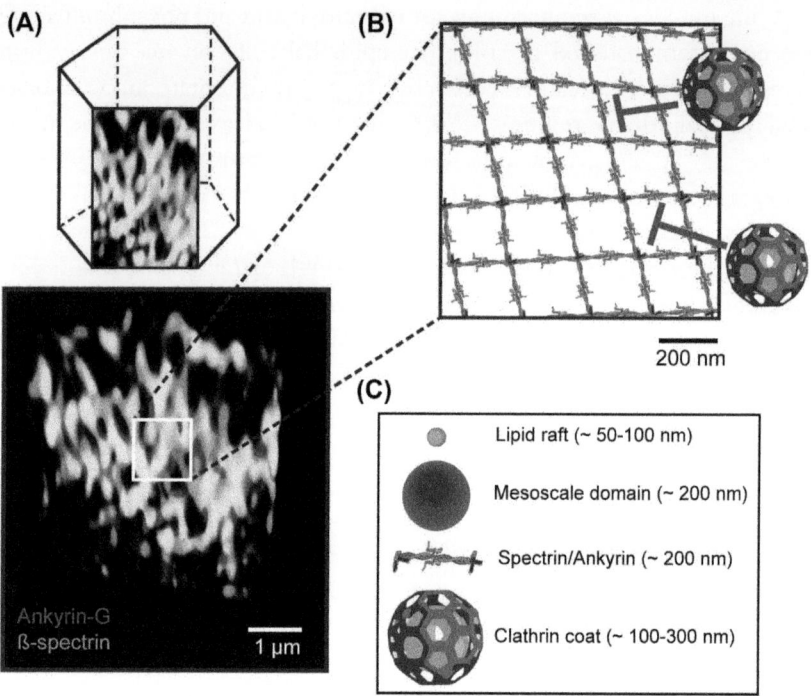

Figure 7 *Ankyrin-G/βII-spectrin-coated microdomains promote lateral membrane height in polarized epithelial cells by opposing clathrin-mediated endocytosis.* (A) Representative deconvolved images of the lateral membrane of Madin—Darby canine kidney cell membranes stained for endogenous ankyrin-G (red) and βII-spectrin (green). (B) Schematic representation of the actin/spectrin/ankyrin lattice on membrane microdomains. (C) Representations of the relative sizes of lipid rafts, mesoscale domains, clathrin coat, and the predicted spectrin/ankyrin network coating membrane microdomains (imaged on the left), drawn at scale (Jenkins, He, & Bennett, 2015). (See color plate)

Several considerations suggest that spectrin/ankyrin microdomains maintain membrane identity and promote membrane domain assembly through opposing endocytosis mediated by clathrin or other mechanisms. Spectrin/ankyrin domains exclude clathrin and clathrin-related membrane cargoes, which are consistent with known dimensions of clathrin-coated pits that are too large to access the plasma membrane through a spectrin—actin network (Jenkins, He, & Bennett, 2015). Moreover, inhibition of spectrin—ankyrin interaction promotes endocytosis of transferrin, while depletion of either protein results in increased endocytosis of bulk membrane lipids (Jenkins, He, & Bennett, 2015). Loss of columnar morphology and collapse of the lateral membrane following depletion of either ankyrin-G or beta-2 spectrin are completely prevented by inhibition of endocytosis with dynasore (Jenkins, He, & Bennett, 2015). Moreover, loss of GABA-A receptors from somatodendritic membrane neurons lacking giant ankyrin-G also is prevented by dynasore (Tseng et al., 2015). In genetic terms, these results demonstrate that dynasore-sensitive endocytosis and ankyrin-G/βII-spectrin-based microdomains are epistatic to each other in the lateral membrane assembly pathway, and that inhibition of endocytosis functions to suppress effects of loss of ankyrin-G/βII-spectrin.

These findings combined with essential roles of membrane lipids discussed above support a dynamic physical model for epithelial lateral membranes and neuronal plasma membranes where micron-scale compartments comprised of lipid-dependent spectrin and ankyrin-G actively patrol the membrane through a microtubule-dependent mechanism, analogous to "window washers," and prevent both receptor-mediated and bulk endocytosis. In addition to microtubule-dependent movements, microdomains are likely engaged in rapid assembly/disassembly through rapidly reversible protein palmitoylation (Fukata et al., 2013), as well as phosphoinositide turnover (Shewan et al., 2011). Spectrin/ankyrin microdomains and their host membranes thus are not permanent structures but instead exist in a dynamic steady state in homeostatic equilibrium with endocytic processes.

10. SUMMARY AND PERSPECTIVES

This chapter reviews mechanisms for evolution and assembly of plasma membrane domains based on spectrin and ankyrin that coordinate functionally related membrane-spanning proteins in diverse vertebrate cells and tissues (Table 1). Membrane transporters and cell adhesion proteins (15 protein families identified so far) associate through independently evolved interactions of intrinsically disordered regions of their cytoplasmic domains (Table 2; Figures 1–3) with a highly conserved peptide-binding groove formed by the ANK repeat solenoid. Ankyrins in turn are coupled to spectrins, which are flexible, highly elongated organelle-scale proteins that form extended arrays through interaction with specialized configurations of actin that are distinct from cortical actin filaments and require accessory proteins such as adducin and protein 4.1.

Spectrin—actin networks likely were present in early metazoans (Baines, 2010; Bennett & Lorenzo, 2013), but around 500 million years ago, bilaterian organisms greatly amplified their capacity to coordinate membrane-spanning proteins through partnership of spectrin, ankyrin, and cell adhesion molecules such as L1CAMs and dystroglycan. Ankyrin may have initially been recruited to plasma membranes from more ancient roles in organelle transport independent of beta-spectrin, which are preserved in vertebrate ankyrin-B (Lorenzo et al., 2014) and invertebrate ankyrins (Koch et al., 2008; Pielage et al., 2008; Stephan et al., 2015). Vertebrates greatly expanded the basic bilaterian spectrin—ankyrin toolkit through gene duplications combined with variation in unstructured intramolecular regulatory sequences as well as independent evolution of ankyrin-binding activity by diverse ion transporters including voltage-gated sodium channels. In addition, giant vertebrate ankyrins with specialized roles in axons acquired completely new coding sequences from an I-connectin/titin-related gene (Bennett & Walder, 2015; Jenkins, Kim, et al., 2015).

Evolution of physiological functions of vertebrate membrane domains involved independent and sequential acquisition of ankyrin-binding activity by cell adhesion molecules and membrane transporters while ankyrins themselves remained relatively unchanged (Hill et al., 2008). We speculate that proto-domains such as early axon initial segments that were based on spectrin—ankyrin-cell adhesion molecule assemblies served as "incubators," where ion transporters independently acquired ankyrin-binding activity through positive selection (Figure 4).

Cellular assembly of spectrin/ankyrin domains depends on membrane lipids in addition to protein interactions. For example, formation of columnar epithelial lateral membranes requires localized palmitoylation of ankyrin-G by DHHC 5 and 8 palmitoyltransferases, and interaction of beta-spectrin with phosphoinositide lipids. High-resolution three-dimensional imaging of columnar epithelial cells has revealed locally dynamic spectrin/ankyrin microdomains covering a portion of the lateral membrane surface area that exclude clathrin and clathrin-dependent membrane cargoes (Jenkins, He, & Bennett, 2015). These spectrin/ankyrin microdomains are distinct from lipid rafts based on their size, microtubule-dependent motility, and requirements for protein interactions. They have escaped notice in part because they do not exist in typical unpolarized cultured cells such as fibroblasts. One function of spectrin/ankyrin domains is to regulate membrane identity through preventing endocytosis of bulk lipids as well as specific proteins such as GABA-A receptors (Jenkins, Kim et al., 2015; Tseng et al., 2015). Spectrin/ankyrin microdomains of epithelial cells are highly enriched in DHHC 5 and 8 palmitoyltransferases (He et al., 2014), which have substrates in addition to ankyrin-G, and may function as signaling platforms similar to proposals for lipid rafts (Lingwood & Simons, 2010).

Finally, on a personal note, the initial impetus for discovery of ankyrin was provided by the eloquently framed fluid mosaic model for plasma membrane structure posed by Singer and Nicolson (1972). It seemed highly unlikely, based on histology of vertebrate tissues known to all first year medical students, that long-range order and mechanical resilience evident in vertebrate plasma membranes could be achieved simply by a bilayer with embedded membrane-spanning proteins. Clearly, peripheral membrane proteins also played an important role, but their identity and how they coupled to membranes needed to be discovered. Spectrin and ankyrin provide one mechanism, as outlined here, but there are certainly many other players such as proteins of the FERM and spectraplakin families. Some general lessons for future research in vertebrate membrane biology include the importance and challenge of studying cells in their native environment. Just as liposomes and artificial bilayers do not fully reproduce biological membranes, undifferentiated cells in two-dimensional culture are not sufficient to understand how plasma membranes are organized in tissues such as the heart and nervous system. Resolving membrane structure at high resolution in three dimensions is not feasible with TIRF or current superresolution microscopy, and it will be critical to develop methods to resolve nanometer-scale three-dimensional structure in tissues.

Evolution has turned out to be much more dynamic and inventive than most investigators expected. As a result, model organisms such as flies and nematodes are complex and interesting in their own right, but are not necessarily relevant for detailed understanding of vertebrate cells. Going forward, humans promise to provide a useful model organism with a wealth of DNA sequence and phenotype information, which can be combined with induced pluripotential stem cell-derived organoids and high-resolution three-dimensional microscopy. The early investigators who began with erythrocyte membranes can only marvel at the potential for discovery that lies ahead.

ACKNOWLEDGMENTS

The authors were supported by the Howard Hughes Medical Institute and the George Barth Geller Fund.

REFERENCES

Abdi, K. M., & Bennett, V. (2008). Adducin promotes micrometer-scale organization of beta2-spectrin in lateral membranes of bronchial epithelial cells. *Molecular Biology of the Cell, 19*(2), 536–545.

Abdi, K. M., Mohler, P. J., Davis, J. Q., & Bennett, V. (2006). Isoform specificity of ankyrin-B: a site in the divergent C-terminal domain is required for intramolecular association. *Journal of Biological Chemistry, 281*(9), 5741–5749.

Agre, P., Casella, J. F., Zinkham, W. H., McMillan, C., & Bennett, V. (1985). Partial deficiency of erythrocyte spectrin in hereditary spherocytosis. *Nature, 314*(6009), 380–383.

Agre, P., Orringer, E. P., & Bennett, V. (1982). Deficient red-cell spectrin in severe, recessively inherited spherocytosis. *The New England Journal of Medicine, 306*(19), 1155–1161.

Akin, E. J., Solé, L., Dib-Hajj, S. D., Waxman, S. G., & Tamkun, M. M. (2015). Preferential targeting of $Na_v1.6$ voltage-gated Na^+ channels to the axon initial segment during development. *PLoS One, 10*(4), e0124397.

Ayalon, G., Davis, J. Q., Scotland, P. B., & Bennett, V. (2008). An ankyrin-based mechanism for functional organization of dystrophin and dystroglycan. *Cell, 135*(7), 1189–1200.

Ayalon, G., Hostettler, J. D., Hoffman, J., Kizhatil, K., Davis, J. Q., & Bennett, V. (2011). Ankyrin-B interactions with spectrin and dynactin-4 are required for dystrophin-based protection of skeletal muscle from exercise injury. *Journal of Biological Chemistry, 286*(9), 7370–7378.

Baines, A. J. (2010). The spectrin-ankyrin-4.1-adducin membrane skeleton: adapting eukaryotic cells to the demands of animal life. *Protoplasma, 244*, 99–131.

Baines, A. J., Lu, H. C., & Bennett, P. M. (February 2014). The Protein 4.1 family: hub proteins in animals for organizing membrane proteins. *Biochimica et Biophysica Acta, 1838*(2), 605–619.

Banuelos, S., Saraste, M., & Carugo, K. D. (1998). Structural comparisons of calponin homology domains: implications for actin binding. *Structure, 6*, 1419–1431.

Barker, J. E., Deveau, S., & Wandersee, N. J. (2000). Amelioration of severe hereditary spherocytosis in nonablated adult mice by marrow transplantation. *Experimental Hematology, 28*(8), 985–992.

Barry, J., Gu, Y., Jukkola, P., O'Neill, B., Gu, H., Mohler, P. J., et al. (2014). Ankyrin-G directly binds to kinesin-1 to transport voltage-gated Na^+ channels into axons. *Developmental Cell, 28*(2), 117–131.

Bennett, V., & Chen, L. (2001). Ankyrins and cellular targeting of diverse membrane proteins to physiological sites. *Current Opinion in Cell Biology, 13*(1), 61–67.

Bennett, V., Davis, J., & Fowler, W. E. (1982). Brain spectrin, a membrane-associated protein related in structure and function to erythrocyte spectrin. *Nature, 299*(5879), 126–131.

Bennett, V., Gardner, K., & Steiner, J. P. (1988). Brain adducin: a protein kinase C substrate that may mediate site-directed assembly at the spectrin-actin junction. *Journal of Biological Chemistry, 263*(12), 5860–5869.

Bennett, V., & Healy, J. (2009). Membrane domains based on ankyrin and spectrin associated with cell–cell interactions. *Cold Spring Harbor Perspectives in Biology, 1*(6), a003012.

Bennett, V., & Lorenzo, D. N. (2013). Spectrin- and ankyrin-based membrane domains and the evolution of vertebrates. *Current Topics in Membranes, 72*, 1–37.

Bennett, V., & Stenbuck, P. J. (1979a). The membrane attachment protein for spectrin is associated with band 3 in human erythrocyte membranes. *Nature, 280*(5722), 468–473.

Bennett, V., & Stenbuck, P. J. (April 10, 1979b). Identification and partial purification of ankyrin, the high affinity membrane attachment site for human erythrocyte spectrin. *Journal of Biological Chemistry, 254*(7), 2533–2541.

Bennett, V., & Stenbuck, P. J. (1980a). Association between ankyrin and the cytoplasmic domain of band 3 isolated from the human erythrocyte membrane. *Journal of Biological Chemistry, 255*(13), 6424–6432.

Bennett, V., & Stenbuck, P. J. (1980b). Human erythrocyte ankyrin. Purification and properties. *Journal of Biological Chemistry, 255*, 2540–2548.

Bennett, V., & Walder, K. (2015). Evolution in action: giant ankyrins awake. *Developmental Cell, 33*(1), 1—2.

Berghs, S., Aggujaro, D., Dirkx, R., Jr., Maksimova, E., Stabach, P., Hermel, J. M., et al. (2000). βIV spectrin, a new spectrin localized at axon initial segments and nodes of ranvier in the central and peripheral nervous system. *Journal of Cell Biology, 151*, 985—1002.

Bodine, D. M., 4th, Birkenmeier, C. S., & Barker, J. E. (1984). Spectrin deficient inherited hemolytic anemias in the mouse: characterization by spectrin synthesis and mRNA activity in reticulocytes. *Cell, 37*(3), 721—729.

Brown, C. J., Johnson, A. K., Dunker, A. K., & Daughdrill, G. W. (2011). Evolution and disorder. *Current Opinion in Structural Biology, 21*(3), 441—446.

Byers, T. J., & Branton, D. (1985). Visualization of the protein associations in the erythrocyte membrane skeleton. *Proceedings of the National Academy of Sciences of the United States of America, 82*(18), 6153—6157.

Cai, X., & Zhang, Y. (March 2006). Molecular evolution of the ankyrin gene family. *Molecular Biology and Evolution, 23*(3), 550—558.

Carugo, K. D., Banuelos, S., & Saraste, M. (1997). Crystal structure of a calponin homology domain. *Nature Structural Biology, 4*, 175—179.

Chan, W., Kordeli, E., & Bennett, V. (1993). 440-kD ankyrinB: structure of the major developmentally regulated domain and selective localization in unmyelinated axons. *Journal of Cell Biology, 123*, 1463—1473.

Chang, K. J., Zollinger, D. R., Susuki, K., Sherman, D. L., Makara, M. A., Brophy, P. J., et al. (2014). Glial ankyrins facilitate paranodal axoglial junction assembly. *Nature Neuroscience, 17*(12), 1673—1681.

Chang, S. H., & Low, P. S. (2003). Identification of a critical ankyrin-binding loop on the cytoplasmic domain of erythrocyte membrane band 3 by crystal structure analysis and site-directed mutagenesis. *Journal of Biological Chemistry, 278*, 6879—6884.

Chen, L., Ong, B., & Bennett, V. (2001). LAD-1, the *Caenorhabditis elegans* L1CAM homologue, participates in embryonic and gonadal morphogenesis and is a substrate for fibroblast growth factor receptor pathway-dependent phosphotyrosine-based signaling. *Journal of Cell Biology, 154*, 841—855.

Cunha, S. R., Le Scouarnec, S., Schott, J. J., & Mohler, P. J. (2008). Exon organization and novel alternative splicing of the human ANK2 gene: implications for cardiac function and human cardiac disease. *Journal of Molecular and Cellular Cardiology, 45*, 724—734.

Cunha, S. R., & Mohler, P. J. (2008). Obscurin targets ankyrin-B and protein phosphatase 2A to the cardiac M-line. *Journal of Biological Chemistry, 283*, 31968—31980.

Curran, J., Makara, M. A., Little, S. C., Musa, H., Liu, B., Wu, X., et al. (2014). EHD3-dependent endosome pathway regulates cardiac membrane excitability and physiology. *Circulation Research, 115*(1), 68—78.

Das, A., Base, C., Manna, D., Cho, W., & Dubreuil, R. R. (2008). Unexpected complexity in the mechanisms that target assembly of the spectrin cytoskeleton. *Journal of Biological Chemistry, 283*(18), 12643—12653.

Daumke, O., Lundmark, R., Vallis, Y., Martens, S., Butler, P. J., & McMahon, H. T. (2007). Architectural and mechanistic insights into an EHD ATPase involved in membrane remodelling. *Nature, 449*, 923—927.

Davis, J., & Bennett, V. (1983). Brain spectrin. Isolation of subunits and formation of hybrids with erythrocyte spectrin subunits. *Journal of Biological Chemistry, 258*, 7757—7766.

Davis, J. Q., & Bennett, V. (1984). Brain ankyrin. A membrane-associated protein with binding sites for spectrin, tubulin, and the cytoplasmic domain of the erythrocyte anion channel. *Journal of Biological Chemistry, 259*, 13550—13559.

Davis, J. Q., Lambert, S., & Bennett, V. (1996). Molecular composition of the node of Ranvier: identification of ankyrin-binding cell adhesion molecules neurofascin (mucin+/third FNIII domain—) and NrCAM at nodal axon segments. *Journal of Cell Biology, 135,* 1355—1367.

Davis, L., Abdi, K., Machius, M., Brautigam, C., Tomchick, D. R., Bennett, V., et al. (2009). Localization and structure of the ankyrin-binding site on beta2- spectrin. *Journal of Biological Chemistry, 284,* 6982—6987.

Davis, L. H., Davis, J. Q., & Bennett, V. (1992). Ankyrin regulation: an alternatively spliced segment of the regulatory domain functions as an intramolecular modulator. *Journal of Biological Chemistry, 267,* 18966—18972.

De Cuevas, M., Lee, J. K., & Spradling, A. C. (1996). Alpha-spectrin is required for germline cell division and differentiation in the Drosophila ovary. *Development, 122,* 3959—3968.

De Rubeis, S., He, X., Goldberg, A. P., Poultney, C. S., Samocha, K., Cicek, A. E., et al. (2014). Synaptic, transcriptional and chromatin genes disrupted in autism. *Nature, 515*(7526), 209—215.

Djinovic-Carugo, K., Young, P., Gautel, M., & Saraste, M. (1999). Structure of the alpha-actinin rod: molecular basis for cross-linking of actin filaments. *Cell, 98,* 537—546.

Dyson, H. J., & Wright, P. E. (2005). Intrinsically unstructured proteins and their functions. *Nature Reviews Molecular Cell Biology, 6,* 197—208.

Dzhashiashvili, Y., Zhang, Y., Galinska, J., Lam, I., Grumet, M., & Salzer, J. L. (2007). Nodes of Ranvier and axon initial segments are ankyrin G-dependent domains that assemble by distinct mechanisms. *Journal of Cell Biology, 177,* 857—870.

Eber, S., & Lux, S. E. (2004). Hereditary spherocytosis—defects in proteins that connect the membrane skeleton to the lipid bilayer. *Seminars in Hematology, 41,* 118—141.

Enneking, E. M., Kudumala, S. R., Moreno, E., Stephan, R., Boerner, J., Godenschwege, T. A., et al. (2013). Transsynaptic coordination of synaptic growth, function, and stability by the L1-type CAM Neuroglian. *PLoS Biology, 11,* e1001537.

Flucher, B. E., & Daniels, M. P. (1989). Distribution of Na$^+$ channels and ankyrin in neuromuscular junctions is complementary to that of acetylcholine receptors and the 43 kd protein. *Neuron, 3*(2), 163—175.

Fowler, V. M. (2013). The human erythrocyte plasma membrane: a Rosetta Stone for decoding membrane-cytoskeleton structure. *Current Topics in Membranes, 72,* 39—88.

Fowler, V. M., & Bennett, V. (1978). Association of spectrin with its membrane attachment site restricts lateral mobility of human erythrocyte integral membrane proteins. *Journal of Supramolecular Structure, 8,* 215—221.

Fukata, Y., Dimitrov, A., Boncompain, G., Vielemeyer, O., Perez, F., & Fukata, M. (2013). Local palmitoylation cycles define activity-regulated postsynaptic subdomains. *Journal of Cell Biology, 202*(1), 145—161.

Fukuzawa, A., Shimamura, J., Takemori, S., Kanzawa, N., Yamaguchi, M., Sun, P., et al. (2001). Invertebrate connectin spans as much as 3.5 μm in the giant sarcomeres of crayfish claw muscle. *EMBO Journal, 20*(17), 4826—4835.

Funderburk, S. F., Wang, Q. J., & Yue, Z. (2010). The Beclin 1-VPS34 complex—at the crossroads of autophagy and beyond. *Trends in Cell Biology, 20*(6), 355—362.

Gao, Y., Perkins, E. M., Clarkson, Y. L., Tobia, S., Lyndon, A. R., Jackson, M., et al. (2011). β-III spectrin is critical for development of purkinje cell dendritic tree and spine morphogenesis. *Journal of Neuroscience, 31,* 16581—16590.

Garcia-Parajo, M. F., Cambi, A., Torreno-Pina, J. A., Thompson, N., & Jacobson, K. (2014). Nanoclustering as a dominant feature of plasma membrane organization. *Journal of Cell Science, 127*(23), 4995—5005.

Gardner, K., & Bennett, V. (1987). Modulation of spectrin-actin assembly by erythrocyte adducin. *Nature, 328*(6128), 359—362.

Garrido, J. J., Giraud, P., Carlier, E., Fernandes, F., Moussif, A., Fache, M. P., et al. (2003). A targeting motif involved in sodium channel clustering at the axonal initial segment. *Science, 300*(5628), 2091−2094.

Genetet, S., Ripoche, P., Le Van Kim, C., Colin, Y., & Lopez, C. (2015). Evidence of a structural and functional ammonium transporter RhBG/anion exchanger 1/ankyrin-G complex in kidney epithelial cells. *Journal of Biological Chemistry, 290*(11), 6925−6936.

Godenschwege, T. A., Kristiansen, L. V., Uthaman, S. B., Hortsch, M., & Murphey, R. K. (2006). A conserved role for Drosophila Neuroglian and human L1-CAM in central-synapse formation. *Current Biology, 16*, 12−23.

Golan, D. E., & Veatch, W. (1980). Lateral mobility of band 3 in the human erythrocyte membrane studied by fluorescence photobleaching recovery: evidence for control by cytoskeletal interactions. *Proceedings of the National Academy of Sciences of the United States of America, 77*(5), 2537−2541.

Grey, J. L., Kodippili, G. C., Simon, K., & Low, P. S. (2012). Identification of contact sites between ankyrin and band 3 in the human erythrocyte membrane. *Biochemistry, 51*, 6838−6846.

Gudmundsson, H., Hund, T. J., Wright, P. J., Kline, C. F., Snyder, J. S., Qian, L., et al. (2010). EH domain proteins regulate cardiac membrane protein targeting. *Circulation Research, 107*, 84−95.

Gudmundsson, H., Curran, J., Kashef, F., Snyder, J. S., Smith, S. A., Vargas-Pinto, P., et al. (2012). Differential regulation of EHD3 in human and mammalian heart failure. *Journal of Molecular and Cellular Cardiology, 52*, 1183−1190.

Hall, T. G., & Bennett, V. (1987). Regulatory domains of erythrocyte ankyrin. *Journal of Biological Chemistry, 262*, 10537−10545.

Hammarlund, M., Jorgensen, E. M., & Bastiani, M. J. (2007). Axons break in animals lacking beta-spectrin. *Journal of Cell Biology, 176*, 269−275.

Hayes, N. V., Scott, C., Heerkens, E., Ohanian, V., Maggs, A. M., Pinder, J. C., et al. (2000). Identification of a novel C-terminal variant of beta II spectrin: two isoforms of beta II spectrin have distinct intracellular locations and activities. *Journal of Cell Science, J113*, 2023−2034.

He, M., Abdi, K. M., & Bennett, V. (July 21, 2014). Ankyrin-G palmitoylation and βII-spectrin binding to phosphoinositide lipids drive lateral membrane assembly. *Journal of Cell Biology, 206*(2), 273−288.

He, M., Jenkins, P., & Bennett, V. (2012). Cysteine 70 of ankyrin-G is S-palmitoylated and is required for function of ankyrin-G in membrane domain assembly. *Journal of Biological Chemistry, 287*(52), 43995−44005.

He, M., Tseng, W. C., & Bennett, V. (2013). A single divergent exon inhibits ankyrin-B association with the plasma membrane. *Journal of Biological Chemistry, 288*(21), 14769−14779.

Hedstrom, K. L., Ogawa, Y., & Rasband, M. N. (2008). AnkyrinG is required for maintenance of the axon initial segment and neuronal polarity. *Journal of Cell Biology, 183*, 635−640.

Hill, A. S., Nishino, A., Nakajo, K., Zhang, G., Fineman, J. R., Selzer, M. E., et al. (2008). Ion channel clustering at the axon initial segment and node of Ranvier evolved sequentially in early chordates. *PLoS Genetics, 4*(12), e1000317.

Hoock, T. C., Peters, L. L., & Lux, S. E. (1997). Isoforms of ankyrin-3 that lack the NH2-terminal repeats associate with mouse macrophage lysosomes. *Journal of Cell Biology, 136*, 1059−1070.

Hopitzan, A. A., Baines, A. J., & Kordeli, E. (2006). Molecular evolution of ankyrin: gain of function in vertebrates by acquisition of an obscurin/titin-binding-related domain. *Molecular Biology and Evolution, 23*, 46−55.

Hopitzan, A. A., Baines, A. J., Ludosky, M. A., Recouvreur, M., & Kordeli, E. (2005). Ankyrin-G in skeletal muscle: tissue-specific alternative splicing contributes to the complexity of the sarcolemmal cytoskeleton. *Experimental Cell Research, S309*, 86—98.

Hortsch, M., Nagaraj, K., & Godenschwege, T. A. (2009). The interaction between L1-type proteins and ankyrins—a master switch for L1-type CAM function. *Cellular and Molecular Biology Letters, 14*, 57—69.

Hu, R. J., Watanabe, M., & Bennett, V. (1992). Characterization of human brain cDNA encoding the general isoform of beta-spectrin. *Journal of Biological Chemistry, 267*, 18715—18722.

Huber, A. H., Stewart, D. B., Laurents, D. V., Nelson, W. J., & Weis, W. I. (2001). The cadherin cytoplasmic domain is unstructured in the absence of beta-catenin. A possible mechanism for regulating cadherin turnover. *Journal of Biological Chemistry, 276*, 12301—12309.

Hund, T. J., Koval, O. M., Li, J., Wright, P. J., Qian, L., Snyder, J. S., et al. (2010). A β(IV)-spectrin/CaMKII signaling complex is essential for membrane excitability in mice. *Journal of Clinical Investigation, 120*, 3508—3519.

Hund, T. J., Snyder, J. S., Wu, X., Glynn, P., Koval, O. M., Onal, B., et al. (2014). β(IV)-Spectrin regulates TREK-1 membrane targeting in the heart. *Cardiovascular Research, 102*(1), 166—175.

Hyvönen, M., Macias, M. J., Nilges, M., Oschkinat, H., Saraste, M., & Wilmanns, M. (1995). Structure of the binding site for inositol phosphates in a PH domain. *EMBO Journal, 14*, 4676—4685.

Ikeda, Y., Dick, K. A., Weatherspoon, M. R., Gincel, D., Armbrust, K. R., Dalton, J. C., et al. (2006). Spectrin mutations cause spinocerebellar ataxia type 5. *Nature Genetics, 38*, 184—190.

Iossifov, I., O'Roak, B. J., Sanders, S. J., Ronemus, M., Krumm, N., Levy, D., et al. (2014). The contribution of de novo coding mutations to autism spectrum disorder. *Nature, 515*(7526), 216—221.

Ipsaro, J. J., Harper, S. L., Messick, T. E., Marmorstein, R., Mondragón, A., & Speicher, D. W. (2010). Crystal structure and functional interpretation of the erythrocyte spectrin tetramerization domain complex. *Blood, 115*, 4843—4852.

Ipsaro, J. J., Huang, L., & Mondragón, A. (2009). Structures of the spectrin-ankyrin interaction binding domains. *Blood, 113*, 5385—5393.

Ipsaro, J. J., & Mondragón, A. (2010). Structural basis for spectrin recognition by ankyrin. *Blood, 115*, 4093—4101.

Iqbal, Z., Vandeweyer, G., van der Voet, M., Waryah, A. M., Zahoor, M. Y., Besseling, J. A., et al. (2013). Homozygous and heterozygous disruptions of ANK3: at the crossroads of neurodevelopmental and psychiatric disorders. *Human Molecular Genetics, 22*, 1960—1970.

Jenkins, P. M., He, M., & Bennett, V. (Sep 11, 2015). Motile spectrin/ankyrin-G microdomains promote lateral membrane assembly by opposing endocytosis. *Sci Adv, 1*(8), e1500301.

Jenkins, P. M., Kim, N., Jones, S. L., Tseng, W. C., Svitkina, T. M., Yin, H. H., et al. (2015). Giant ankyrin-G: a critical innovation in vertebrate evolution of fast and integrated neuronal signaling. *Proceedings of the National Academy of Sciences of the United States of America, 112*(4), 957—964.

Jenkins, P. M., Vasavda, C., Hostettler, J., Davis, J. Q., Abdi, K., & Bennett, V. (2013). E-cadherin polarity is determined by a multifunction motif mediating lateral membrane retention through ankyrin-G and apical-lateral transcytosis through clathrin. *Journal of Biological Chemistry, 288*(20), 14018—14031.

Jenkins, S. M., & Bennett, V. (2001). Ankyrin-G coordinates assembly of the spectrin-based membrane skeleton, voltage-gated sodium channels, and L1 CAMs at Purkinje neuron initial segments. *Journal of Cell Biology, 155*, 739—746.

Jones, S. L., Korobova, F., & Svitkina, T. (2014). Axon initial segment cytoskeleton comprises a multiprotein submembranous coat containing sparse actin filaments. *Journal of Cell Biology, 205*(1), 67–81.

Kizhatil, K., Baker, S. A., Arshavsky, V. Y., & Bennett, V. (2009). Ankyrin-G promotes cyclic nucleotide-gated channel transport to rod photoreceptor sensory cilia. *Science, 323*, 1614–1617.

Kizhatil, K., & Bennett, V. (2004). Lateral membrane biogenesis in human bronchial epithelial cells requires 190 kDa Ankyrin-G. *Journal of Biological Chemistry, 279*, 16706–16714.

Kizhatil, K., Davis, J. Q., Davis, L., Hoffman, J., Hogan, B. L., & Bennett, V. (2007). Ankyrin-G is a molecular partner of E-cadherin in epithelial cells and early embryos. *Journal of Biological Chemistry, 282*, 26552–26561.

Kizhatil, K., Sandhu, N., Peachy, N. S., & Bennett, V. (2009). Ankyrin-B is required for coordinated expression of beta-2 spectrin, the Na/K ATPase and the Na/Ca exchanger in the inner segment of rod photoreceptors. *Experimental Eye Research, 88*, 59–64.

Kizhatil, K., Yoon, W., Mohler, P. J., Davis, L. H., Hoffman, J. A., & Bennett, V. (2007). Ankyrin-G and beta2-spectrin collaborate in biogenesis of lateral membrane of human bronchial epithelial cells. *Journal of Biological Chemistry, 282*, 2029–2037.

Kline, C. F., Scott, J., Curran, J., Hund, T. J., & Mohler, P. J. (2014). Ankyrin-B regulates Cav2.1 and Cav2.2 channel expression and targeting. *Journal of Biological Chemistry, 289*(8), 5285–5295.

Koch, I., Schwarz, H., Beuchle, D., Goellner, B., Langegger, M., & Aberle, H. (2008). Drosophila ankyrin 2 is required for synaptic stability. *Neuron, 58*, 210–222.

Komada, M., & Soriano, P. (January 21, 2002). βIV-spectrin regulates sodium channel clustering through ankyrin-G at axon initial segments and nodes of Ranvier. *Journal of Cell Biology, 156*(2), 337–348.

Kordeli, E., Lambert, S., & Bennett, V. (1995). AnkyrinG. A new ankyrin gene with neural-specific isoforms localized at the axonal initial segment and node of Ranvier. *Journal of Biological Chemistry, 270*, 2352–2359.

Kordeli, E., Ludosky, M. A., Deprette, C., Frappier, T., & Cartaud, J. (1998). AnkyrinG is associated with the postsynaptic membrane and the sarcoplasmic reticulum in the skeletal muscle fiber. *Journal of Cell Science, 111*(Pt 15), 2197–2207.

Krieger, C. C., An, X., Tang, H. Y., Mohandas, N., Speicher, D. W., & Discher, D. E. (2011). Cysteine shotgun-mass spectrometry (CS-MS) reveals dynamic sequence of protein structure changes within mutant and stressed cells. *Proceedings of the National Academy of Sciences of the United States of America, 108*, 8269–8274.

Kuhlman, P. A., Hughes, C. A., Bennett, V., & Fowler, V. M. (1996). A new function for adducin. Calcium/calmodulin-regulated capping of the barbed ends of actin filaments. *Journal of Biological Chemistry, 271*(14), 7986–7991.

Kumar, S., Park, S. H., Cieply, B., Schupp, J., Killiam, E., Zhang, F., et al. (October 2011). A pathway for the control of anoikis sensitivity by E-cadherin and epithelial-to-mesenchymal transition. *Molecular and Cellular Biology, 31*(19), 4036–4051.

Kunimoto, M. (1995). A neuron-specific isoform of brain ankyrin, 440-kD ankyrinB, is targeted to the axons of rat cerebellar neurons. *Journal of Cell Biology, 131*, 1821–1829.

Kunimoto, M., Otto, E., & Bennett, V. (1991). A new 440-kD isoform is the major ankyrin in neonatal rat brain. *Journal of Cell Biology, 115*, 1319–1331.

Kusumi, A., Fujiwara, T. K., Chadda, R., Xie, M., Tsunoyama, T. A., Kalay, Z., et al. (2012). Dynamic organizing principles of the plasma membrane that regulate signal transduction: commemorating the fortieth anniversary of Singer and Nicolson's fluid-mosaic model. *Annual Review of Cell and Developmental Biology, 28*, 215–250.

Lee, G., Abdi, K., Jiang, Y., Michaely, P., Bennett, V., & Marszalek, P. E. (2006). Nanospring behaviour of ankyrin repeats. *Nature, 440*, 246–249.

Lemaillet, G., Walker, B., & Lambert, S. (2003). Identification of a conserved ankyrin-binding motif in the family of sodium channel alpha subunits. *Journal of Biological Chemistry, 278*(30), 27333—27339.

Li, J., Kline, C. F., Hund, T. J., Anderson, M. E., & Mohler, P. J. (2010). Ankyrin-B regulates Kir6.2 membrane expression and function in heart. *Journal of Biological Chemistry, 285*(37), 28723—28730.

Li, J., Mahajan, A., & Tsai, M. D. (2006). Ankyrin repeat: a unique motif mediating protein-protein interactions. *Biochemistry, 45*(51), 15168—15178.

Li, X., Matsuoka, Y., & Bennett, V. (1998). Adducin preferentially recruits spectrin to the fast growing ends of actin filaments in a complex requiring the MARCKS-related domain and a newly defined oligomerization domain. *Journal of Biological Chemistry, 273*(30), 19329—19338.

Lighthouse, D. V., Buszczak, M., & Spradling, A. C. (2008). New components of the Drosophila fusome suggest it plays novel roles in signaling and transport. *Developmental Biology, 317*, 59—71.

Lin, H., Yue, L., & Spradling, A. C. (1994). The Drosophila fusome, a germline-specific organelle, contains membrane skeletal proteins and functions in cyst formation. *Development, 120*, 947—956.

Lingwood, D., & Simons, K. (2010). Lipid rafts as a membrane-organizing principle. *Science, 327*(5961), 46—50.

Lise, S., Clarkson, Y., Perkins, E., Kwasniewska, A., Sadighi Akha, E., Schnekenberg, R. P., et al. (2012). Recessive mutations in SPTBN2 implicate β-III spectrin in both cognitive and motor development. *PLoS Genetics, 8*(12), e1003074.

Lopez, C., Metral, S., Eladari, D., Drevensek, S., Gane, P., Chambrey, R., et al. (2005). The ammonium transporter RhBG requirement of a tyrosine-based signal and ankyrin-G for basolateral targeting and membrane anchorage in polarized kidney epithelial cells. *Journal of Biological Chemistry, 280*, 8221—8228.

Lorenzo, D. N., Badea, A., Davis, J., Hostettler, J., He, J., Zhong, G., et al. (2014). A PIK3C3-ankyrin-B-dynactin pathway promotes axonal growth and multiorganelle transport. *Journal of Cell Biology, 207*(6), 735—752.

Lorenzo, D. N., Li, M. G., Mische, S. E., Armbrust, K. R., Ranum, L. P., & Hays, T. S. (2010). Spectrin mutations that cause spinocerebellar ataxia type 5 impair axonal transport and induce neurodegeneration in Drosophila. *Journal of Cell Biology, 189*, 143—158.

Lowe, J. S., Palygin, O., Bhasin, N., Hund, T. J., Boyden, P. A., Shibata, E., et al. (2008). Voltage-gated Nav channel targeting in the heart requires an ankyrin-G dependent cellular pathway. *Journal of Cell Biology, 180*(1), 173—186.

Lux, S. E., John, K. M., & Bennett, V. (1990). Analysis of cDNA for human erythrocyte ankyrin indicates a repeated structure with homology to tissue- differentiation and cell cycle control proteins. *Nature, 344*, 36—42.

Macias, M. J., Musacchio, A., Ponstingl, H., Nilges, M., Saraste, M., & Oschkinat, H. (1994). Structure of the pleckstrin homology domain from beta-spectrin. *Nature, 369*, 675—677.

Makara, M. A., Curran, J., Little, S. C., Musa, H., Polina, I., Smith, S. A., et al. (2014). Ankyrin-G coordinates intercalated disc signaling platform to regulate cardiac excitability in vivo. *Circulation Research, 115*(11), 929—938.

Malhotra, J. D., Koopmann, M. C., Kazen-Gillespie, K. A., Fettman, N., Hortsch, M., & Isom, L. L. (2002). Structural requirements for interaction of sodium channel beta 1 subunits with ankyrin. *Journal of Biological Chemistry, 277*(29), 26681—26688.

Maniar, T. A., Kaplan, M., Wang, G. J., Shen, K., Wei, L., Shaw, J. E., et al. (2011). UNC-33 (CRMP) and ankyrin organize microtubules and localize kinesin to polarize axon-dendrite sorting. *Nature Neuroscience, 15*(1), 48—56.

Martin, M. W., Grazhdankin, D. V., Bowring, S. A., Evans, D. A., Fedonkin, M. A., & Kirschvink, J. L. (2000). Age of Neoproterozoic bilatarian body and trace fossils, White Sea, Russia: implications for metazoan evolution. *Science, 288*, 841—845.

Martin-Belmonte, F., Gassama, A., Datta, A., Yu, W., Rescher, U., Gerke, V., et al. (2007). PTEN-mediated apical segregation of phosphoinositides controls epithelial morphogenesis through Cdc42. *Cell, 128*, 383—397.

Matsuoka, Y., Li, X., & Bennett, V. (2000). Adducin: structure, function and regulation. *Cellular and Molecular Life Sciences, 57*(6), 884—895.

Mehboob, S., Song, Y., Witek, M., Long, F., Santarsiero, B. D., Johnson, M. E., et al. (2010). Crystal structure of the nonerythroid alpha-spectrin tetramerization site reveals differences between erythroid and nonerythroid spectrin tetramer formation. *Journal of Biological Chemistry, 285*, 14572—14584.

Michaely, P., Tomchick, D. R., Machius, M., & Anderson, R. G. (2002). Crystal structure of a 12 ANK repeat stack from human ankyrinR. *EMBO Journal, 21*, 6387—6396.

Mohler, P. J., Davis, J. Q., & Bennett, V. (2005). Ankyrin-B coordinates the Na/K ATPase, Na/Ca exchanger, and InsP$_3$ receptor in a cardiac T-tubule/SR microdomain. *PLoS Biology, 3*, e423.

Mohler, P. J., Gramolini, A. O., & Bennett, V. (2002). The ankyrin-B C-terminal domain determines activity of ankyrin-B/G chimeras in rescue of abnormal inositol 1,4,5-trisphosphate and ryanodine receptor distribution in ankyrin-B $(-/-)$ neonatal cardiomyocytes. *Journal of Biological Chemistry, 277*, 10599—10607.

Mohler, P. J., Rivolta, I., Napolitano, C., LeMaillet, G., Lambert, S., Priori, S. G., et al. (2004). Nav1.5 E1053K mutation causing Brugada syndrome blocks binding to ankyrin-G and expression of Nav1.5 on the surface of cardiomyocytes. *Proceedings of the National Academy of Sciences of the United States of America, 101*, 17533—17538.

Mohler, P. J., Schott, J. J., Gramolini, A. O., Dilly, K. W., Guatimosim, S., duBell, W. H., et al. (2003). Ankyrin-B mutation causes type 4 long-QT cardiac arrhythmia and sudden cardiac death. *Nature, 421*, 634—639.

Mohler, P. J., Yoon, W., & Bennett, V. (2004). Ankyrin-B targets beta2-spectrin to an intracellular compartment in neonatal cardiomyocytes. *Journal of Biological Chemistry, 279*, 40185—40193.

Nelson, W. J., & Hammerton, R. W. (1989). A membrane-cytoskeletal complex containing Na$^+$,K$^+$-ATPase, ankyrin, and fodrin in Madin-Darby canine kidney (MDCK) cells: implications for the biogenesis of epithelial cell polarity. *Journal of Cell Biology, 108*(3), 893—902.

Nelson, W. J., Shore, E. M., Wang, A. Z., & Hammerton, R. W. (February 1990). Identification of a membrane-cytoskeletal complex containing the cell adhesion molecule uvomorulin (E-cadherin), ankyrin, and fodrin in Madin-Darby canine kidney epithelial cells. *Journal of Cell Biology, 110*(2), 349—357.

Nematollahi, L. A., Garza-Garcia, A., Bechara, C., Esposito, D., Morgner, N., Robinson, C. V., et al. (2015). Flexible stoichiometry and asymmetry of the PIDDosome core complex by heteronuclear NMR spectroscopy and mass spectrometry. *Journal of Molecular Biology, 427*(4), 737—752.

Ohara, O., Ohara, R., Yamakawa, H., Nakajima, D., & Nakayama, M. (1998). Characterization of a new beta-spectrin gene which is predominantly expressed in brain. *Brain Research Molecular Brain Research, 57*, 181—192.

Otsuka, A. J., Franco, R., Yang, B., Shim, K. H., Tang, L. Z., Zhang, Y. Y., et al. (1995). An ankyrin-related gene (unc-44) is necessary for proper axonal guidance in *Caenorhabditis elegans. Journal of Cell Biology, 129*, 1081—1092.

Otto, E., Kunimoto, M., McLaughlin, T., & Bennett, V. (1991). Isolation and characterization of cDNAs encoding human brain ankyrins reveal a family of alternatively-spliced genes. *Journal of Cell Biology, 114*, 241—253.

Pan, Z., Kao, T., Horvath, Z., Lemos, J., Sul, J. Y., Cranstoun, S. D., et al. (2006). A common ankyrin-G-based mechanism retains KCNQ and NaV channels at electrically active domains of the axon. *Journal of Neuroscience, 26*(10), 2599–2613.

Parkinson, N. J., Olsson, C. L., Hallows, J. L., McKee-Johnson, J., Keogh, B. P., Noben-Trauth, K., et al. (September 2001). Mutant beta-spectrin 4 causes auditory and motor neuropathies in quivering mice. *Nature Genetics, 29*(1), 61–65.

Patthy, L. (1999). Genome evolution and the evolution of exon-shuffling—a review. *Gene, 238*(1), 103–114.

Perkins, E. M., Clarkson, Y. L., Sabatier, N., Longhurst, D. M., Millward, C. P., Jack, J., et al. (2010). Loss of beta-III spectrin leads to Purkinje cell dysfunction recapitulating the behavior and neuropathology of spinocerebellar ataxia type 5 in humans. *Journal of Neuroscience, 30*, 4857–4867.

Peters, L. L., John, K. M., Lu, F. M., Eicher, E. M., Higgins, A., Yialamas, M., et al. (1995). Ank3 (epithelial ankyrin), a widely distributed new member of the ankyrin gene family and the major ankyrin in kidney, is expressed in alternatively spliced forms, including forms that lack the repeat domain. *Journal of Cell Biology, 130*, 313–330.

Pielage, J., Cheng, L., Fetter, R. D., Carlton, P. M., Sedat, J. W., & Davis, G. W. (2008). A presynaptic giant ankyrin stabilizes the NMJ through regulation of presynaptic microtubules and transsynaptic cell adhesion. *Neuron, 58*(2), 195–209.

Pielage, J., Fetter, R. D., & Davis, G. W. (2005). Presynaptic spectrin is essential for synapse stabilization. *Current Biology, 15*(10), 918–928.

Plückthun, A. (2015). Designed ankyrin repeat proteins (DARPins): binding proteins for research, diagnostics, and therapy. *Annual Review of Pharmacology and Toxicology, 55*, 489–511.

Pocock, R., Bénard, C. Y., Shapiro, L., & Hobert, O. (2008). Functional dissection of the C. elegans cell adhesion molecule SAX-7, a homologue of human L1. *Molecular and Cellular Neurosciences, 37*, 56–68.

Rief, M., Pascual, J., Saraste, M., & Gaub, H. E. (1999). Single molecule force spectroscopy of spectrin repeats: low unfolding forces in helix bundles. *Journal of Molecular Biology, 286*, 553–561.

Saka, S. K., Honigmann, A., Eggeling, C., Hell, S. W., Lang, T., & Rizzoli, S. O. (2014). Multi-protein assemblies underlie the mesoscale organization of the plasma membrane. *Nature Communications, 5*, 4509.

Salomao, M., An, X., Guo, X., Gratzer, W. B., Mohandas, N., & Baines, A. J. (2006). Mammalian alpha I-spectrin is a neofunctionalized polypeptide adapted to small highly deformable erythrocytes. *Proceedings of the National Academy of Sciences of the United States of America, 103*(3), 643–648.

Scotland, P., Zhou, D., Benveniste, H., & Bennett, V. (1998). Nervous system defects of ankyrin-B (−/−) mice suggest functional overlap between the cell adhesion molecule L1 and 440 kD ankyrin-B in premyelinated axons. *Journal of Cell Biology, 143*, 1305–1315.

Sheetz, M. P., Schindler, M., & Koppel, D. E. (1980). Lateral mobility of integral membrane proteins is increased in spherocytic erythrocytes. *Nature, 285*(5765), 510–511.

Shewan, A., Eastburn, D. J., & Mostov, K. (2011). Phosphoinositides in cell architecture. *Cold Spring Harbor Perspectives in Biology, 3*, a004796.

Shotton, D. M., Burke, B. E., & Branton, D. (1979). The molecular structure of human erythrocyte spectrin. Biophysical and electron microscopic studies. *Journal of Molecular Biology, 131*(2), 303–329.

Siegenthaler, D., Enneking, E. M., Moreno, E., & Pielage, J. (2015). L1CAM/Neuroglian controls the axon-axon interactions establishing layered and lobular mushroom body architecture. *Journal of Cell Biology, 208*(7), 1003–1018.

Singer, S. J., & Nicolson, G. L. (1972). The fluid mosaic model of the structure of cell membranes. *Science, 175*(4023), 720–731.

Smith, K. R., Kopeikina, K. J., Fawcett-Patel, J. M., Leaderbrand, K., Gao, R., Schürmann, B., et al. (2014). Psychiatric risk factor ANK3/ankyrin-G nanodomains regulate the structure and function of glutamatergic synapses. *Neuron, 84*(2), 399–415.

Sobotzik, J. M., Sie, J. M., Politi, C., Del Turco, D., Bennett, V., Deller, T., et al. (2009). AnkyrinG is required to maintain axo-dendritic polarity in vivo. *Proceedings of the National Academy of Sciences of the United States of America, 106,* 17564–17569.

Stabach, P. R., Simonović, I., Ranieri, M. A., Aboodi, M. S., Steitz, T. A., Simonović, M., et al. (2009). The structure of the ankyrin-binding site of beta-spectrin reveals how tandem spectrin-repeats generate unique ligand-binding properties. *Blood, 113,* 5377–5384.

Stankewich, M. C., Tse, W. T., Peters, L. L., Ch'ng, Y., John, K. M., Stabach, P. R., et al. (1998). A widely expressed betaIII spectrin associated with Golgi and cytoplasmic vesicles. *Proceedings of the National Academy of Sciences of the United States of America, 95,* 14158–14163.

Staufenbiel, M. (1987). Ankyrin-bound fatty acid turns over rapidly at the erythrocyte plasma membrane. *Molecular and Cellular Biology, 7,* 2981–2984.

Stephan, R., Goellner, B., Moreno, E., Frank, C. A., Hugenschmidt, T., Genoud, C., et al. (2015). Hierarchical microtubule organization controls axon caliber and transport and determines synaptic structure and stability. *Developmental Cell, 33*(1), 5–21.

Susuki, K., Chang, K. J., Zollinger, D. R., Liu, Y., Ogawa, Y., Eshed-Eisenbach, Y., et al. (2013). Three mechanisms assemble central nervous system nodes of Ranvier. *Neuron, 78,* 469–482.

Tang, Y., Katuri, V., Dillner, A., Mishra, B., Deng, C. X., & Mishra, L. (2003). Disruption of transforming growth factor-beta signaling in ELF beta-spectrin-deficient mice. *Science, 299,* 574–777.

Trunova, S., Baek, B., & Giniger, E. (2011). Cdk5 regulates the size of an axon initial segment-like compartment in mushroom body neurons of the Drosophila central brain. *Journal of Neuroscience, 31*(29), 10451–10462.

Tseng, W. C., Jenkins, P. M., Tanaka, M., Mooney, R., & Bennett, V. (January 27, 2015). Giant ankyrin-G stabilizes somatodendritic GABAergic synapses through opposing endocytosis of GABAA receptors. *Proceedings of the National Academy of Sciences of the United States of America, 112*(4), 1214–1219.

Uemoto, Y., Suzuki, S., Terada, N., Ohno, N., Ohno, S., Yamanaka, S., et al. (2007). Specific role of the truncated betaIV-spectrin Sigma6 in sodium channel clustering at axon initial segments and nodes of Ranvier. *Journal of Biological Chemistry, 282,* 6548–6555.

Vosseller, K., Trinidad, J. C., Chalkley, R. J., Specht, C. G., Thalhammer, A., Lynn, A. J., et al. (2006). O-linked N-acetylglucosamine proteomics of postsynaptic density preparations using lectin weak affinity chromatography and mass spectrometry. *Molecular and Cellular Proteomics, 5,* 923–934.

Wang, C., Wei, Z., Chen, K., Ye, F., Yu, C., Bennett, V., et al. (Nov 10, 2014). Structural basis of diverse membrane target recognitions by ankyrins. *Elife, 3.* http://dx.doi.org/10.7554/eLife.04353.

Wang, C., Yu, C., Ye, F., Wei, Z., & Zhang, M. (2012). Structure of the ZU5-ZU5-UPA-DD tandem of ankyrin-B reveals interaction surfaces necessary for ankyrin function. *Proceedings of the National Academy of Sciences of the United States of America, 109,* 4822–4827.

Wasenius, V. M., Saraste, M., Salvén, P., Erämaa, M., Holm, L., & Lehto, V. P. (1989). Primary structure of the brain alpha-spectrin. *Journal of Cell Biology, 108,* 79–93.

Winkelmann, J. C., Chang, J. G., Tse, W. T., Scarpa, A. L., Marchesi, V. T., & Forget, B. G. (1990). Full-length sequence of the cDNA for human erythroid beta-spectrin. *Journal of Biological Chemistry, 265,* 11827–11832.

Winkelmann, J. C., Costa, F. F., Linzie, B. L., & Forget, B. G. (1990). Beta spectrin in human skeletal muscle. Tissue-specific differential processing of 3' beta spectrin pre-mRNA generates a beta spectrin isoform with a unique carboxyl terminus. *Journal of Biological Chemistry, 265*, 20449—20454.

Wood, S. J., & Slater, C. R. (February 9, 1998). β-Spectrin is colocalized with both voltage-gated sodium channels and ankyrinG at the adult rat neuromuscular junction. *Journal of Cell Biology, 140*(3), 675—684.

Xu, K., Zhong, G., & Zhuang, X. (2013). Actin, spectrin, and associated proteins form a periodic cytoskeletal structure in axons. *Science, 339*(6118), 452—456.

Yan, Y., Winograd, E., Viel, A., Cronin, T., Harrison, S. C., & Branton, D. (1993). Crystal structure of the repetitive segments of spectrin. *Science, 262*, 2027—2030.

Zhang, X., & Bennett, V. (1996). Identification of O-linked N-acetyl glucosamine modification of ankyrinG isoforms targeted to nodes of Ranvier. *Journal of Biological Chemistry, 271*, 31391—31398.

Zhang, X., Davis, J. Q., Carpenter, S., & Bennett, V. (1998). Structural requirements for association of neurofascin with ankyrin. *Journal of Biological Chemistry, 273*, 30785—30794.

Zhang, Y., Resneck, W. G., Lee, P. C., Randall, W. R., Bloch, R. J., & Ursitti, J. A. (2010). Characterization and expression of a heart-selective alternatively spliced variant of alpha II-spectrin, cardi+, during development in the rat. *Journal of Molecular and Cellular Cardiology, 48*, 1050—1059.

Zhong, G., He, J., Zhou, R., Lorenzo, D., Babcock, H. P., Bennett, V., et al. (Dec 23, 2014). Developmental mechanism of the periodic membrane skeleton in axons. *Elife, 3*. http://dx.doi.org/10.7554/eLife.04581.

CHAPTER SIX

The Axon Initial Segment, 50 Years Later: A Nexus for Neuronal Organization and Function

Christophe Leterrier
Aix Marseille Université, CNRS, CRN2M UMR 7286, Marseille, France
E-mail: christophe.leterrier@univ-amu.fr

Contents

Current Topics in Membranes, Volume 77
ISSN 1063-5823
http://dx.doi.org/10.1016/bs.ctm.2015.10.005

Abstract

The axon initial segment is a highly specialized neuronal compartment, identified almost 50 years ago by the pioneers of electron microscopy. Located in the first 50 μm of the axon, it contains unique cytoskeletal features and concentrates a repertoire of specific scaffold and membrane proteins that assembles just after axon determination. The axon initial segment (AIS) supports two crucial physiological functions of the mature neuron: first, it generates and shapes the action potential. Second, it separates the cell body from the axon, preserving the molecular identity of each compartment. In addition to a diffusion barrier restricting membrane proteins and lipids exchange, an intracellular filter has been proposed that could selectively exclude somatodendritic vesicles and recruit axonal cargoes. Finally, the AIS scaffold is capable of morphological plasticity during development or in response to network activity. These changes directly impact the neuron excitability, allowing an adaptive and homeostatic response. These plastic electrogenic properties, as well as the regulation of protein transport to and from the axon, may have important implications in several neuropathological contexts where the AIS structure is altered. Fifty years after its first characterization, the AIS thus emerges as a nexus for both neuronal organization and physiology.

1. INTRODUCTION

The neuron is the epitome of cell polarity. In 1865, Otto Deiters first described the archetypal neuron: a cell body bearing several "protoplasmic processes" (dendrites) and a single "axis cylinder process" (axon) (Craig & Banker, 1994; Deiters & Guillery, 2013). As hypothesized later by Ramon y Cajal as the "law of dynamic polarization" in 1889 (Sotelo, 2003), this asymmetry supports the directionality of signal transfer in the brain: the somatodendritic compartment receives synaptic inputs, and the axonal compartment sends signals toward downstream cells in the form of action potentials (Craig & Banker, 1994). The advent of neuronal cell culture allowed deciphering the respective molecular compositions of the somato-dendritic and axonal compartments (Craig & Banker, 1994) and the mechanisms underlying the emergence of polarity during neuronal development (Caceres, Ye, & Dotti, 2012; Cheng & Poo, 2012). Later studies have focused on how neuronal polarization arises in vivo, where extracellular cues complement cell-autonomous stochastic events (Lewis, Courchet, & Polleux, 2013; Namba et al., 2015). After the initial axon specification, the neuron grows elaborate dendritic and axonal ramifications and maintains this polarized architecture using a multitude of molecular transport, sorting, and diffusing mechanisms (Kapitein & Hoogenraad, 2011; Lasiecka, Yap, Vakulenko, & Winckler, 2009).

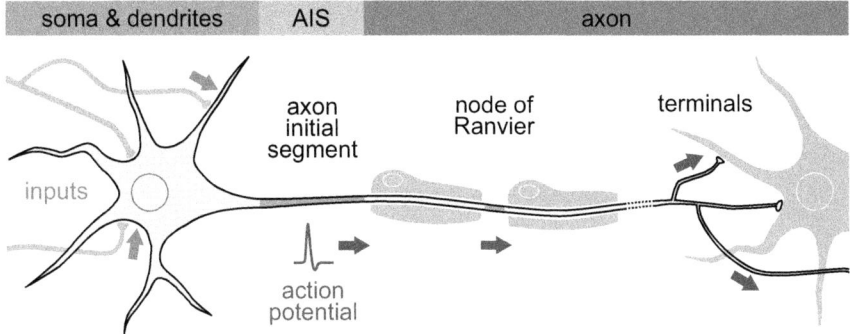

Figure 1 *The axon initial segment (AIS).* The neuron is a polarized cell that integrates synaptic inputs over the cell body (soma) and dendrites (on the left, blue (dark gray in print versions)). The AIS (orange (light gray in print versions)) located along the first 20–60 μm of the axon, realizes the final integration of inputs and the initiation of the action potential (red (darker gray in print versions)) that propagates along the axon toward presynaptic terminals that contact downstream neurons (on the right, red (darker gray in print versions)). The axon can be myelinated by Schwann cells or oligodendrocytes (along the axon in gray), and the action potential is regenerated at nodes of Ranvier between myelin sheets (red (darker gray in print versions)).

Strategically located between the somatodendritic and axonal compartments, the axon initial segment (AIS) occupies the first 20–60 μm of the axon (Figure 1). It is present in almost all vertebrate neuronal types, forming a unique compartment with specific morphological features and protein components (Chang & Rasband, 2013). The AIS ensures two functions that are crucial for the neuron physiology: it generates the action potentials, and separates the axon from the cell body to maintain its molecular identity (Leterrier & Dargent, 2014). I will detail how AIS components and organization were determined along the years, before outlining the mechanisms of its formation and maintenance. After reviewing its role in electrogenesis and polarity maintenance, I will discuss its recently discovered plasticity and alterations in disease.

2. THE AIS MORPHOLOGY: EARLY ELECTRON MICROSCOPY WORKS

The emergence of the axon from the cell body was first described by Kölliker in 1849 and pictured by Deiters in 1865 (Clark, Goldberg, & Rudy, 2009; Deiters & Guillery, 2013). The first precise descriptions from electron microscopy works defined the AIS as the axonal portion immediately following the hillock, usually 20—60 μm long, stopping at the beginning of the myelin sheath in myelinated neurons. The AIS has a 0.5—2 μm diameter and presents unique morphological features: a dense layer undercoating the plasma membrane, a grouping of microtubules into fascicles, and a sharp decrease in the number of ribosomes (Conradi, 1966; Kohno, 1964; Palay, Sotelo, Peters, & Orkand, 1968; Peters, Proskauer, & Kaiserman-Abramof, 1968). On close examination, the undercoat is connected with an external density on the other side of the plasma membrane, suggesting a dense complex of submembrane and membrane proteins (Palay et al., 1968). The undercoat is tripartite, with a 7.5-nm granular layer connecting the plasma membrane to a 7.5-nm-thick lamina that lies above 35 -nm deep filamentous tufts (Chan-Palay, 1972). In pyramidal cortical neurons and cerebellar Purkinje cells, microtubules form 5 to 8 fascicles that group together 3 to 10 microtubules. In a fascicle, microtubules are ~40 nm apart (microtubule center to center) and connected by protein bridges (Kohno, 1964; Palay et al., 1968; Peters et al., 1968) of unknown molecular identity. Finally, most of these early works emphasized the presence of axoaxonic, symmetric synapses: these postsynaptic specializations are devoid of the AIS undercoat and often associated with cisternal organelles (Figure 2(A) and (B)). Besides synaptic appositions, the AIS is also specifically contacted by microglia processes in the cortex (Baalman et al., 2015).

3. THE AIS MOLECULAR COMPONENTS AND THEIR INTERACTIONS

3.1 The Ankyrin G-Bound Scaffold

Over the years, studies have identified molecular components of the AIS. The canonical protein complex in the AIS consists in ion channels and cell adhesion molecules (CAMs) anchored to a specialized ankyrin/spectrin scaffold that is connected to both actin filaments and microtubules. The central interacting hub of this complex is ankyrin G (from the Greek *ankyra*, anchor), which interacts with all other components and is often designated as the AIS "key organizer" (Figure 2(C) and Figure 3(A), Rasband, 2010). Ankyrin G is the third member of the ankyrin family in mammals that also includes ankyrin R (part of the ankyrin/spectrin complex in red blood cells, Bennett & Stenbuck, 1979) and ankyrin B (originally described as the "brain" ankyrin, Davis & Bennett, 1984). Ankyrin G was first detected using anti-ankyrin R antibodies in an ankyrin R-deficient

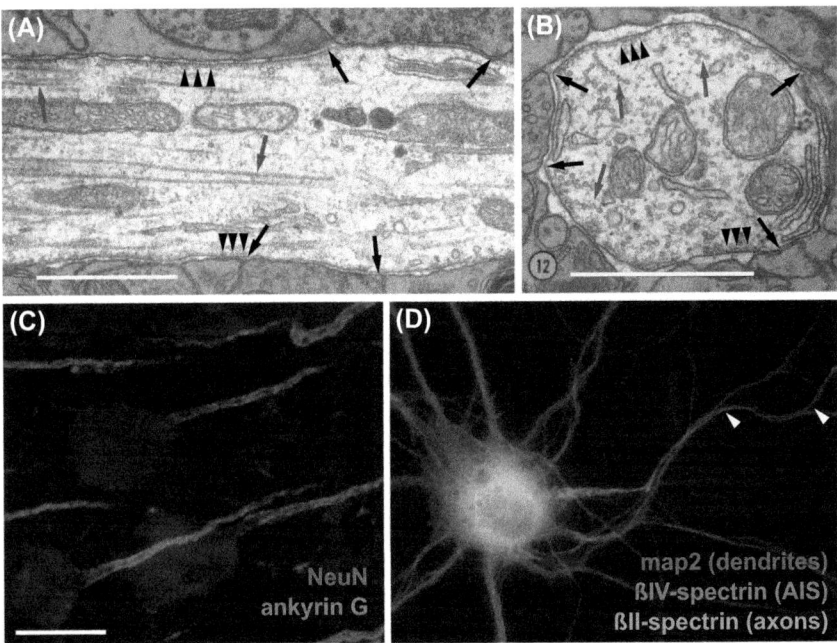

Figure 2 *Morphology of the axon initial segment (AIS).* (A) Electron microscopy image of a longitudinal section through the AIS of a rat hippocampal CA3 neuron. The microtubule fascicles are visible (red arrows) supporting the transport of organelles and vesicles. The dense undercoat lines the plasma membrane (arrowheads) and stops at axoaxonic synaptic contacts present on the AIS (top and bottom membrane, between arrows). Scale bar, 1 μm. (B) Electron microscopy image of a cross section through the AIS of a rat hippocampal CA3 neuron. Microtubule fascicles are seen as string of beads (red arrows). The dense undercoat can sometimes be seen as a row of small dots (arrowheads). The undercoat stops at sites of cisternal organelles apposition (arrows). Scale bar, 1 μm. *((A) and (B) are reproduced from (Kosaka, 1980) with permission from Springer.)* (C) Brain cortical section from an adult rat showing neurons (labeled with NeuN, blue) and their initial segments (ankyrin G, red). Scale bar, 10 μm. (D) Rat hippocampal neuron after 15 days in culture, labeled for the somatodendritic marker map2 (blue) the AIS scaffold βIV-spectrin (red) and the distal axon scaffold βII-spectrin (green, arrowheads). Scale bar, 20 μm. (See color plate)

mice (Kordeli & Bennett, 1991; Kordeli, Davis, Trapp, & Bennett, 1990). The cloning of ankyrin G and the generation of specific antibodies demonstrated ankyrin G localization at the AIS and nodes of Ranvier (Kordeli, Lambert, & Bennett, 1995). By contrast, ankyrin B is found along the whole shaft of unmyelinated axons (Boiko et al., 2007; Chan, Kordeli, & Bennett, 1993) and clusters around paranodes in myelinated fibers (Chang et al.,

2014; Ogawa et al., 2006). The largest isoform of ankyrin G is a 480-kDa protein (4377 amino acids for human ankyrin G) composed of different domains (Bennett & Lorenzo, 2013): a membrane-binding domain that contains 24 ankyrin repeats and binds to ion channels and CAMs in the AIS; a spectrin-binding domain that interacts with β-spectrin; a serine-rich domain; a ~2300 amino acids unstructured tail (sharing homology with i-connexin/titin); and a carboxyterminus containing a death domain. A 270 kDa isoform of ankyrin G is also present in the AIS that results from a splicing event that removes the last 1900 amino acids of the tail. The common domains shared by these two AIS-specific ankyrin G isoforms are the serin-rich domain and the tail, which are missing in the smaller 190 kDa ankyrin G found in the brain, kidney, and heart (Kordeli et al., 1995). The ternary structure of several ankyrin domains has been resolved (Liu, Zhang, & Wang, 2014; Wang, Wei, et al., 2014; Wang, Yu, Ye, Wei, & Zhang, 2012). Ankyrin G labeling is found in close apposition to the plasma membrane (Iwakura, Uchigashima, Miyazaki, Yamasaki, & Watanabe, 2012; Le Bras et al., 2013). This membrane association is important for ankyrin G localization at the AIS, as shown by the role of palmitoylation at cystein 70 in the membrane-binding domain (He, Jenkins, & Bennett, 2012). In contrast, membrane association of distal axon ankyrin B is inhibited by intramolecular interactions (Abdi, Mohler, Davis, & Bennett, 2006; He, Tseng, & Bennett, 2013).

In the AIS, the spectrin that interacts with ankyrin G is βIV-spectrin (Figures 2(D) and 3(A), Berghs et al., 2000). The full-length βIV-spectrin $\sum 1$ isoform is a 280-kDa protein (2564 amino acids for human βIV-spectrin $\sum 1$) that contains: two adjacent aminoterminal calponin-homology domains that bind actin; 17 consecutive triple helical repeats; a charged domain specific to βIV-spectrin; and a carboxyterminus containing a pleckstrin-homology domain (Bennett & Baines, 2001). In rodents and probably in humans, a shorter βIV-spectrin $\sum 6$ isoform (140 kDa) is also present at the AIS: it lacks the aminoterminal half of the protein, including actin-binding sites and the first nine triple helical repeats (Komada & Soriano, 2002). The canonical spectrins exist as heterotetramers of two α-spectrins and two β-spectrins, with lateral antiparallel α-β dimerization combined with end-to-end interaction between α dimers. Although βIV-spectrin can interact with αII-spectrin (Uemoto et al., 2007), no α-spectrin has been specifically identified at the AIS, in contrast to the αII-spectrin/βII-spectrin assembly found along the distal axon and at paranodes (Galiano et al., 2012; Ogawa et al., 2006). Ankyrins interact with the 15th helical repeat of β-spectrins (Davis et al., 2009; Kennedy, Warren, Forget, & Morrow, 1991; Yang, Ogawa, Hedstrom, & Rasband, 2007) via their spectrin-binding domain composed of two ZU5 (*ZO-1 and unc5*) and one UPA (*unc5, PIDD, and ankyrin*) modules (Ipsaro, Huang, & Mondragón, 2009; Mohler, Yoon, & Bennett, 2004; Wang et al., 2012). In their recent work uncovering the specific role of 480-kDa ankyrin G at the AIS, Vann Bennett's group uncovered a potentially phosphorylated serine in the specific tail domain that is critical for βIV-spectrin interaction and recruitment to the AIS (Jenkins et al., 2015). The possible interplay between this tail motif and the spectrin-binding domain remains to be clarified.

Beside its linkage with actin via βIV-spectrin, ankyrin G is also linked to the microtubule cytoskeleton via the *end binding* (EB) proteins EB1 and EB3 (Leterrier, Vacher, et al., 2011). EB proteins are dimers that associate with growing microtubule plus tips in interphase cells (+TIPs) and coordinate the assembly of the others +TIPs such as CLASPs and CLIPs (Akhmanova & Steinmetz, 2010). On electron microscopy images of the AIS, microtubule fascicles often appears apposed to the membrane undercoat, suggested a link between these structures (Nakajima, 1974; Westrum & Gray, 1976). Interestingly, ankyrin B was initially found to bind brain microtubules (Bennett & Davis, 1981; Davis & Bennett, 1984), using tubulin preparations that are rich in EB proteins (Sweet et al., 2011). At the AIS, ankyrin G binds to the hydrophobic pocket of the EB dimer, preventing binding to +TIPs. This results in a high local concentration of EB and its binding along the microtubules, leading to a stable, filamentous labeling (Leterrier, Vacher, et al., 2011; Nakata & Hirokawa, 2003). The EB-binding site on ankyrin G is unknown, but the tail domain of the 480-kDa ankyrin G notably contains six EB-binding SxIP motifs, making it a strong candidate for EB binding (Jiang et al., 2012). It is likely that several other microtubule-associated proteins have a specific role at the AIS: nuclear distribution element-like 1 (Ndel1/Nudel), a cytoskeleton regulator, concentrates at the AIS (Bradshaw et al., 2008; 2011; Pei et al., 2014; Sasaki et al., 2000). Antibodies against phosphorylated IκBα label a microtubule-associated epitope along the AIS (Sanchez-Ponce, Tapia, Muñoz, & Garrido, 2008; Schultz et al., 2006), but the target is not IκBα itself (Buffington, Sobotzik, Schultz, & Rasband, 2012). Finally, an autoantibody found in a patient suffering from paraneoplastic encephalomyelitis results in microtubule-associated labeling along the AIS, but its target is unknown (Shams'ili, de Leeuw, Hulsenboom, Jaarsma, & Smitt, 2009).

3.2 Ankyrin G-Bound Membrane Proteins: CAMs and Ion Channels

As the central hub of the complex, ankyrin G interacts with a range of membrane proteins that concentrate at the AIS (Figure 3(A)). CAMs of the L1-CAM family represent up to 1% of the membrane proteins in the brain (Bennett & Baines, 2001). The 186-kDa isoform of neurofascin (NF-186) as well as the *neuronal cell adhesion molecule* (NrCAM) is specifically concentrated at the AIS and nodes of Ranvier (Davis, Lambert, & Bennett, 1996). Their cytoplasmic tail interacts with several sites along the inner groove of the ankyrin G membrane-binding domain (Michaely & Bennett, 1995; Wang, Wei, et al., 2014). Relatively to the diverse roles of L1-CAMs in cellular interactions and neuronal growth, not much is known so far about the specific role of NF-186 and NrCAM at the AIS. NF-186 recruits several extracellular matrix components such as aggrecan, brevican (Hedstrom et al., 2007), tenascin R, and versican that concentrate in the AIS in vitro and in vivo (Brückner, Szeöke, Pavlica, Grosche, & Kacza, 2006; Frischknecht et al., 2009; John et al., 2006).

Consistent with its role as the action potential initiation site, the AIS contains a high concentration of voltage-gated sodium channels (Catterall, 1981; Wollner & Catterall, 1986). These sodium channels are found at a 5- to 50-fold higher density than in the dendrites or distal axon, based on quantitative immunohistochemical or electrophysiological evidence (Kole et al., 2008; Lorincz & Nusser, 2010; but see Colbert & Pan, 2002; Fleidervish, Lasser-Ross, Gutnick, & Ross, 2010). The dominant isoform of sodium channels at the AIS of mature neurons is Nav1.6 in rodent (Jenkins & Bennett, 2001; Kress, Dowling, Eisenman, & Mennerick, 2010; Lorincz & Nusser, 2008; 2010; Van Wart, Trimmer, & Matthews, 2007) and human neurons (Inda, DeFelipe, & Muñoz, 2006; Tian, Wang, Ke, Guo, & Shu, 2014). Nav1.2 is more abundant in the immature AIS and gradually complemented by Nav1.6 during development (Boiko et al., 2003; Osorio et al., 2010). In mature neurons of the cortex and hippocampus, Nav1.2 localizes in the proximal part of the AIS, and Nav1.6 is more concentrated along the distal part (Hu et al., 2009; Lorincz & Nusser, 2010; Tian et al., 2014; Wimmer, Reid, Mitchell, et al., 2010). Nav1.1 channels have also been detected along the proximal part of the AIS in cortical interneurons (Ogiwara et al., 2007; Tian et al., 2014; Van Wart et al., 2007; Wimmer et al., 2015), motoneurons (Duflocq, Le Bras, Bullier, Couraud, & Davenne, 2008), and cerebellar Purkinje cells (Xiao, Bosch, Nerbonne, & Ornitz, 2013). Nav channel concentration is due to their interaction with ankyrin G (Srinivasan, Elmer, Davis, Bennett, & Angelides, 1988) that occurs between a motif in the intracellular loop II—III of the channels (Garrido et al., 2003; Gasser et al., 2012; Lemaillet, Walker, & Lambert, 2003) and ankyrin repeats in the ankyrin G membrane-binding domain (Srinivasan, Lewallen, & Angelides, 1992; Wang, Wei, et al., 2014).

At the AIS, the Nav1.1/1.2/1.6 α subunits are often associated with auxiliary β subunits (O'Malley & Isom, 2015). These immunoglobulin loops containing CAMs associate with the α pore-forming subunit via noncovalent interaction (for $\beta 1$ and $\beta 3$) or disulfide bonds (for $\beta 2$ and $\beta 4$). Depending on the neuronal type, the AIS can exhibit a concentration of $\beta 1$ (Brackenbury et al., 2010; Wimmer et al., 2015), $\beta 2$ (Chen et al., 2012), or $\beta 4$ (Buffington & Rasband, 2013; Miyazaki et al., 2014) subunits. These subunits have been reported to interact with several other AIS components, including NrCAM, NF-186 (McEwen & Isom, 2004), and ankyrin G (Malhotra et al., 2002). Intracellular homologous factors of fibroblast growth factors FGF12 (Wildburger et al., 2015), FGF13 (Goldfarb et al., 2007), and FGF14 (Laezza et al., 2007; Lou et al., 2005; Xiao et al., 2013) also interact with Nav α subunits at the AIS, where they modulate their electrophysiological properties. In addition to voltage-gated sodium channels, ankyrin G also binds to potassium channels KCNQ2 and KCNQ3 (Kv7.2 and Kv7.3) that are concentrated at the AIS (Devaux, Kleopa, Cooper, & Scherer, 2004; Klinger, Gould, Boehm, & Shapiro, 2011; Pan et al., 2006; Rasmussen et al., 2007; Sanchez-Ponce, DeFelipe, Garrido, & Muñoz, 2012). The ankyrin G-binding motif, present in multiple copies in the tetrameric channel, is homologous to the sodium loop II—III motif (Pan et al., 2006), but is likely to have appeared independently (Hill et al., 2008), and binds to an overlapping site in the ankyrin G membrane-binding domain (Xu & Cooper, 2015).

3.3 AIS Synapses and Cisternal Organelles

As noted in the early electron microscopy works, the AIS often contains postsynaptic specializations characterized by an interruption of the AIS undercoat (Jones & Powell, 1969; Peters et al., 1968) sometimes located on axonic spines (Figure 3(B), Kosaka, 1980; Sloper & Powell, 1979; Westrum, 1970). These are GABAergic synapses and their density and repartition depend on the neuronal types studied: cerebellar Purkinje cells AIS only have rare synapses on the proximal AIS (Iwakura et al., 2012; Somogyi & Hámori, 1976), whereas pyramidal neurons from the cortex

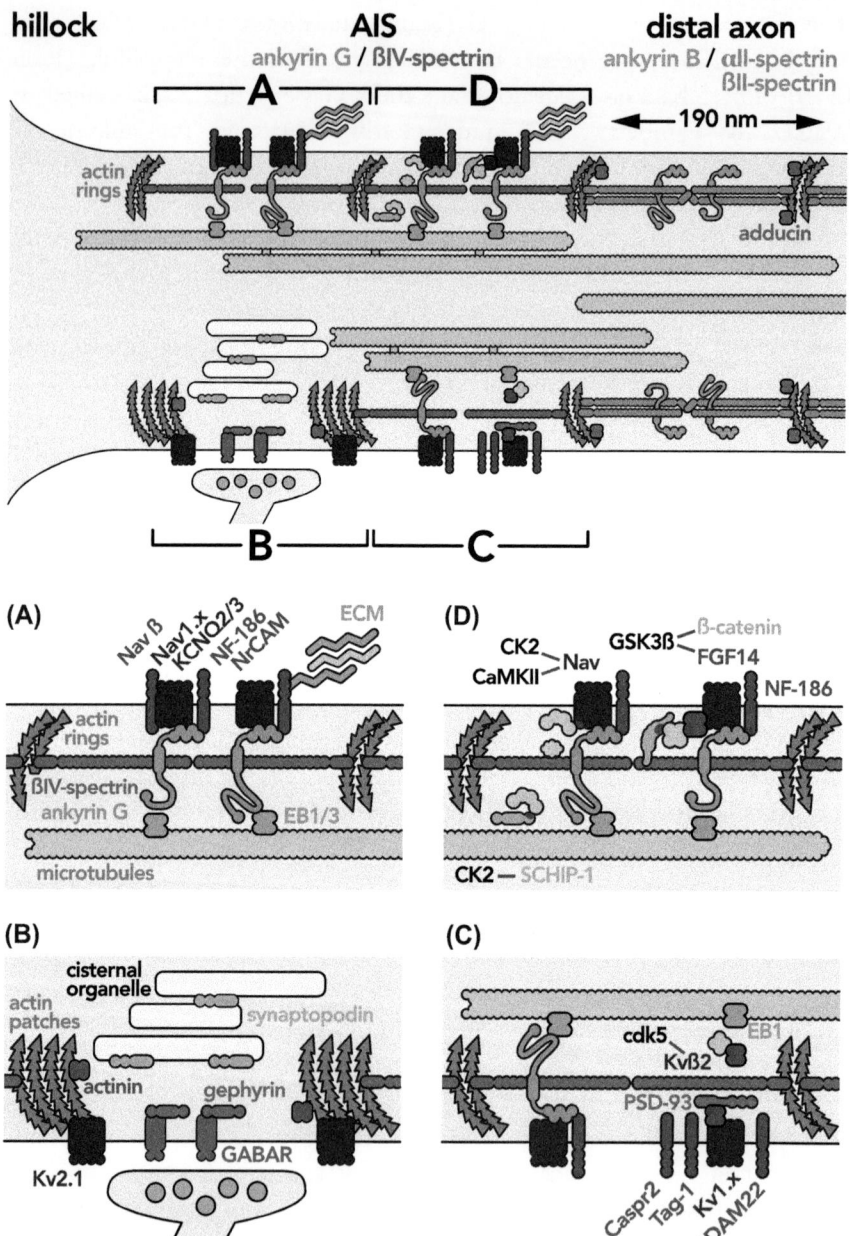

Figure 3 *Components of the AIS and their modulation.* The AIS, characterized by its ankyrin G/βIV-spectrin scaffold sits between the axon hillock (left) and the distal axon (right), which contains an ankyrin B/βII-spectrin scaffold. Actin rings are regularly found (190 nm period) along the AIS and distal axon, capped by adducin (left). (A) The canonical AIS complex. Actin rings are connected by dimers of βIV-spectrin, that

are contacted by 30−200 synapses, with a higher density on the distal AIS (Fish, Hoftman, Sheikh, Kitchens, & Lewis, 2013; Kosaka, 1980; Sloper & Powell, 1979). In the cortex, AIS GABAergic innervation is ensured by specialized axoaxonic interneurons called chandelier cells (DeFelipe, Hendry, Jones, & Schmechel, 1985; Inan et al., 2013; Lewis, 2011; Somogyi, Freund, & Cowey, 1982; Tai, Janas, Wang, & Van Aelst, 2014). GABAergic AIS innervation is also present in hippocampal cultures (Brünig, Scotti, Sidler, & Fritschy, 2002; Burkarth, Kriebel, Kranz, & Volkmer, 2007; Muir & Kittler, 2014). AIS synapses contain GABA $\alpha2$ and to a lesser extent $\alpha1$ receptors (Brünig et al., 2002; Christie & De Blas, 2003; Nusser, Sieghart, Benke, Fritschy, & Somogyi, 1996) anchored to gephyrin (Burkarth et al., 2007; Panzanelli et al., 2011; Tretter et al., 2008). Concentration of NF-186 is important for proper innervation of the AIS in Purkinje cells (Ango et al., 2004) and hippocampal neurons (Burkarth et al., 2007; Kriebel et al., 2011).

Peculiar cisternal organelles, similar to the spine apparatus found in a subset of dendritic spines, are often seen associated with AIS synapses on electron microscopy images (Kosaka, 1980; Peters et al., 1968). These cisternal organelles are calcium stores (Benedeczky, Molnár, & Somogyi, 1994) enriched in synaptopodin, inositol 1,4,5-triphosphate receptor 1 (IP(3)R1), α-actinin, and annexin 6 (Figure 3(B), Antón-Fernández, Rubio-Garrido, DeFelipe, & Muñoz, 2015; Bas Orth, Schultz, Müller,

---

interact with the ankyrin G scaffold. Nav α subunits and KCNQ2/3 ion channels are concentrated by their anchoring to ankyrin G. Also anchored to ankyrin G are CAMs (NF-186 and NrCAM) that bind ECM components (aggrecan, brevican, tenascin R, versican), whereas Nav β subunits are linked to their α counterpart. Ankyrin G is in turn bound to microtubules via the EB end binding proteins EB1 and EB3. (B) The AIS synapse. GABAergic synapses contain $\alpha2$ and $\alpha1$ GABA receptors (GABAR), anchored by gephyrin. Apposed to the synapse are cisternal organelles rich in synaptopodin. The synapses are often associated with actin patches rich in actinin, and membrane clusters of Kv2.1. (C) The distal AIS complex. In certain neuronal types, potassium channels Kv1.2 and Kv1.1 are concentrated along the distal portion of the AIS. These are associated with the CAMs Caspr2 and Tag-1, the metalloprotease ADAM22, and the PSD-93 scaffold. Kv$\beta2$ subunits associate with Kv1 channels, and regulate their delivery via their phosphorylation by cdk2/5 that detach them from the microtubule-associated EB1. (D) Modulation of component interactions by phosphorylation: a number of kinases can phosphorylate AIS components to regulate their interactions. CK2 phosphorylates Nav channels, modulating their binding to ankyrin G. CK2 also phosphorylates SCHIP-1 to modulate its interaction with ankyrin G. CaMKII is bound to βIV-spectrin and also phosphorylates Nav channels. GSK3β phosphorylates both FGF14 (a Nav channel partner) and β-catenin at the AIS. (See color plate)

Frotscher, & Deller, 2007; Sanchez-Ponce, Blázquez-Llorca, DeFelipe, Garrido, & Muñoz, 2012; Sanchez-Ponce, DeFelipe, Garrido, & Muñoz, 2011). Clustering of potassium channels Kv2.1 at the AIS occur in proximity to these cisternal organelles (King, Manning, & Trimmer, 2014; Sarmiere, Weigle, & Tamkun, 2008). Interestingly, these sites of Kv2.1 clustering have been implicated in endoplasmic reticulum—plasma membrane junction (Fox et al., 2015), suggesting a protein trafficking role for the actin-rich assembly of GABAergic synapses and cisternal organelles. These complexes correspond to gaps in the ankyrin G/βIV-spectrin undercoat (King et al., 2014; Peters et al., 1968). This does not rule out a role for the ankyrin-based scaffold in their assembly, as perisynaptic 480-kDa ankyrin G organizes GABAergic synapses around the cell body and AIS of mature neurons (Tseng, Jenkins, Tanaka, Mooney, & Bennett, 2015). This is done via interaction with GABARAP, which was shown to concentrate at the AIS when expressed in a transgenic mouse model (Koike et al., 2013).

3.4 The Distal AIS Complex and Other AIS Components

Another distinct, but less well-characterized complex is found along the distal AIS (Figure 3(C)). This complex is found in cortical neurons and motoneurons, but not in cerebellar Purkinje cells (Lorincz & Nusser, 2008) and includes voltage-gated potassium channels Kv1.1 and Kv1.2 (Dodson, Barker, & Forsythe, 2002; Duflocq, Chareyre, Giovannini, Couraud, & Davenne, 2011; Goldberg et al., 2008; Inda et al., 2006; Kirizs, Kerti-Szigeti, Lorincz, & Nusser, 2014; Lorincz & Nusser, 2008; Sanchez-Ponce, DeFelipe, et al., 2012; Trimmer, 2015). They are associated with the CAMs Caspr2 (*contactin-associated protein-like 2*) and Tag-1 (*transient axonal glycoprotein-1*, Ogawa et al., 2008), as well as the metalloprotease ADAM22 (*A disintegrin and metalloprotease domain 22*, Ogawa et al., 2010). These proteins lack known interactions with ankyrin G, but Kv1.1/1.2 channels bind to PSD-93 (Ogawa et al., 2008) as well as the cytoplasmic subunit Kvβ2 in the distal AIS (Vacher et al., 2011).

Over the years, several other membrane and cytoplasmic proteins have been reported to localize or act at the AIS. Voltage-gated calcium channels Cav2.1 (P/Q-type current) and Cav2.2 (N-type) concentrate along the AIS of cortical neurons (Yu, Maureira, Liu, & McCormick, 2010). Modulation of the action potential occurs at the AIS via Cav3.1/3.2 (T-type) channels (Bender & Trussell, 2009; Bender, Ford, & Trussell, 2010; Bender, Uebele, Renger, & Trussell, 2012), but no AIS immunolocalization has been reported. Serotoninergic receptors 5HT1a have been found to localize along the AIS of cortical neurons in humans (Cruz, Eggan, Azmitia, & Lewis, 2004; Czyrak, Czepiel, Maćkowiak, Chocyk, & Wędzony, 2003; DeFelipe, Arellano, Gómez, Azmitia, & Muñoz, 2001) and could inhibit action potential generation via spillover from soma-liberated serotonin (Cotel, Exley, Cragg, & Perrier, 2013). Regarding cytoplasmic proteins, SCHIP1 (*Schwannomin interacting protein 1*) has been found to localize at the AIS, thanks to its interaction with ankyrin G (Martin et al., 2008; Papandréou et al., 2015). The endocytic protein amphiphysin II/BIN1 was found to localize very specifically to the AIS, but this finding was not replicated or pursued (Butler et al., 1997). Finally, an antibody against a phosphorylated form of β-catenin was found to preferentially label the AIS of neurons in cultures or brain sections (Tapia et al., 2013), as well as antibodies against the E3 ubiquitin ligase MDM2 (*murine double minute 2*, Zhao, Wang, Xu, & Zhu, 2014).

4. THE MOLECULAR STRUCTURE OF THE AIS

To precisely understand the AIS functions, it is necessary to determine the precise arrangement of AIS components. Super-resolution microscopy now allows studying the organization of protein complexes down to a few tens of nanometers in situ (Maglione & Sigrist, 2013) and has provided decisive insights into the organization of the axon and AIS. The flexible nature of helical repeats allows spectrin to act like a spring, with length of tetramers varying between 60 and 200 nm depending on biological contexts (Brown et al., 2015). In erythrocytes, a hexagonal lattice is formed by spectrins with a 160-nm radial extent (Byers & Branton, 1985; Liu, Derick, & Palek, 1987; Nans, Mohandas, & Stokes, 2011). In axons, extension of the αII/βII tetramer connected to actin on both sides results in a striking periodic lattice along the axon plasma membrane, with longitudinal spectrins connecting rings of adducin-capped actin filaments (Xu, Zhong, & Zhuang, 2013). Actin rings and spectrin epitopes are thus regularly spaced every 190 nm (Xu et al., 2013; Zhong et al., 2014), a period corresponding to the purified spectrin tetramer length observed by electron microscopy (Bennett, Davis, & Fowler, 1982; Glenney, Glenney, & Weber, 1982). Accordingly, ankyrin B (that interacts with the center of the spectrin tetramer) is found as regularly spaced clusters between the actin rings (Xu et al., 2013). This regular actin/βII-spectrin submembrane lattice could provide both flexibility and resistance to the axon, with rigid rings connected by flexible linkers, like a vacuum cleaner hose (Brown et al., 2015; Lai & Cao, 2014; Morris, 2001). In support to this hypothesis,

Caenorhabditis elegans axonal β-spectrin/UNC-70 forms a prestressed lattice important for touch sensation(Krieg, Dunn, & Goodman, 2014), and mutants lacking β-spectrin/UNC-70 have fragile axons that break when the animal moves (Hammarlund, Jorgensen, & Bastiani, 2007).

In the AIS, early electron microscopy work detailed the membrane undercoat as being made of three radial layers, and presenting a longitudinal periodicity of ~70 nm (Chan-Palay, 1972). Similar to the actin/βII-spectrin lattice in the distal axon, super-resolution microscopy resolves βIV-spectrin and actin as regularly spaced clusters with a 190-nm period (Figure 3, D'Este, Kamin, Göttfert, El-Hady, & Hell, 2015; Leterrier et al., 2015; Xu et al., 2013; Zhong et al., 2014). βIV-spectrin aminoterminus localizes along actin bands, and βIV-spectrin carboxyterminus bands are found in the void between actin bands, showing that the AIS undercoat is formed by head-to-head dimers of βIV-spectrin connecting rings of actin filaments (Leterrier et al., 2015). Whether these filaments are associated with adducin is unsure, given the low adducin immunolabeling in the AIS (Jones, Korobova, & Svitkina, 2014). Sodium channels are also periodically clustered between actin rings, and it was hypothesized that this could have an influence on action potential generation at the AIS (Rasband, 2013; Xu et al., 2013). NF-186 clusters also localize between actin rings (D'Este et al., 2015; Leterrier et al., 2015). Scanning electron microscopy coupled to immunogold labeling was recently used to visualize individual proteins inside the AIS and showed ankyrin G and βIV-spectrin were visible as elongated rods (120–160 and 40–80 nm long, respectively) that formed a dense coat around AIS microtubules (Jones et al., 2014). The discrepancy between the ~60 nm length of individual βIV-spectrin proteins seen by electron microscopy, and the 190-nm periodicity of βIV-spectrin observed by super-resolution microscopy could be due to the presence of the $\sum$6 isoform, that is shorter and does not bind actin, or to refolding of spectrins during sample processing for electron microscopy.

Sodium channels, NF-186, and the carboxyterminal side of βIV-spectrin all bind to the aminoterminal part of ankyrin G, which is also periodically organized in the submembrane complex (Leterrier et al., 2015; Zhong et al., 2014). Strikingly, no periodic pattern is detected for the carboxyterminus of ankyrin G (Leterrier et al., 2015; Zhong et al., 2014). This could be due to the presence of the 480- and 270-kDa isoforms that have tails of different lengths, or to a localization of the ankyrin G carboxyterminus away from the periodic submembrane lattice. Radial mapping of the ankyrin

G epitopes using super-resolution microscopy supports the latter hypothesis, showing that the ankyrin G carboxyterminus localizes below the submembrane complex, 35 nm deeper than its aminoterminal part (Leterrier et al., 2015).

Along the distal axon, 220- and 440-kDa isoforms of ankyrin B are likely to have distinct roles, with the shorter 220 kDa ankyrin B primarily needed for axonal transport (Lorenzo et al., 2014). Finally, actin rings may not be the sole form of actin present in the AIS and axon: ~1-μm actin patches have also been described in the AIS (Jones et al., 2014; Watanabe et al., 2012), and a system of actin "hot spots" and labile longitudinal filaments named "actin trails" have been identified along the whole axon (Ganguly et al., 2015). Actin patches have been proposed to help filtering vesicular transport through the axon (Watanabe et al., 2012; see below), or assigned to AIS synapses (D'Este et al., 2015). Orientation of actin filaments is crucial to the filtering mechanism and is still debated, with different results depending on the technique used (Jensen et al., 2014; Jones et al., 2014). Importantly, disruption of actin impacts the nascent AIS (Song et al., 2009; Xu & Shrager, 2005), but the mature AIS scaffold is resistant to actin depolymerization (Jones et al., 2014; Li et al., 2011; Sanchez-Ponce, DeFelipe, et al., 2011).

5. AIS ASSEMBLY AND MAINTENANCE
5.1 Timing and Relation to Axon Formation

Accumulation of ankyrin G and βIV-spectrin is detected in cortical neurons right after birth in rodents (Galiano et al., 2012; Jenkins & Bennett, 2001) and primates (Cruz, Weaver, Lovallo, Melchitzky, & Lewis, 2009), with reports of AIS-positive labeling detected as soon as E19-E20 in rats (Berghs et al., 2000; Gutzmann et al., 2014). Ankyrin G is first detected at the distal part of the AIS and accumulates toward the soma in cortical neurons (Galiano et al., 2012). In mouse motoneurons, ankyrin G restriction to the AIS begins at E11.5, after being detected along the whole axon (Le Bras et al., 2013). In hippocampal culture, ankyrin G and spectrin appear at the AIS after around two days in culture, are detected in all neurons after 4–6 days (Boiko et al., 2003; Hedstrom et al., 2007; Yang et al., 2007), and progressively accumulate over the weeks (Jones et al., 2014). In all cases, AIS assembly occurs after axon specification and is not required for the initiation of neuronal polarity, which occurs normally in ankyrin

G–depleted neurons (Galiano et al., 2012; Hedstrom et al., 2007; Jenkins & Bennett, 2001; Zhou et al., 1998).

5.2 Ankyrin—Spectrin Trafficking and Clustering

Time course analysis of protein concentration at the AIS in wild-type or component-depleted neurons have delineated the hierarchical "inside-out" sequence of AIS assembly. Ankyrin G is the main AIS organizer: it appears early at the AIS and is responsible for the recruitment of its partners. Depletion of ankyrin G results in the complete absence of an AIS (Dzhashiashvili et al., 2007; Hedstrom et al., 2007; Jenkins & Bennett, 2001; Sobotzik et al., 2009; Zhou et al., 1998). The initial concentration of ankyrin G in the AIS depends on the interplay between the proximal ankyrin G/βIV-spectrin and the distal ankyrin B/αII/βII-spectrin scaffolds (Galiano et al., 2012). In cortical neurons, ankyrin B/αII/βII-spectrin are first expressed and invade the distal part of the axon, due to transport via a kinesin-2 (KIF3/KAP3) complex (Galiano et al., 2012; Takeda et al., 2000). Ankyrin G is expressed later and accumulates along the plasma membrane not yet occupied by ankyrin B, resulting in a concentration along the proximal axon (Galiano et al., 2012). Ankyrin G is transported into the axon by a kinesin-1 motor (KIF5) that binds its membrane-binding domain (Barry et al., 2014). In addition, the specific domains found in the 270- and 480-kDa isoforms (serin-rich domain and tail) help restricting ankyrin G expression to the proximal axon (Zhang & Bennett, 1998).

Ankyrin B could also have a role in βII-spectrin transport to the axon, as its 220-kDa isoform is necessary for axonal transport (Lorenzo et al., 2014), and neurons lacking ankyrin B have more βII-spectrin in dendrites than in the axon (Zhong et al., 2014). The βII-spectrin periodic scaffold, as well as the AIS morphology, are not altered in ankB−/− neurons (Lorenzo et al., 2014). This suggests that ankyrin B is more important for transport than for the assembly of submembrane scaffolds in distal axons, in contrast to ankyrin G in the AIS. Accordingly, the distal periodic scaffold would primarily be formed by βII-spectrin. Interestingly, βII-spectrin seems to be present as a periodic lattice in the proximal axon and nascent AIS, before being gradually replaced by βIV-spectrin (Zhong et al., 2014). βIV-spectrin is recruited by ankyrin G in the AIS (Yang et al., 2007), and is not necessary for AIS formation (Hedstrom et al., 2007; Lacas-Gervais et al., 2004; Yang, Lacas-Gervais, Morest, Solimena, & Rasband, 2004). However, prolonged absence of βIV-spectrin leads to a progressive destabilization of the AIS in vivo, with a decreased clustering of ankyrin G and ion channels

(Lacas-Gervais et al., 2004; Uemoto et al., 2007; Yang et al., 2004). Similarly, knockdown of the EB1/EB3 link between ankyrin G and microtubules does not prevent AIS formation, but progressively decreases AIS components concentration (Leterrier, Vacher, et al., 2011).

As both ankyrin G (480/270 kDa) and βIV-spectrin ($\sum 1/\sum 6$) have long and short isoforms at the AIS, an interesting question is the respective role of these isoforms in the AIS formation and maintenance. Recently, Vann Bennett's group has developed tools to address the specific roles of 480- and 270-kDa ankyrin G at the AIS (Jenkins et al., 2015). Using knockout mice lacking both isoforms coupled with rescue by transfection in cultured neurons, they showed that 480-kDa ankyrin G is necessary and sufficient for AIS formation. One should thus consider 480-kDa ankyrin G as the main AIS ankyrin G isoform, the shorter isoform being unable to concentrate on its own in the AIS (Jenkins et al., 2015). The shorter βIV-spectrin $\sum 6$ isoform has been shown to be necessary for AIS maintenance in addition to the $\sum 1$ isoform (Uemoto et al., 2007), but whether the two isoforms have a different role at the molecular level is unknown.

5.3 Recruitment of Membrane Proteins and Role in Maintenance

Shortly after the concentration of ankyrin G and βIV-spectrin, membrane protein partners (CAMs and ion channels) start to accumulate at the AIS (Hedstrom et al., 2007; Jenkins & Bennett, 2001). Ankyrin G is directly responsible for the recruitment of most of its membrane protein partners: Nav α and β subunits, KCNQ2/3 channels, NF–186 (Ango et al., 2004; Buffington & Rasband, 2013; Hedstrom et al., 2007; Jenkins & Bennett, 2001; Zhou et al., 1998). Several mechanisms have been proposed to drive the accumulation of AIS membrane proteins: (1) nonselective insertion followed by selective endocytosis from outside the AIS, (2) nonselective insertion followed by diffusion and capture by the AIS scaffold, and (3) direct insertion in the AIS, with eventual cotransport of scaffolds and membrane proteins (Leterrier, Brachet, Dargent, & Vacher, 2011).

Nonspecific delivery followed by selective endocytosis from the somato-dendritic compartment was originally put forth to explain the AIS localization of the Nav channels, based on the study of CD4 fused to intracellular domains of Nav1.2 (Fache et al., 2004; Garrido et al., 2001). A first endocytosis motif in the Nav1.2 carboxyterminus drives axonal expression by triggering somatodendritic-selective endocytosis (Garrido et al., 2001). A second one, located near to the ankyrin-binding motif in the intracellular loop II–III, is necessary for restricting surface localization to the AIS (Fache et al., 2004). Recently, the use of full-length Nav1.6 constructs confirmed that the ankyrin-binding motif was necessary and sufficient for AIS concentration (Gasser et al., 2012). The diffusion-trapping mechanism is supported by the ankyrin G-dependent immobilization of Kv-Nav chimeras and full-length channels at the AIS (Akin, Solé, Dib-Hajj, Waxman, & Tamkun, 2015; Brachet et al., 2010). However, pulse-chase surface labeling experiments suggest that insertion of the full-length Nav1.6 occurs directly at the AIS, with fewer nonselective delivery events followed by endocytosis or diffusion (Akin et al., 2015). This direct insertion is consistent with the proposed cotransport of Nav channels together with ankyrin G by kinesin-1 motors (Barry et al., 2014). It is possible that both selective endocytosis/diffusion trapping and direct insertion can occur, depending on the type of sodium channel and the developmental stage, Nav1.2 being expressed earlier than Nav1.6. Axonal targeting of KCNQ2/3 channels depends on multiple motifs in the carboxyterminal tail (Chung, Jan, & Jan, 2006), and an ankyrin G-binding motif homologous to the Nav channel motif directs their AIS concentration (Pan et al., 2006; Rasmussen et al., 2007). Less is known about Kv1 trafficking to the distal AIS complex: axonal targeting occurs via the Kvβ2 cytoplasmic subunit that binds the Kv1 T1 domain (Gu, Jan, & Jan, 2003; Rivera, Chu, & Arnold, 2005), kinesin-2, and EB1 (Gu et al., 2006; Vacher et al., 2011). Kv1 are thought to be anchored by PSD-93 at the AIS, but PSD-93 depletion impairs targeting only in vitro, not in vivo (Duflocq et al., 2011; Ogawa et al., 2008; 2010), and the accumulation of Kv1 depends on ankyrin G (Sanchez-Ponce, DeFelipe, et al., 2012). Nonspecific delivery followed by selective endocytosis is also an axonal targeting mechanism for CAMs like L1/NgCAM (*neuron-glia cell adhesion molecule*, Wisco et al., 2003) and has been shown to target NF-186 to the AIS via a doublecortin-dependent mechanism (Yap et al., 2012). Downstream immobilization of NF-186 at the AIS depends on ankyrin G binding (Boiko et al., 2007; Garver, Ren, Tuvia, & Bennett, 1997; Zhang, Davis, Carpenter, & Bennett, 1998).

Once formed, the AIS is strikingly stable, as shown by the slow decrease of protein content after shRNA transfection (Hedstrom, Ogawa, & Rasband, 2008), the persistence of live-cell NF-186 labeling over 10 days (Schafer et al., 2009), and the negligible recovery of Nav1.6 in FRAP experiments (Akin et al., 2015). The low turnover of AIS membrane proteins seems in turn to stabilize the whole compartment. NF-186 is not necessary for AIS formation (Dzhashiashvili et al., 2007; Hedstrom et al., 2007; Zonta et al., 2011), but conditional ablation of NF-186 suggests that it is recruiting NrCAM to the AIS, and helps maintaining the AIS integrity over time (Zonta et al., 2011). It is not known if Nav channels have a role in stabilizing the AIS assembly, but in cultured motoneurons where the AIS appears rapidly (within 2 days), Nav depletion hinders its assembly (Xu & Shrager, 2005).

5.4 Modulation of Component Interactions

Protein interactions within the AIS are dynamically modulated via multiple phosphorylation events (Figure 3(D)). The first example of such a mechanism is the regulation of the NF-186 binding to ankyrin G: an unknown tyrosine kinase phosphorylates NF-186 at its FIGQY motif, resulting in decreased binding to ankyrin G and decreased anchoring at the AIS (Boiko et al., 2007; Garver et al., 1997). Another example is the regulation of the interaction between Nav channels and ankyrin G by protein kinase CK2: CK2 enhances Nav channels binding to ankyrin G by phosphorylating the binding motif on Nav channels in vitro and in vivo (Bréchet et al., 2008; Hien et al., 2014). This strengthened interaction helps sodium channels accumulation at the AIS (Bréchet et al., 2008; Gasser et al., 2012) by enhancing their immobilization (Brachet et al., 2010). The analogous KCNQ2/3 ankyrin-binding motif can also be phosphorylated by CK2, strengthening ankyrin binding (Xu & Cooper, 2015), whereas Nav1.8 (which is normally not expressed in the CNS) has a phosphomimetic motif that natively binds to ankyrin G with a high affinity (Montersino et al., 2014). CK2 is concentrated in the AIS in a Nav-dependent manner (Bréchet et al., 2008; Hien et al., 2014), and CK2 inhibition inhibits AIS formation and maintenance (Sanchez-Ponce, Muñoz, & Garrido, 2011). CK2 also phosphorylate SCHIP1, regulating its binding to ankyrin G (Papandréou et al., 2015). It would be interesting to determine if CK2 directly phosphorylates ankyrin G to modulate its function, as it was shown for ankyrin R in erythrocytes (Ghosh, Dorsey, & Cox, 2002) and for the *Drosophila* neuronal ankyrin 2 (Bulat, Rast, & Pielage, 2014). Another kinase that is likely to regulate the AIS structure is cyclin-dependent kinase 5 (cdk5). In mammalian neurons, cdk5 is present in the AIS and modulates the transport of potassium channels Kvβ2 subunits by regulating their interaction with EB1 (Vacher et al., 2011). In *Drosophila*, cdk5 controls the length of an AIS-like complex enriched in ANK1, actin and ion channels (Trunova, Baek, & Giniger, 2011). Another kinase, Ca^{2+}/calmodulin-dependent protein kinase CaMKII, is present in the AIS where it associates with βIV-spectrin and phosphorylates Nav channels (Hund et al., 2010). Glycogen synthase kinase GSK3β participates in sodium channel targeting to the AIS by regulating its interaction with FGF14 (Shavkunov et al., 2013), and phosphorylates β–catenin, helping AIS maintenance (Tapia et al., 2013).

6. ELECTROGENESIS AT THE AIS

In the next two parts, I will focus on the physiological functions of the AIS. The role of the AIS in neuronal electrogenesis has been thoroughly reviewed (Bender & Trussell, 2012; Clark et al., 2009; Kole & Stuart, 2012; Kress & Mennerick, 2009) and will only be briefly summarized here. Early experimental work (Araki & Otani, 1955; Coombs, Curtis, & Eccles, 1957; Edwards & Ottoson, 1958; Fatt, 1957), as well as modeling studies (Dodge & Cooley, 1973; Mainen, Joerges, Huguenard, & Sejnowski, 1995), suggested the proximal axon as the site for action potential generation. Over the last decades, accumulating evidence has confirmed this for a repertoire of neuronal types: excitatory neurons in the neocortex (Hu et al., 2009; Kole, Letzkus, & Stuart, 2007; Meeks & Mennerick, 2007; Palmer & Stuart, 2006; Popovic, Foust, McCormick, & Zecevic, 2011; Shu, Duque, Yu, Haider, & McCormick, 2007), hippocampus (Kress et al., 2010; Kress, Dowling, Meeks, & Mennerick, 2008; Schmidt-Hieber, Jonas, & Bischofberger, 2008), and basal ganglia (Atherton, Wokosin, Ramanathan, & Bevan, 2008); cerebellar Purkinje cells (Foust, Popovic, Zecevic, & McCormick, 2010; Khaliq & Raman, 2006; Palmer et al., 2010); and parvalbumin and somatostatin-expressing inhibitory interneurons of the neocortex (Li et al., 2014). In addition to the Nav channels density, the low action potential threshold at the AIS could be due to specific biophysical properties of the channels (Colbert & Pan, 2002; Fleidervish et al., 2010; Hu et al., 2009), with a possible role for their interaction with ankyrin G (Shirahata et al., 2006). Importantly, differences in AIS localization, sodium channels density, and subtype content influence the neuron intrinsic excitability (Kress et al., 2010; Kuba & Ohmori, 2009; Kuba, Ishii, & Ohmori, 2006; Wang, Wang, Yu, & Chen, 2011). Notably, the localization of the AIS emergence from a basal dendrite rather than the cell body, which occurs in a subset of hippocampal principal cells, specifically modifies the sensitivity to inputs from this dendrite (Thome et al., 2014). Other ion channels at the AIS modulate the action potential initiation and shape. KCNQ2/3 channels generate the I_M current, restraining intrinsic firing and neuronal excitability (Brown & Passmore, 2009; Guan, Higgs, Horton, Spain, & Foehring, 2011; Shah, Migliore, Valencia, Cooper, & Brown, 2008; Tran, Gavrilis, Cooper, & Kole, 2014). Kv1 channels can modulate the firing threshold (Cudmore, Fronzaroli-Molinieres, Giraud, & Debanne, 2010; Kole et al., 2007) and are important for shaping the action potential waveform (Bialowas et al., 2015; Dodson et al., 2002; Kole et al., 2007; Rowan, Tranquil, & Christie, 2014; Shu, Yu, Yang, & McCormick, 2007) and dampening excitability in interneurons (Campanac et al., 2013; Goldberg et al., 2008). Regarding calcium channels, Cav2.1 (P/Q-type) and Cav2.2 (N-type) channels have been shown to allow calcium entry into the AIS during electrical activity in cerebellar Purkinje cells (Callewaert, Eilers, & Konnerth, 1996) and cortical neurons (Yu et al., 2010). In auditory brainstem interneurons, Cav3.1/3.2 channels (T-type) modulated by dopamine release regulate complex spiking patterns (Bender et al., 2010; 2012; Bender & Trussell, 2009).

Another aspect of electrogenesis is the role of GABAergic synapses along the AIS. It was hypothesized that these synapses could exert a strong inhibitory control on action potential generation (Jones & Powell, 1969). It was later found that GABAergic axoaxonic synapses could have an excitatory effect on cortical pyramidal cells due to an increased local chloride concentration (Khirug et al., 2008; Molnár et al., 2008; Szabadics et al., 2006; Woodruff, Xu, Anderson, & Yuste, 2009). However, the issue is still debated, with numerous examples of inhibitory action of these synapses (Dugladze, Schmitz, Whittington, Vida, & Gloveli, 2012; Glickfeld, Roberts, Somogyi, & Scanziani, 2009; Veres, Nagy, Vereczki, Andrási, & Hájos, 2014; Wang, Hooks, & Sun, 2014) suggesting that AIS synapses could either have an inhibitory or excitatory effect depending on the neuronal type and network state.

7. THE AIS AND THE MAINTENANCE OF AXONAL IDENTITY

7.1 Maintenance of Axonal Identity by the AIS

The other important role of the AIS is to separate the somatodendritic compartment from the axon, maintaining the axonal molecular identity and thus the overall neuronal polarity. In fact, this polarity maintenance function is likely to have appeared early during evolution, before the clustering of ion channels via ankyrin-binding and the electro-genic properties (Rolls & Jegla, 2015). Accordingly, concentration of ankyrin in the proximal axon has been described in *Drosophila*, forming a proto-AIS (Trunova et al., 2011). In *C. elegans*, there is no clustering of the ankyrin homolog UNC-44, but it has a role in regulating proper microtubule organization and axonal sorting (Maniar et al., 2012), and organelles access to the axon is regulated (Edwards et al., 2013). Mamma-lian neurons lacking an AIS progressively acquire dendritic properties along the axon (Rasband, 2010). After AIS disassembly triggered by depletion of ankyrin G, the proximal axon of cultured neurons becomes positive for the somatodendritic marker map2, with the formation of ectopic spines bearing excitatory postsynaptic specializations (Hedstrom et al., 2008). Furthermore, membrane proteins that are restricted to the somatodendritic domain such as β1-integrin can enter the axon in AIS-deficient neurons (Franssen et al., 2015). In vivo, mice lacking ankyrin G in Purkinje cells also show ectopic spines and a perturbation of the proximal axon identity (Sobotzik et al., 2009).

The location of the AIS immediately suggests that it maintains the molecular identity of the axon by physically separating it from the somato-dendritic compartment. Indeed, the AIS regulates protein mobility between the soma and axon by two distinct mechanisms: on the one hand, the AIS assembles a surface diffusion barrier that keeps the somatodendritic and axonal membrane proteins from mixing. On the other hand, the AIS acts as an intracellular traffic filter that exclude somatodendritic protein from entering the axon, and could help to recruit axonal proteins. The diffusion barrier mechanism is well established, and is consistent with the progressive proximo-distal perturbation of axonal identity in AIS-defective neurons (Rasband, 2010). By contrast, the intracellular filter was more recently proposed (Song et al., 2009) and is still hotly debated (Jenkins et al., 2015; Leterrier & Dargent, 2014).

7.2 The AIS Membrane Diffusion Barrier

In the AIS, membrane proteins and lipids do not freely diffuse through the AIS, but are slowed down or immobilized, resulting in the segregation of membrane components. Initially proposed by Dotti and Simons (Dotti & Simons, 1990), this hypothesis was first demonstrated by showing that fluorescent lipids could not diffuse from the axon to the cell body

(Kobayashi, Storrie, Simons, & Dotti, 1992). This result was challenged by later studies, either theoretical (Futerman, Khanin, & Segel, 1993) or experimental (Winckler & Poo, 1996). However, confirmation that the AIS formed a diffusion barrier for axonal proteins was firmly established by measuring the tractability of beads attached to different transmembrane proteins (Winckler, Forscher, & Mellman, 1999). The capacity of the diffusion barrier to also immobilize membrane lipids was conclusively confirmed by video-rate tracking of fluorescent or nanogold-tagged lipids (Nakada et al., 2003). Although a few FRAP studies have been inconclusive (Fukano, Hama, & Miyawaki, 2004; Howarth et al., 2003), the existence of the diffusion barrier has since been confirmed using single-particle tracking of proteins labeled with quantum dots: VAMP2 (Song et al., 2009) or nonpolarized ion channel constructs (Brachet et al., 2010).

An intact AIS structure is necessary for the existence of the diffusion barrier. Nakada et al. showed correlation between the AIS formation and the immobilization of lipids in the proximal axon (Nakada et al., 2003). Furthermore, AIS disassembly via ankyrin G depletion leads to the disappearance of the diffusion barrier (Song et al., 2009). The diffusion barrier is also impaired by actin disruption (Nakada et al., 2003; Winckler et al., 1999). According to the classical "fences and pickets" model, the submembrane lattice (fences) coupled with its high density of bound transmembrane proteins (pickets) is able to slow all lipids and proteins diffusion (Kusumi et al., 2005). In the AIS, the picket role for membrane proteins bound to ankyrin G is confirmed by the immobilization of ion channels and CAMs at the AIS (Angelides, Elmer, Loftus, & Elson, 1988; Boiko et al., 2007; Brachet et al., 2010; Nakada et al., 2003). However, the periodic AIS lattice is a particular case, as the submembrane actin rings (fences) and membrane proteins anchored by ankyrin G (pickets) are separated by the 90 nm length of the βIV-spectrin protein (Xu et al., 2013). It is likely that ankyrin G-bound ion channels and CAMs (rather than the actin/βIV-spectrin lattice) are the primary source of the diffusion barrier because a similar periodic actin/βII-spectrin lattice is present along the distal axon, where surface diffusion is unhindered. An interesting result in this regard is the role of βIV-spectrin, which selectively restricts the diffusion of L1 at the AIS compared to the distal axon. This could suggest a protein-specific component to the diffusion barrier, or a partial compensation by βII-spectrin in βIV-spectrin knockout mice (Nishimura, Akiyama, Komada, & Kamiguchi, 2007).

7.3 Intracellular Filter: Role of Ankyrin G and Actin

In addition to the surface diffusion barrier, it has been proposed that the AIS also assembles an intracellular filter for both cytoplasmic diffusion and vesicular transport. In 2009, Moo-Ming Poo's lab showed that cytoplasmic protein diffusion was hindered at the AIS in a size-dependent manner (Song et al., 2009). In addition, vesicular transport through the AIS was found to be slower than in the distal axon and successful axonal entry to depend on the high transport efficacy of axonal kinesins. Both processes depend on actin and ankyrin G, leading the authors to propose an intracellular "molecular sieve" model. The sieve (average pore size of 13 nm) would sterically slow diffusion and transport in the AIS, helping in the maintenance of neuronal polarity (Song et al., 2009). The diffusion barrier for intracellular proteins has been confirmed (Sun, Wu, Gu, Liu, et al., 2014) and its disruption could be the basis of the Map2 invasion of the proximal axon in AIS-depleted neurons (Hedstrom et al., 2008; Jenkins et al., 2015). In contrast, the existence and mechanism of a vesicular transport filter are currently under question. Several subsequent reports did not detect a slowing down of axonal vesicles through the AIS (Jenkins et al., 2015; Petersen, Kaech, & Banker, 2014), questioning the validity of the physical sieve model. Although accumulation of mitochondria at the AIS entrance has been described (Li et al., 2004), the generally unhindered passage of large organelles into the axon also argues against this model.

It is nonetheless accepted that a selective transport occurs at the axon entrance, with specific exclusion of somatodendritic proteins (Al-Bassam, Xu, Wandless, & Arnold, 2012; Burack, Silverman, & Banker, 2000; Petersen et al., 2014). However, a specific role of the AIS scaffold in this process is disputed. Somatodendritic vesicles and organelles did not significantly enter the axon in ankyrin G-depleted neurons (Jenkins et al., 2015). Regarding a distinct role for AIS actin in this process, Don Arnold's lab has demonstrated the targeting of somatodendritic proteins via myosin V binding (Lewis, Mao, Svoboda, & Arnold, 2009). Myosin VI, which travels in the opposite direction on actin filaments, conversely helps axonal targeting (Lewis, Mao, & Arnold, 2011). This leads to an "AIS actin filter" model where oriented actin patches and filaments in the AIS can stop vesicular transport of somatodendritic vesicles harboring myosin V motors (Al-Bassam et al., 2012; Watanabe et al., 2012). In addition to submembrane actin rings, it is clear that the AIS and distal axon can contain intracellular actin filaments and patches (D'Este et al., 2015; Ganguly et al., 2015; Watanabe et al., 2012). In the AIS, the interplay between this actin-based filter and the ankyrin G/βIV-spectrin scaffold remains to be determined. Others have argued that the main sorting mechanism occurs upstream in the cell body, when vesicles are packaged and bud from the Golgi apparatus (Farías et al., 2012; Petersen et al., 2014), with actin disruption resulting in disorganized packaging and mistargeting of proteins in all compartments (Petersen et al., 2014). It is likely that both mechanisms are complementary, the evolution of these partially redundant processes allowing a low rate of mistargeted proteins.

7.4 Microtubule and Vesicular Transport through the AIS

Long-range vesicular transport is ensured by microtubules in neuronal cells (Coles & Bradke, 2015; Kapitein & Hoogenraad, 2015; Prokop, 2013). In the axon, the uniform polarity of microtubules (plus-end out) leads to specific directional transport by kinesins (anterograde) and dynein (retrograde) (Baas & Lin, 2011; Kevenaar & Hoogenraad, 2015). As kinesins can transport both somatodendritic and axonal cargoes, there must be some cues that allow motors to specifically find and walk along microtubules that enter the axon (Hirokawa & Tanaka, 2015). The kinesin-1 (KIF5) head is selectively targeted to axons (Jacobson, Schnapp, & Banker, 2006; Nakata & Hirokawa, 2003; Nakata, Niwa, Okada, Perez, & Hirokawa, 2011) and its nonprocessive mutant accumulates selectively along microtubules in the axon hillock, suggesting the existence of a specific recruitment cue (Nakata & Hirokawa, 2003). Microtubules in the AIS and proximal axon are thought to be stable (Baas, Ahmad, Pienkowski, Brown, & Black, 1993; Hammond et al., 2010) and they are enriched in posttranslational modifications such as acetylation, detyrosination, polyglutamylation, and possibly polyamination (Hammond et al., 2010; Konishi & Setou, 2009; Song et al., 2013; Tapia, Wandosell, & Garrido, 2010). Kinesin-1 preferably binds to highly modified, stabilized MTs (Cai, McEwen, Martens, Meyhofer, & Verhey, 2009; Dunn et al., 2008; Reed et al., 2006), suggesting that these modifications drive the recruitment of kinesin-1 to the AIS. Stabilizing all neuronal microtubules using taxol abolishes kinesin-1 targeting to the axon (Hammond et al., 2010; Nakata & Hirokawa, 2003), but no consensus exist yet on whether acetylation (Reed et al., 2006; Tapia et al., 2010), detyrosination (Konishi & Setou, 2009), or a combination of multiple modifications (Hammond et al., 2010) have a specific impact on kinesin-1 binding. The concentration of the microtubule minus-end associated protein CAMSAP2 at the proximal AIS could also play a role in the selective recruitment of axonal cargoes (Yau et al., 2014).

An interesting alternative is the possible role of GTP remnants along AIS MTs, resulting from incomplete hydrolysis of tubulin–GTP along microtubules (Dimitrov et al., 2008). Interestingly, these remnants are enriched along the AIS microtubules and could trigger the recruitment of kinesin-1 (Nakata et al., 2011). There could be an interplay between these GTP remnants and EB proteins concentrated in the AIS (Leterrier, Vacher, et al., 2011), as EB proteins preferentially interact with GTP-tubulin when interacting with the dynamic plus ends (Maurer, Fourniol, Bohner, Moores, & Surrey, 2012). Furthermore, the link between ankyrin G and microtubules via EB proteins or other partners could help organize microtubules along the AIS, indirectly driving the recruitment of axonal vesicles. Microtubule organization by the AIS scaffold is further suggested by the lack of fascicles in ankyrin G-deficient Purkinje cells (Sobotzik et al., 2009) and is consistent with the recent discovery of *Drosophila* giant ankyrins that organize microtubules in synapses (Koch et al., 2008; Pielage et al., 2008) and the axonal shaft (Bennett & Walder, 2015; Stephan et al., 2015).

8. AIS MORPHOLOGICAL PLASTICITY AND ALTERATION IN DISEASE

8.1 Developmental and Activity-Dependent Plasticity

AIS components have been identified over the years, uncovering an intricate assembly of cytoskeleton, scaffold, and membrane proteins, analogous in some ways to the postsynaptic scaffold (Sheng & Hoogenraad, 2007). In line with this analogy, recent work has unraveled the morphological plasticity of the AIS, with its position and content regulated by developmental, physiological, and pathological processes (Yoshimura & Rasband, 2014). These changes profoundly influence the AIS electrogenic properties. AIS length and position vary during development (Cruz, Lovallo, Stockton, Rasband, & Lewis, 2009; Galiano et al., 2012) and this can tune sensitivity to inputs, as exemplified by the relationship between AIS morphology and the development of the visual sensory cortex (Gutzmann et al., 2014) or avian auditory neurons (Kuba, Adachi, & Ohmori, 2014). Moreover, two studies have uncovered how that neuronal activity can change the AIS location as a new mechanism of slow homeostatic adaptation (Grubb & Burrone, 2010; Kuba, Oichi, & Ohmori, 2010). Depriving inputs to chick brainstem auditory neurons leads to a longer AIS associated with enhanced excitability (Kuba et al., 2010). Elevating neuronal activity in cultured hippocampal neurons leads to a slow displacement of the AIS toward the distal axon, which leads to a lower excitability (Grubb & Burrone, 2010), a finding recently confirmed using channelrhodopsin-expressing neurons in hippocampal slices (Wefelmeyer, Cattaert, & Burrone, 2015). Interestingly, GABAergic synapses stay close to the cell body, resulting in increased GABA receptors mobility (Muir & Kittler, 2014) and increased inhibition of action potential firing (Wefelmeyer et al., 2015). In inhibitory dopaminergic interneurons, this homeostatic plasticity is reversed: elevated activity leads to longer initial segments that are closer to the soma, resulting in increased excitability (Chand, Galliano, Chesters, & Grubb, 2015).

8.2 Potential Mechanisms for AIS Plasticity and Alteration

How does a compartment as stable as the AIS profoundly rearrange or disassemble in a matter of hours? The proposed mechanisms for the AIS morphological plasticity are diverse and incomplete, but calcium signaling seems to be the common denominator. In line with the presence of calcium stores (cisternal organelles) and calcium channels at the AIS, calcium-dependent pathways can have a profound impact on the AIS stability. Calmodulin binding drives the assembly of KCNQ2/3 heteromers and thus their targeting to the AIS (Liu & Devaux, 2014). AIS relocalization in response to elevated activity depends on Cav1 (L-type)-mediated calcium entry and on the activation of calcineurin, a calcium- and calmodulin-dependent phosphatase implicated in synaptic plasticity and memory (Evans et al., 2013). Finally, fast AIS disassembly in response to ischemia depends on scaffold proteolysis by calcium-dependent cystein protease calpain, which degrades ankyrin G and βIV-spectrin (Schafer et al., 2009). Interestingly, calpain is also involved in AIS partial disassembly triggered by ATP-activated purinergic P2X7 receptors leading to calcium entry (Del Puerto et al., 2015).

8.3 Alterations of the AIS in Disease

Beyond physiological adaptation, the AIS is also affected in a vast range of pathological conditions. Complete disassembly has been observed in different in vivo and in vitro models of neuronal injury (Schafer et al., 2009). Subtle alterations in AIS length have been detected after stroke (Hinman, Rasband, & Carmichael, 2013), and small changes in AIS location have been measured in mouse and rat models of epilepsy (Harty et al., 2013). Axotomy caused by mild traumatic brain injury preferentially happens at the AIS (Greer, Hånell, McGinn, & Povlishock, 2013), and less impacted neurons have a lengthier AIS (Baalman, Cotton, Rasband, & Rasband, 2013). A mouse model for Angelman syndrome (caused by deletion of the ubiquitin ligase UBE3A gene) exhibits upregulated ankyrin G and Nav1.6 channels resulting in longer initial segments, an alteration that can be rescued by genetically reducing the Na/K-ATPase α1 (Kaphzan, Buffington, Jung, Rasband, & Klann, 2011; Kaphzan et al., 2013). AIS innervation by GABAergic synapses and associated presence of 5HT1a receptors is perturbed in the cortex of schizophrenic patients (Cruz et al., 2004; Cruz, Lovallo, 2009; Woo, Whitehead, Melchitzky, & Lewis, 1998). AIS alteration could be a common feature in nervous disease and injury and a possible target for therapeutic intervention.

A detailed account of the pathologies associated with defects or mutations in AIS components is beyond the present scope and this subject has been reviewed recently (Buffington & Rasband, 2011; Hsu, Nilsson, & Laezza, 2014), so I will only briefly summarize them and outline most recent works. Due to the role of AIS ion channels in the initiation of action potentials, a large number of mutations in sodium or potassium channels present at the AIS (Nav1.1, Nav1.2, Nav1.6, KCNQ2/3) result in epileptic conditions in humans (Wimmer, Reid, So, Berkovic, & Petrou, 2010). Ankyrin G has been repeatedly associated with risks of schizophrenia and bipolar disorder (Leussis, Madison, & Petryshen, 2012), and some mechanistic insights were recently obtained. AnkG downregulation has behavioral effects in mice (Leussis et al., 2013): variation in ankyrin G mRNA expression was found in bipolar patients (Wirgenes et al., 2014) and could result from the deregulation of the miR-34 microRNA (Bavamian et al., 2015). Identification of patients with heterologous disruption of ankyrin isoforms or homozygous truncation in 480-kDa ankyrin G revealed a range of intellectual disabilities (Iqbal et al., 2013). However, it is not known if these are the consequences of ankyrin G role at the AIS or caused by alternative mechanisms such as GABAergic synapse organization by 480-kDa ankyrin G (Tseng et al., 2015), or synapse maintenance (Smith et al., 2014) and control of neurogenesis (Durak et al., 2015; Paez-Gonzalez et al., 2011) by the smaller 190-kDa isoform.

The AIS could also be the place of axonal transport perturbations that are present in several neurodegenerative diseases (Encalada & Goldstein, 2014; Millecamps & Julien, 2013). Ankyrin G has been diversely implicated in Alzheimer's disease (León-Espinosa, DeFelipe, & Muñoz, 2012; Sanchez-Mut et al., 2013; Santuccione et al., 2012). In particular, mouse models of Alzheimer's disease have decreased ankyrin G levels at the AIS due to upregulation of the miR-342-5p microRNA (Sun, Wu, Gu, & Zhang, 2014). This downregulation of ankyrin G results in an altered intracellular diffusion filter, with easier axonal entry of large cytoplasmic proteins

(Sun, Wu, Gu, Liu, et al., 2014). By contrast, motor neurons of amyotrophic lateral sclerosis (ALS) mouse models show an accumulation of neurofilaments and mitochondria in the AIS (Sasaki, Warita, Abe, & Iwata, 2005), consistent with organelle accumulation and proximal axon swelling in ALS patients (Sasaki & Iwata, 1996). Finally, a recent work suggests that demyelinating diseases such as multiple sclerosis implicate defects in axonal transport resulting in gray matter damage (Calabrese et al., 2015; Dutta et al., 2011; Hares et al., 2013; Sorbara et al., 2014). Subtle alterations in the AIS morphology were observed in demyelinated cortical neurons (Hamada & Kole, 2015), and it would be interesting to examine the possible AIS alterations occurring in multiple sclerosis.

9. CONCLUSION

The AIS is a unique neuronal subcompartment that serves crucial neurophysiological functions. Recognized for a long time as the site of action potential initiation site, it was merely seen as a static assembly of scaffold and channels. Similar to what happened for the synapse in the past two decades, the AIS has now been revealed as a dynamic assembly, capable of morphologic plasticity in physiological and pathological conditions, resulting in profound effects on the neuron excitability. Moreover, cell biology studies uncovered the essential role of the AIS in the maintenance of axonal identity, with insights into its regulation of protein mobility and trafficking. Several key questions remain: What is the exact nature of the vesicular sorting that occurs at the AIS? What are the structural mechanisms of the AIS morphological plasticity? Is the organizing role of ankyrin G at the AIS implicated in neuropsychiatric disorders where it is recognized as a major risk factor? Thanks to the growing interest for the AIS among neurobiologists, the coming years will undoubtedly be very exciting, providing numerous answers and even more numerous new questions.

ACKNOWLEDGMENTS

I would like to thank Bénédicte Dargent for discussions and support; Subhojit Roy, Francis Castets, Marie-Jeanne Papandreou, Jean-Marc Goaillard, and Mickaël Zbili for discussions and careful reading of the manuscript.

REFERENCES

Abdi, K. M., Mohler, P. J., Davis, J. Q., & Bennett, V. (2006). Isoform specificity of ankyrin-B: a site in the divergent C-terminal domain is required for intramolecular association. *The Journal of Biological Chemistry, 281*(9), 5741–5749.

Akhmanova, A., & Steinmetz, M. O. (2010). Microtubule +TIPs at a glance. *Journal of Cell Science, 123*(Pt 20), 3415–3419.

Akin, E. J., Solé, L., Dib-Hajj, S. D., Waxman, S. G., & Tamkun, M. M. (2015). Preferential targeting of Nav1.6 voltage-gated Na$^+$ channels to the axon initial segment during development. *PloS One, 10*(4), e0124397.

Al-Bassam, S., Xu, M., Wandless, T. J., & Arnold, D. B. (2012). Differential trafficking of transport vesicles contributes to the localization of dendritic proteins. *Cell Reports, 2*(1), 89–100.

Angelides, K. J., Elmer, L., Loftus, D., & Elson, E. (1988). Distribution and lateral mobility of voltage-dependent sodium channels in neurons. *The Journal of Cell Biology, 106*(6), 1911–1925.

Ango, F., di Cristo, G., Higashiyama, H., Bennett, V., Wu, P., & Huang, Z. (2004). Ankyrin-based subcellular gradient of neurofascin, an immunoglobulin family protein, directs GABAergic innervation at Purkinje axon initial segment. *Cell, 119*(2), 257–272.

Antón-Fernández, A., Rubio-Garrido, P., DeFelipe, J., & Muñoz, A. (2015). Selective presence of a giant saccular organelle in the axon initial segment of a subpopulation of layer V pyramidal neurons. *Brain Structure and Function, 220*(2), 869–884.

Araki, T., & Otani, T. (1955). Response of single motoneurons to direct stimulation in toad's spinal cord. *Journal of Neurophysiology, 18*(5), 472–485.

Atherton, J. F., Wokosin, D. L., Ramanathan, S., & Bevan, M. D. (2008). Autonomous initiation and propagation of action potentials in neurons of the subthalamic nucleus. *The Journal of Physiology, 586*(Pt 23), 5679–5700.

Baalman, K. L., Cotton, R. J., Rasband, S. N., & Rasband, M. N. (2013). Blast wave exposure impairs memory and decreases axon initial segment length. *Journal of Neurotrauma, 30*(9), 741–751.

Baalman, K. L., Marin, M. A., Ho, T. S.-Y., Godoy, M., Cherian, L., Robertson, C., et al. (2015). Axon initial segment-associated microglia. *The Journal of Neuroscience, 35*(5), 2283–2292.

Baas, P. W., & Lin, S. (2011). Hooks and comets: the story of microtubule polarity orientation in the neuron. *Developmental Neurobiology, 71*(6), 403–418.

Baas, P. W., Ahmad, F. J., Pienkowski, T., Brown, A. E., & Black, M. M. (1993). Sites of microtubule stabilization for the axon. *The Journal of Neuroscience, 13*(5), 2177–2185.

Barry, J., Gu, Y., Jukkola, P., O'Neill, B., Gu, H., Mohler, P. J., et al. (2014). Ankyrin-G directly binds to kinesin-1 to transport voltage-gated Na$^+$ channels into axons. *Developmental Cell, 28*(2), 117–131.

Bas Orth, C., Schultz, C., Müller, C., Frotscher, M., & Deller, T. (2007). Loss of the cisternal organelle in the axon initial segment of cortical neurons in synaptopodin-deficient mice. *Journal of Comparative Neurology, 504*(5), 441–449.

Bavamian, S., Mellios, N., Lalonde, J., Fass, D. M., Wang, J., Sheridan, S. D., et al. (2015). Dysregulation of miR-34a links neuronal development to genetic risk factors for bipolar disorder. *Molecular Psychiatry, 20*(5), 573–584.

Bender, K. J., & Trussell, L. O. (2009). Axon initial segment Ca^{2+} channels influence action potential generation and timing. *Neuron, 61*(2), 259–271.

Bender, K. J., & Trussell, L. O. (2012). The physiology of the axon initial segment. *Annual Review of Neurosciences, 35,* 249–265.

Bender, K. J., Ford, C. P., & Trussell, L. O. (2010). Dopaminergic modulation of axon initial segment calcium channels regulates action potential initiation. *Neuron, 68*(3), 500–511.

Bender, K. J., Uebele, V. N., Renger, J. J., & Trussell, L. O. (2012). Control of firing patterns through modulation of axon initial segment T-type calcium channels. *The Journal of Physiology, 590*(Pt 1), 109–118.

Benedeczky, I., Molnár, E., & Somogyi, P. (1994). The cisternal organelle as a Ca(2+)-storing compartment associated with GABAergic synapses in the axon initial segment of hippocampal pyramidal neurones. *Experimental Brain Research (Experimentelle Hirnforschung. Expérimentation Cérébrale), 101*(2), 216–230.

Bennett, V., & Baines, A. J. (2001). Spectrin and ankyrin-based pathways: metazoan inventions for integrating cells into tissues. *Physiological Reviews, 81*(3), 1353–1392.

Bennett, V., & Davis, J. Q. (1981). Erythrocyte ankyrin: immunoreactive analogues are associated with mitotic structures in cultured cells and with microtubules in brain. *Proceedings of the National Academy of Sciences of the United States of America, 78*(12), 7550–7554.

Bennett, V., & Lorenzo, D. N. (2013). Spectrin- and ankyrin-based membrane domains and the evolution of vertebrates. *Current Topics in Membranes, 72*, 1–37.

Bennett, V., & Stenbuck, P. J. (1979). Identification and partial purification of ankyrin, the high affinity membrane attachment site for human erythrocyte spectrin. *The Journal of Biological Chemistry, 254*(7), 2533–2541.

Bennett, V., & Walder, K. (2015). Evolution in action: giant ankyrins awake. *Developmental Cell, 33*(1), 1–2.

Bennett, V., Davis, J. Q., & Fowler, W. E. (1982). Brain spectrin, a membrane-associated protein related in structure and function to erythrocyte spectrin. *Nature, 299*(5879), 126–131.

Berghs, S., Aggujaro, D., Dirkx, R., Maksimova, E., Stabach, P., Hermel, J. M., et al. (2000). betaIV spectrin, a new spectrin localized at axon initial segments and nodes of ranvier in the central and peripheral nervous system. *The Journal of Cell Biology, 151*(5), 985–1002.

Bialowas, A., Rama, S., Zbili, M., Marra, V., Fronzaroli-Molinieres, L., Ankri, N., et al. (2015). Analog modulation of spike-evoked transmission in CA3 circuits is determined by axonal Kv1.1 channels in a time-dependent manner. *The European Journal of Neuroscience, 41*(3), 293–304.

Boiko, T., Van Wart, A., Caldwell, J. H., Levinson, S. R., Trimmer, J. S., & Matthews, G. (2003). Functional specialization of the axon initial segment by isoform-specific sodium channel targeting. *The Journal of Neuroscience, 23*(6), 2306–2313.

Boiko, T., Vakulenko, M., Ewers, H., Yap, C. C. C., Norden, C., & Winckler, B. (2007). Ankyrin-dependent and -independent mechanisms orchestrate axonal compartmentalization of L1 family members neurofascin and L1/neuron-glia cell adhesion molecule. *The Journal of Neuroscience, 27*(3), 590–603.

Brachet, A., Leterrier, C., Irondelle, M., Fache, M.-P., Racine, V., Sibarita, J.-B., et al. (2010). Ankyrin G restricts ion channel diffusion at the axonal initial segment before the establishment of the diffusion barrier. *The Journal of Cell Biology, 191*(2), 383–395.

Brackenbury, W. J., Calhoun, J. D., Chen, C., Miyazaki, H., Nukina, N., Oyama, F., et al. (2010). Functional reciprocity between Na^+ channel Nav1.6 and beta1 subunits in the coordinated regulation of excitability and neurite outgrowth. *Proceedings of the National Academy of Sciences of the United States of America, 107*(5), 2283–2288.

Bradshaw, N. J., Ogawa, F., Antolin-Fontes, B., Chubb, J. E., Carlyle, B. C., Christie, S., et al. (2008). DISC1, PDE4B, and NDE1 at the centrosome and synapse. *Biochemical and Biophysical Research Communications, 377*(4), 1091–1096.

Bradshaw, N. J., Soares, D. C., Carlyle, B. C., Ogawa, F., Davidson-Smith, H., Christie, S., et al. (2011). PKA phosphorylation of NDE1 is DISC1/PDE4 dependent and modulates its interaction with LIS1 and NDEL1. *The Journal of Neuroscience, 31*(24), 9043–9054.

Bréchet, A., Fache, M.-P., Brachet, A., Ferracci, G., Baude, A., Irondelle, M., et al. (2008). Protein kinase CK2 contributes to the organization of sodium channels in axonal membranes by regulating their interactions with ankyrin G. *The Journal of Cell Biology, 183*(6), 1101–1114.

Brown, D. A., & Passmore, G. M. (2009). Neural KCNQ (Kv7) channels. *British Journal of Pharmacology, 156*(8), 1185–1195.

Brown, J. W., Bullitt, E., Sriswasdi, S., Harper, S., Speicher, D. W., & McKnight, C. J. (2015). The physiological molecular shape of spectrin: a compact supercoil resembling a Chinese finger trap. *PLoS Computational Biology, 11*(6), e1004302.

Brückner, G., Szeöke, S., Pavlica, S., Grosche, J., & Kacza, J. (2006). Axon initial segment ensheathed by extracellular matrix in perineuronal nets. *Neuroscience, 138*(2), 365–375.

Brünig, I., Scotti, E., Sidler, C., & Fritschy, J.-M. (2002). Intact sorting, targeting, and clustering of gamma-aminobutyric acid A receptor subtypes in hippocampal neurons in vitro. *Journal of Comparative Neurology, 443*(1), 43–55.

Buffington, S. A., & Rasband, M. N. (2011). The axon initial segment in nervous system disease and injury. *The European Journal of Neuroscience, 34*(10), 1609–1619.

Buffington, S. A., & Rasband, M. N. (2013). Na^+ channel-dependent recruitment of $Nav\beta4$ to axon initial segments and nodes of ranvier. *The Journal of Neuroscience, 33*(14), 6191–6202.

Buffington, S. A., Sobotzik, J. M., Schultz, C., & Rasband, M. N. (2012). IκBα is not required for axon initial segment assembly. *Molecular and Cellular Neurosciences, 50*(1), 1–9.

Bulat, V., Rast, M., & Pielage, J. (2014). Presynaptic CK2 promotes synapse organization and stability by targeting Ankyrin2. *The Journal of Cell Biology, 204*(1), 77–94.

Burack, M., Silverman, M., & Banker, G. A. (2000). The role of selective transport in neuronal protein sorting. *Neuron, 26*(2), 465–472.

Burkarth, N., Kriebel, M., Kranz, E., & Volkmer, H. (2007). Neurofascin regulates the formation of gephyrin clusters and their subsequent translocation to the axon hillock of hippocampal neurons. *Molecular and Cellular Neurosciences, 36*(1), 59–70.

Butler, M. H., David, C., Ochoa, G. C., Freyberg, Z., Daniell, L., Grabs, D., et al. (1997). Amphiphysin II (SH3P9; BIN1), a member of the amphiphysin/Rvs family, is concentrated in the cortical cytomatrix of axon initial segments and nodes of ranvier in brain and around T tubules in skeletal muscle. *The Journal of Cell Biology, 137*(6), 1355–1367.

Byers, T. J., & Branton, D. (1985). Visualization of the protein associations in the erythrocyte membrane skeleton. *Proceedings of the National Academy of Sciences of the United States of America, 82*(18), 6153–6157.

Caceres, A., Ye, B., & Dotti, C. G. (2012). Neuronal polarity: demarcation, growth and commitment. *Current Opinion in Cell Biology, 24*(4), 547–553.

Cai, D., McEwen, D. P., Martens, J. R., Meyhofer, E., & Verhey, K. J. (2009). Single molecule imaging reveals differences in microtubule track selection between Kinesin motors. *PLoS Biology, 7*(10), e1000216.

Calabrese, M., Magliozzi, R., Ciccarelli, O., Geurts, J. J. G., Reynolds, R., & Martin, R. (2015). Exploring the origins of grey matter damage in multiple sclerosis. *Nature Reviews Neurosciences, 16*(3), 147–158.

Callewaert, G., Eilers, J., & Konnerth, A. (1996). Axonal calcium entry during fast "sodium" action potentials in rat cerebellar Purkinje neurones. *The Journal of Physiology, 495*(Pt 3), 641–647.

Campanac, E., Gasselin, C., Baude, A., Rama, S., Ankri, N., & Debanne, D. (2013). Enhanced intrinsic excitability in basket cells maintains excitatory-inhibitory balance in hippocampal circuits. *Neuron, 77*(4), 712–722.

Catterall, W. A. (1981). Localization of sodium channels in cultured neural cells. *The Journal of Neuroscience, 1*(7), 777–783.

Chan, W., Kordeli, E., & Bennett, V. (1993). 440-kD ankyrin B: structure of the major developmentally regulated domain and selective localization in unmyelinated axons. *The Journal of Cell Biology, 123*(6 Pt 1), 1463–1473.

Chand, A. N., Galliano, E., Chesters, R. A., & Grubb, M. S. (2015). A distinct subtype of dopaminergic interneuron displays inverted structural plasticity at the axon initial segment. *The Journal of Neuroscience, 35*(4), 1573–1590.

Chang, K.-J., & Rasband, M. N. (2013). Excitable domains of myelinated nerves: axon initial segments and nodes of ranvier. *Current Topics in Membranes, 72*, 159—192.

Chang, K.-J., Zollinger, D. R., Susuki, K., Sherman, D. L., Makara, M. A., Brophy, P. J., et al. (2014). Glial ankyrins facilitate paranodal axoglial junction assembly. *Nature Neuroscience, 17*(12), 1673—1681.

Chan-Palay, V. (1972). The tripartite structure of the undercoat in initial segments of Purkinje cell axons. *Zeitschrift Für Anatomie Und Entwicklungsgeschichte, 139*(1), 1—10.

Chen, C., Calhoun, J. D., Zhang, Y., Lopez-Santiago, L., Zhou, N., Davis, T. H., et al. (2012). Identification of the cysteine residue responsible for disulfide linkage of Na^+ channel α and β_2 subunits. *The Journal of Biological Chemistry, 287*(46), 39061—39069.

Cheng, P.-L., & Poo, M.-M. (2012). Early events in axon/dendrite polarization. *Annual Review of Neurosciences, 35*, 181—201.

Christie, S. B., & De Blas, A. L. (2003). GABAergic and glutamatergic axons innervate the axon initial segment and organize GABA(A) receptor clusters of cultured hippocampal pyramidal cells. *Journal of Comparative Neurology, 456*(4), 361—374.

Chung, H. J., Jan, Y.-N., & Jan, L. Y. (2006). Polarized axonal surface expression of neuronal KCNQ channels is mediated by multiple signals in the KCNQ2 and KCNQ3 C-terminal domains. *Proceedings of the National Academy of Sciences of the United States of America, 103*(23), 8870—8875.

Clark, B. D., Goldberg, E. M., & Rudy, B. (2009). Electrogenic tuning of the axon initial segment. *The Neuroscientist, 15*(6), 651—668.

Colbert, C. M., & Pan, E. (2002). Ion channel properties underlying axonal action potential initiation in pyramidal neurons. *Nature Neuroscience, 5*(6), 533—538.

Coles, C. H., & Bradke, F. (2015). Coordinating neuronal actin-microtubule dynamics. *Current Biology, 25*(15), R677—R691.

Conradi, S. (1966). Ultrastructural specialization of the initial axon segment of cat lumbar motoneurons. Preliminary observations. *Acta Societatis Medicorum Upsaliensis, 71*(5), 281—284.

Coombs, J. S., Curtis, D. R., & Eccles, J. C. (1957). The generation of impulses in motoneurones. *The Journal of Physiology, 139*(2), 232—249.

Cotel, F., Exley, R., Cragg, S. J., & Perrier, J.-F. (2013). Serotonin spillover onto the axon initial segment of motoneurons induces central fatigue by inhibiting action potential initiation. *Proceedings of the National Academy of Sciences of the United States of America, 110*(12), 4774—4779.

Craig, A. M., & Banker, G. A. (1994). Neuronal polarity. *Annual Review of Neurosciences, 17*, 267—310.

Cruz, D. A., Eggan, S. M., Azmitia, E. C., & Lewis, D. A. (2004). Serotonin1A receptors at the axon initial segment of prefrontal pyramidal neurons in schizophrenia. *The American Journal of Psychiatry, 161*(4), 739—742.

Cruz, D. A., Lovallo, E. M., Stockton, S., Rasband, M., & Lewis, D. A. (2009). Postnatal development of synaptic structure proteins in pyramidal neuron axon initial segments in monkey prefrontal cortex. *Journal of Comparative Neurology, 514*(4), 353—367.

Cruz, D. A., Weaver, C. L., Lovallo, E. M., Melchitzky, D. S., & Lewis, D. A. (2009). Selective alterations in postsynaptic markers of chandelier cell inputs to cortical pyramidal neurons in subjects with schizophrenia. *Neuropsychopharmacology: Official Publication of the American College of Neuropsychopharmacology, 34*(9), 2112—2124.

Cudmore, R. H., Fronzaroli-Molinieres, L., Giraud, P., & Debanne, D. (2010). Spike-time precision and network synchrony are controlled by the homeostatic regulation of the D-type potassium current. *The Journal of Neuroscience, 30*(38), 12885—12895.

Czyrak, A., Czepiel, K., Maćkowiak, M., Chocyk, A., & Wędzony, K. (2003). Serotonin 5-HT1A receptors might control the output of cortical glutamatergic neurons in rat cingulate cortex. *Brain Research, 989*(1), 42—51.

Davis, J. Q., & Bennett, V. (1984). Brain ankyrin. A membrane-associated protein with binding sites for spectrin, tubulin, and the cytoplasmic domain of the erythrocyte anion channel. *The Journal of Biological Chemistry, 259*(21), 13550–13559.

Davis, J. Q., Lambert, S., & Bennett, V. (1996). Molecular composition of the node of Ranvier: identification of ankyrin-binding cell adhesion molecules neurofascin (mucin+/third FNIII domain-) and NrCAM at nodal axon segments. *The Journal of Cell Biology, 135*(5), 1355–1367.

Davis, L. H., Abdi, K. M., Machius, M., Brautigam, C., Tomchick, D. R., Bennett, V., et al. (2009). Localization and structure of the ankyrin-binding site on beta2-spectrin. *The Journal of Biological Chemistry, 284*(11), 6982–6987.

DeFelipe, J., Hendry, S. H., Jones, E. G., & Schmechel, D. (1985). Variability in the terminations of GABAergic chandelier cell axons on initial segments of pyramidal cell axons in the monkey sensory-motor cortex. *Journal of Comparative Neurology, 231*(3), 364–384.

DeFelipe, J., Arellano, J. I., Gómez, A., Azmitia, E. C., & Muñoz, A. (2001). Pyramidal cell axons show a local specialization for GABA and 5-HT inputs in monkey and human cerebral cortex. *Journal of Comparative Neurology, 433*(1), 148–155.

Deiters, V. S., & Guillery, R. W. (2013). Otto Friedrich Karl Deiters (1834–1863). *Journal of Comparative Neurology, 521*(9), 1929–1953.

Del Puerto, A., Fronzaroli-Molinieres, L., Perez-Alvarez, M. J., Giraud, P., Carlier, E., Wandosell, F., et al. (2015). ATP-P2X7 receptor modulates axon initial segment composition and function in physiological conditions and brain injury. *Cerebral Cortex, 25*(8), 2282–2294.

D'Este, E., Kamin, D., Göttfert, F., El-Hady, A., & Hell, S. W. (2015). STED nanoscopy reveals the ubiquity of subcortical cytoskeleton periodicity in living neurons. *Cell Reports, 10*(8), 1246–1251.

Devaux, J. J., Kleopa, K. A., Cooper, E. C., & Scherer, S. S. (2004). KCNQ2 is a nodal K$^+$ channel. *The Journal of Neuroscience, 24*(5), 1236–1244.

Dimitrov, A., Quesnoit, M., Moutel, S., Cantaloube, I., Poüs, C., & Perez, F. (2008). Detection of GTP-tubulin conformation in vivo reveals a role for GTP remnants in microtubule rescues. *Science, 322*(5906), 1353–1356.

Dodge, F. A., & Cooley, J. W. (1973). Action potential of the motor neuron. *IBM Journal of Research and Development, 17*(3), 219–229.

Dodson, P. D., Barker, M. C., & Forsythe, I. D. (2002). Two heteromeric Kv1 potassium channels differentially regulate action potential firing. *The Journal of Neuroscience, 22*(16), 6953–6961.

Dotti, C. G., & Simons, K. (1990). Polarized sorting of viral glycoproteins to the axon and dendrites of hippocampal neurons in culture. *Cell, 62*(1), 63–72.

Duflocq, A., Le Bras, B., Bullier, E., Couraud, F., & Davenne, M. (2008). Nav1.1 is predominantly expressed in nodes of Ranvier and axon initial segments. *Molecular and Cellular Neurosciences, 39*(2), 180–192.

Duflocq, A., Chareyre, F., Giovannini, M., Couraud, F., & Davenne, M. (2011). Characterization of the axon initial segment (AIS) of motor neurons and identification of a para-AIS and a juxtapara-AIS, organized by protein 4.1B. *BMC Biology, 9, 66.*

Dugladze, T., Schmitz, D., Whittington, M. A., Vida, I., & Gloveli, T. (2012). Segregation of axonal and somatic activity during fast network oscillations. *Science, 336*(6087), 1458–1461.

Dunn, S., Morrison, E. E., Liverpool, T., Molina-París, C., Cross, R., Alonso, M., et al. (2008). Differential trafficking of Kif5c on tyrosinated and detyrosinated microtubules in live cells. *Journal of Cell Science, 121*(Pt 7), 1085–1095.

Durak, O., de Anda, F. C., Singh, K. K., Leussis, M. P., Petryshen, T. L., Sklar, P., et al. (2015). Ankyrin-G regulates neurogenesis and Wnt signaling by altering the subcellular localization of β-catenin. *Molecular Psychiatry, 20*(3), 388–397.

Dutta, R., Chang, A., Doud, M. K., Kidd, G. J., Ribaudo, M. V., Young, E. A., et al. (2011). Demyelination causes synaptic alterations in hippocampi from multiple sclerosis patients. *Annals of Neurology, 69*(3), 445—454.

Dzhashiashvili, Y., Zhang, Y., Galinska, J., Lam, I., Grumet, M., & Salzer, J. L. (2007). Nodes of Ranvier and axon initial segments are ankyrin G-dependent domains that assemble by distinct mechanisms. *The Journal of Cell Biology, 177*(5), 857—870.

Edwards, C., & Ottoson, D. (1958). The site of impulse initiation in a nerve cell of a crustacean stretch receptor. *The Journal of Physiology, 143*(1), 138—148.

Edwards, S. L., Yu, S.-C., Hoover, C. M., Phillips, B. C., Richmond, J. E., & Miller, K. G. (2013). An organelle gatekeeper function for *Caenorhabditis elegans* UNC-16 (JIP3) at the axon initial segment. *Genetics, 194*(1), 143—161.

Encalada, S. E., & Goldstein, L. S. B. (2014). Biophysical challenges to axonal transport: motor-cargo deficiencies and neurodegeneration. *Annual Review of Biophysics, 43*, 141—169.

Evans, M. D., Sammons, R. P., Lebron, S., Dumitrescu, A. S., Watkins, T. B. K., Uebele, V. N., et al. (2013). Calcineurin signaling mediates activity-dependent relocation of the axon initial segment. *The Journal of Neuroscience, 33*(16), 6950—6963.

Fache, M.-P., Moussif, A., Fernandes, F., Giraud, P., Garrido, J. J., & Dargent, B. (2004). Endocytotic elimination and domain-selective tethering constitute a potential mechanism of protein segregation at the axonal initial segment. *The Journal of Cell Biology, 166*(4), 571—578.

Farías, G. G., Cuitino, L., Guo, X., Ren, X., Jarnik, M., Mattera, R., et al. (2012). Signal-mediated, AP-1/clathrin-dependent sorting of transmembrane receptors to the somatodendritic domain of hippocampal neurons. *Neuron, 75*(5), 810—823.

Fatt, P. (1957). Sequence of events in synaptic activation of a motoneurone. *Journal of Neurophysiology, 20*(1), 61—80.

Fish, K. N., Hoftman, G. D., Sheikh, W., Kitchens, M., & Lewis, D. A. (2013). Parvalbumin-containing chandelier and basket cell boutons have distinctive modes of maturation in monkey prefrontal cortex. *The Journal of Neuroscience, 33*(19), 8352—8358.

Fleidervish, I. A., Lasser-Ross, N., Gutnick, M. J., & Ross, W. N. (2010). Na^+ imaging reveals little difference in action potential-evoked Na^+ influx between axon and soma. *Nature Neuroscience, 13*(7), 852—860.

Foust, A. J., Popovic, M., Zecevic, D., & McCormick, D. A. (2010). Action potentials initiate in the axon initial segment and propagate through axon collaterals reliably in cerebellar Purkinje neurons. *The Journal of Neuroscience, 30*(20), 6891—6902.

Fox, P. D., Haberkorn, C. J., Akin, E. J., Seel, P. J., Krapf, D., & Tamkun, M. M. (2015). Induction of stable ER-plasma-membrane junctions by Kv2.1 potassium channels. *Journal of Cell Science, 128*(11), 2096—2105.

Franssen, E. H. P., Zhao, R.-R., Koseki, H., Kanamarlapudi, V., Hoogenraad, C. C., Eva, R., et al. (2015). Exclusion of integrins from CNS axons is regulated by Arf6 activation and the AIS. *The Journal of Neuroscience, 35*(21), 8359—8375.

Frischknecht, R., Heine, M., Perrais, D., Seidenbecher, C. I., Choquet, D., & Gundelfinger, E. D. (2009). Brain extracellular matrix affects AMPA receptor lateral mobility and short-term synaptic plasticity. *Nature Neuroscience, 12*(7), 897—904.

Fukano, T., Hama, H., & Miyawaki, A. (2004). Similar diffusibility of membrane proteins across the axon-soma and dendrite-soma boundaries revealed by a novel FRAP technique. *Journal of Structural Biology, 147*(1), 12—18.

Futerman, A. H., Khanin, R. R., & Segel, L. A. L. (1993). Lipid diffusion in neurons. *Nature, 362*(6416), 119.

Galiano, M. R., Jha, S., Ho, T. S.-Y., Zhang, C., Ogawa, Y., Chang, K.-J., et al. (2012). A distal axonal cytoskeleton forms an intra-axonal boundary that controls axon initial segment assembly. *Cell, 149*(5), 1125—1139.

Ganguly, A., Tang, Y., Wang, L., Ladt, K., Loi, J., Dargent, B., et al. (2015). A dynamic formin-dependent deep F-actin network in axons. *The Journal of Cell Biology, 104*(51), 20576—21417.

Garrido, J. J., Fernandes, F., Giraud, P., Mouret, I., Pasqualini, E., Fache, M.-P., et al. (2001). Identification of an axonal determinant in the C-terminus of the sodium channel Na(v) 1.2. *The EMBO Journal, 20*(21), 5950—5961.

Garrido, J. J., Giraud, P., Carlier, E., Fernandes, F., Moussif, A., Fache, M.-P., et al. (2003). A targeting motif involved in sodium channel clustering at the axonal initial segment. *Science, 300*(5628), 2091—2094.

Garver, T. D., Ren, Q., Tuvia, S., & Bennett, V. (1997). Tyrosine phosphorylation at a site highly conserved in the L1 family of cell adhesion molecules abolishes ankyrin binding and increases lateral mobility of neurofascin. *The Journal of Cell Biology, 137*(3), 703—714.

Gasser, A., Ho, T. S.-Y., Cheng, X., Chang, K.-J., Waxman, S. G., Rasband, M. N., et al. (2012). An ankyrin G-binding motif is necessary and sufficient for targeting Na$_v$1.6 sodium channels to axon initial segments and nodes of Ranvier. *The Journal of Neuroscience, 32*(21), 7232—7243.

Ghosh, S., Dorsey, F. C., & Cox, J. V. (2002). CK2 constitutively associates with and phosphorylates chicken erythroid ankyrin and regulates its ability to bind to spectrin. *Journal of Cell Science, 115*(Pt 21), 4107—4115.

Glenney, J. R., Glenney, P., & Weber, K. (1982). F-actin-binding and cross-linking properties of porcine brain fodrin, a spectrin-related molecule. *The Journal of Biological Chemistry, 257*(16), 9781—9787.

Glickfeld, L. L., Roberts, J. D., Somogyi, P., & Scanziani, M. (2009). Interneurons hyperpolarize pyramidal cells along their entire somatodendritic axis. *Nature Neuroscience, 12*(1), 21—23.

Goldberg, E. M., Clark, B. D., Zagha, E., Nahmani, M., Erisir, A., & Rudy, B. (2008). K$^+$ channels at the axon initial segment Dampen near-threshold excitability of neocortical fast-spiking GABAergic interneurons. *Neuron, 58*(3), 387—400.

Goldfarb, M., Schoorlemmer, J., Williams, A., Diwakar, S., Wang, Q., Huang, X., et al. (2007). Fibroblast growth factor homologous factors control neuronal excitability through modulation of voltage-gated sodium channels. *Neuron, 55*(3), 449—463.

Greer, J. E., Hånell, A., McGinn, M. J., & Povlishock, J. T. (2013). Mild traumatic brain injury in the mouse induces axotomy primarily within the axon initial segment. *Acta Neuropathologica, 126*(1), 59—74.

Grubb, M. S., & Burrone, J. (2010). Activity-dependent relocation of the axon initial segment fine-tunes neuronal excitability. *Nature, 465*(7301), 1070—1074.

Gu, J., Jan, Y.-N., & Jan, L. Y. (2003). A conserved domain in axonal targeting of Kv1 (Shaker) voltage-gated potassium channels. *Science, 301*(5633), 646—649.

Gu, C., Zhou, W., Puthenveedu, M. A., Xu, M., Jan, Y.-N., & Jan, L. Y. (2006). The microtubule plus-end tracking protein EB1 is required for Kv1 voltage-gated K$^+$ channel axonal targeting. *Neuron, 52*(5), 803—816.

Guan, D., Higgs, M. H., Horton, L. R., Spain, W. J., & Foehring, R. C. (2011). Contributions of Kv7-mediated potassium current to sub- and suprathreshold responses of rat layer II/III neocortical pyramidal neurons. *Journal of Neurophysiology, 106*(4), 1722—1733.

Gutzmann, A., Ergül, N., Grossmann, R., Schultz, C., Wahle, P., & Engelhardt, M. (2014). A period of structural plasticity at the axon initial segment in developing visual cortex. *Frontiers in Neuroanatomy, 8*, 11.

Hamada, M. S., & Kole, M. H. P. (2015). Myelin loss and axonal ion channel adaptations associated with gray matter neuronal hyperexcitability. *The Journal of Neuroscience, 35*(18), 7272—7286.

Hammarlund, M., Jorgensen, E. M., & Bastiani, M. J. (2007). Axons break in animals lacking beta-spectrin. *The Journal of Cell Biology, 176*(3), 269—275.

Hammond, J. W., Huang, C.-F., Kaech, S., Jacobson, C., Banker, G. A., & Verhey, K. J. (2010). Posttranslational modifications of tubulin and the polarized transport of kinesin-1 in neurons. *Molecular Biology of the Cell, 21*(4), 572—583.

Hares, K., Kemp, K., Rice, C., Gray, E., Scolding, N., & Wilkins, A. (2013). Reduced axonal motor protein expression in non-lesional grey matter in multiple sclerosis. *Multiple Sclerosis.*

Harty, R. C., Kim, T. H., Thomas, E. A., Cardamone, L., Jones, N. C., Petrou, S., et al. (2013). Axon initial segment structural plasticity in animal models of genetic and acquired epilepsy. *Epilepsy Research, 105*(3), 272—279.

He, M., Jenkins, P. M., & Bennett, V. (2012). Cysteine 70 of ankyrin-G is S-palmitoylated and is required for function of ankyrin-G in membrane domain assembly. *The Journal of Biological Chemistry, 287*(52), 43995—44005.

He, M., Tseng, W.-C., & Bennett, V. (2013). A single divergent exon inhibits ankyrin-B association with the plasma membrane. *The Journal of Biological Chemistry, 288*(21), 14769—14779.

Hedstrom, K., Xu, X., Ogawa, Y., Frischknecht, R., Seidenbecher, C., Shrager, P., et al. (2007). Neurofascin assembles a specialized extracellular matrix at the axon initial segment. *The Journal of Cell Biology, 178*(5), 875—886.

Hedstrom, K., Ogawa, Y., & Rasband, M. N. (2008). Ankyrin G is required for maintenance of the axon initial segment and neuronal polarity. *The Journal of Cell Biology, 183*(4), 635—640.

Hien, Y. E., Montersino, A., Castets, F., Leterrier, C., Filhol, O., Vacher, H., et al. (2014). CK_2 accumulation at the axon initial segment depends on sodium channel Nav1. *FEBS Letters, 588*(18), 3403—3408.

Hill, A. S., Nishino, A., Nakajo, K., Zhang, G., Fineman, J. R., Selzer, M. E., et al. (2008). Ion channel clustering at the axon initial segment and node of ranvier evolved sequentially in early chordates. *PLoS Genetics, 4*(12), e1000317.

Hinman, J. D., Rasband, M. N., & Carmichael, S. T. (2013). Remodeling of the axon initial segment after focal cortical and white matter stroke. *Stroke, 44*(1), 182—189.

Hirokawa, N., & Tanaka, Y. (2015). Kinesin superfamily proteins (KIFs): various functions and their relevance for important phenomena in life and diseases. *Experimental Cell Research, 334*(1), 16—25.

Howarth, M., Spiller, D., Reed, J., McNamee, C., White, M., & Moss, D. (2003). No barrier to diffusion between cell soma and neurite membranes in sympathetic neurons for a GPI-anchored glycoprotein. *Molecular and Cellular Neurosciences, 24*(2), 296—306.

Hsu, W.-C. J., Nilsson, C. L., & Laezza, F. (2014). Role of the axonal initial segment in psychiatric disorders: function, dysfunction, and intervention. *Frontiers in Psychiatry, 5,* 109.

Hu, W., Tian, C., Li, T., Yang, M., Hou, H., & Shu, Y. (2009). Distinct contributions of Na(v)1.6 and Na(v)1.2 in action potential initiation and backpropagation. *Nature Neuroscience, 12*(8), 996—1002.

Hund, T. J., Koval, O. M., Li, J., Wright, P. J., Qian, L., Snyder, J. S., et al. (2010). A β(IV)-spectrin/CaMKII signaling complex is essential for membrane excitability in mice. *The Journal of Clinical Investigation, 120*(10), 3508—3519.

Inan, M., Blázquez-Llorca, L., Merchán-Pérez, A., Anderson, S. A., DeFelipe, J., & Yuste, R. (2013). Dense and overlapping innervation of pyramidal neurons by chandelier cells. *The Journal of Neuroscience, 33*(5), 1907—1914.

Inda, M., DeFelipe, J., & Muñoz, A. (2006). Voltage-gated ion channels in the axon initial segment of human cortical pyramidal cells and their relationship with chandelier cells. *Proceedings of the National Academy of Sciences of the United States of America, 103*(8), 2920—2925.

Ipsaro, J. J., Huang, L., & Mondragón, A. (2009). Structures of the spectrin–ankyrin interaction binding domains. *Blood, 113*(22), 5385–5393.

Iqbal, Z., Vandeweyer, G., van der Voet, M., Waryah, A. M., Zahoor, M. Y., Besseling, J. A., et al. (2013). Homozygous and heterozygous disruptions of ANK3: at the crossroads of neurodevelopmental and psychiatric disorders. *Human Molecular Genetics, 22*(10), 1960–1970.

Iwakura, A., Uchigashima, M., Miyazaki, T., Yamasaki, M., & Watanabe, M. (2012). Lack of molecular-anatomical evidence for GABAergic influence on axon initial segment of cerebellar Purkinje cells by the pinceau formation. *The Journal of Neuroscience, 32*(27), 9438–9448.

Jacobson, C., Schnapp, B., & Banker, G. A. (2006). A change in the selective translocation of the Kinesin-1 motor domain marks the initial specification of the axon. *Neuron, 49*(6), 797–804.

Jenkins, S. M., & Bennett, V. (2001). Ankyrin-G coordinates assembly of the spectrin-based membrane skeleton, voltage-gated sodium channels, and L1 CAMs at Purkinje neuron initial segments. *The Journal of Cell Biology, 155*(5), 739–746.

Jenkins, P. M., Kim, N., Jones, S. L., Tseng, W.-C., Svitkina, T. M., Yin, H. H., et al. (2015). Giant ankyrin-G: a critical innovation in vertebrate evolution of fast and integrated neuronal signaling. *Proceedings of the National Academy of Sciences of the United States of America, 112*(4), 957–964.

Jensen, C. S., Watanabe, S., Rasmussen, H. B., Schmitt, N., Olesen, S.-P., Frost, N. A., et al. (2014). Specific sorting and post-Golgi trafficking of dendritic potassium channels in living neurons. *The Journal of Biological Chemistry, 289*(15), 10566–10581.

Jiang, K., Toedt, G., Montenegro Gouveia, S., Davey, N. E., Hua, S., van der Vaart, B., et al. (2012). A proteome-wide screen for mammalian SxIP motif-containing microtubule plus-end tracking proteins. *Current Biology, 22*(19), 1800–1807.

John, N., Krügel, H., Frischknecht, R., Smalla, K., Schultz, C., Kreutz, M., et al. (2006). Brevican-containing perineuronal nets of extracellular matrix in dissociated hippocampal primary cultures. *Molecular and Cellular Neurosciences, 31*(4), 774–784.

Jones, E. G., & Powell, T. P. (1969). Synapses on the axon hillocks and initial segments of pyramidal cell axons in the cerebral cortex. *Journal of Cell Science, 5*(2), 495–507.

Jones, S. L., Korobova, F., & Svitkina, T. (2014). Axon initial segment cytoskeleton comprises a multiprotein submembranous coat containing sparse actin filaments. *The Journal of Cell Biology, 2*(1), 89.

Kaphzan, H., Buffington, S. A., Jung, J. I., Rasband, M. N., & Klann, E. (2011). Alterations in intrinsic membrane properties and the axon initial segment in a mouse model of Angelman syndrome. *The Journal of Neuroscience, 31*(48), 17637–17648.

Kaphzan, H., Buffington, S. A., Ramaraj, A. B., Lingrel, J. B., Rasband, M. N., Santini, E., et al. (2013). Genetic reduction of the α1 subunit of Na/K-ATPase corrects multiple hippocampal phenotypes in Angelman syndrome. *Cell Reports, 4*(3), 405–412.

Kapitein, L. C., & Hoogenraad, C. C. (2011). Which way to go? Cytoskeletal organization and polarized transport in neurons. *Molecular and Cellular Neurosciences, 46*(1), 9–20.

Kapitein, L. C., & Hoogenraad, C. C. (2015). Building the neuronal microtubule cytoskeleton. *Neuron, 87*(3), 492–506.

Kennedy, S. P., Warren, S. L., Forget, B. G., & Morrow, J. S. (1991). Ankyrin binds to the 15th repetitive unit of erythroid and nonerythroid beta-spectrin. *The Journal of Cell Biology, 115*(1), 267–277.

Kevenaar, J. T., & Hoogenraad, C. C. (2015). The axonal cytoskeleton: from organization to function. *Frontiers in Molecular Neuroscience, 8*, 44.

Khaliq, Z. M., & Raman, I. M. (2006). Relative contributions of axonal and somatic Na channels to action potential initiation in cerebellar Purkinje neurons. *The Journal of Neuroscience, 26*(7), 1935–1944.

Khirug, S., Yamada, J., Afzalov, R., Voipio, J., Khiroug, L., & Kaila, K. (2008). GABAergic depolarization of the axon initial segment in cortical principal neurons is caused by the Na-K-2Cl cotransporter NKCC1. *The Journal of Neuroscience, 28*(18), 4635—4639.

King, A. N., Manning, C. F., & Trimmer, J. S. (2014). A unique ion channel clustering domain on the axon initial segment of mammalian neurons. *Journal of Comparative Neurology, 522*(11), 2594—2608.

Kirizs, T., Kerti-Szigeti, K., Lorincz, A., & Nusser, Z. (2014). Distinct axo-somato-dendritic distributions of three potassium channels in CA1 hippocampal pyramidal cells. *The European Journal of Neuroscience, 39*(11), 1771—1783.

Klinger, F., Gould, G., Boehm, S., & Shapiro, M. S. (2011). Distribution of M-channel subunits KCNQ2 and KCNQ3 in rat hippocampus. *NeuroImage, 58*(3), 761—769.

Kobayashi, T., Storrie, B., Simons, K., & Dotti, C. G. (1992). A functional barrier to movement of lipids in polarized neurons. *Nature, 359*(6396), 647—650.

Koch, I., Schwarz, H., Beuchle, D., Goellner, B., Langegger, M., & Aberle, H. (2008). *Drosophila* Ankyrin 2 is required for synaptic stability. *Neuron, 58*(2), 210—222.

Kohno, K. (1964). Neurotubules contained within the dendrite and axon of Purkinje cell of frog. *The Bulletin of Tokyo Medical and Dental University, 11*(4), 411—442.

Koike, M., Tanida, I., Nanao, T., Tada, N., Iwata, J.-I., Ueno, T., et al. (2013). Enrichment of GABARAP relative to LC3 in the axonal initial segments of neurons. *PloS One, 8*(5), e63568.

Kole, M. H. P., & Stuart, G. J. (2012). Signal processing in the axon initial segment. *Neuron, 73*(2), 235—247.

Kole, M. H. P., Letzkus, J. J., & Stuart, G. J. (2007). Axon initial segment Kv1 channels control axonal action potential waveform and synaptic efficacy. *Neuron, 55*(4), 633—647.

Kole, M. H. P., Ilschner, S., Kampa, B., Williams, S., Ruben, P., & Stuart, G. (2008). Action potential generation requires a high sodium channel density in the axon initial segment. *Nature Neuroscience, 11*(2), 178—186.

Komada, M., & Soriano, P. (2002). [Beta]IV-spectrin regulates sodium channel clustering through ankyrin-G at axon initial segments and nodes of Ranvier. *The Journal of Cell Biology, 156*(2), 337—348.

Konishi, Y., & Setou, M. (2009). Tubulin tyrosination navigates the kinesin-1 motor domain to axons. *Nature Neuroscience, 12*(5), 559—567.

Kordeli, E., & Bennett, V. (1991). Distinct ankyrin isoforms at neuron cell bodies and nodes of Ranvier resolved using erythrocyte ankyrin-deficient mice. *The Journal of Cell Biology, 114*(6), 1243—1259.

Kordeli, E., Davis, J. Q., Trapp, B., & Bennett, V. (1990). An isoform of ankyrin is localized at nodes of Ranvier in myelinated axons of central and peripheral nerves. *The Journal of Cell Biology, 110*(4), 1341—1352.

Kordeli, E., Lambert, S., & Bennett, V. (1995). Ankyrin G. A new ankyrin gene with neural-specific isoforms localized at the axonal initial segment and node of Ranvier. *The Journal of Biological Chemistry, 270*(5), 2352—2359.

Kosaka, T. (1980). The axon initial segment as a synaptic site: ultrastructure and synaptology of the initial segment of the pyramidal cell in the rat hippocampus (CA3 region). *Journal of Neurocytology, 9*(6), 861—882.

Kress, G. J., & Mennerick, S. (2009). Action potential initiation and propagation: upstream influences on neurotransmission. *Neuroscience, 158*(1), 211—222.

Kress, G. J., Dowling, M. J., Meeks, J., & Mennerick, S. (2008). High threshold, proximal initiation, and slow conduction velocity of action potentials in dentate granule neuron mossy fibers. *Journal of Neurophysiology, 100*(1), 281.

Kress, G. J., Dowling, M. J., Eisenman, L. N., & Mennerick, S. (2010). Axonal sodium channel distribution shapes the depolarized action potential threshold of dentate granule neurons. *Hippocampus, 20*(4), 558—571.

Kriebel, M., Metzger, J., Trinks, S., Chugh, D., Harvey, R. J., Harvey, K., et al. (2011). The cell adhesion molecule neurofascin stabilizes axo-axonic GABAergic terminals at the axon initial segment. *The Journal of Biological Chemistry, 286*(27), 24385–24393.

Krieg, M., Dunn, A. R., & Goodman, M. B. (2014). Mechanical control of the sense of touch by β-spectrin. *Nature Cell Biology, 16*(3), 224–233.

Kuba, H., & Ohmori, H. (2009). Roles of axonal sodium channels in precise auditory time coding at nucleus magnocellularis of the chick. *The Journal of Physiology, 587*(Pt 1), 87–100.

Kuba, H., Ishii, T. M., & Ohmori, H. (2006). Axonal site of spike initiation enhances auditory coincidence detection. *Nature, 444*(7122), 1069–1072.

Kuba, H., Oichi, Y., & Ohmori, H. (2010). Presynaptic activity regulates Na(+) channel distribution at the axon initial segment. *Nature, 465*(7301), 1075–1078.

Kuba, H., Adachi, R., & Ohmori, H. (2014). Activity-dependent and activity-independent development of the axon initial segment. *The Journal of Neuroscience, 34*(9), 3443–3453.

Kusumi, A., Nakada, C., Ritchie, K., Murase, K., Suzuki, K. G. N., Murakoshi, H., et al. (2005). Paradigm shift of the plasma membrane concept from the two-dimensional continuum fluid to the partitioned fluid: high-speed single-molecule tracking of membrane molecules. *Annual Review of Biophysics and Biomolecular Structure, 34*, 351–378.

Lacas-Gervais, S., Guo, J., Strenzke, N., Scarfone, E., Kolpe, M., Jahkel, M., et al. (2004). BetaIVSigma1 spectrin stabilizes the nodes of Ranvier and axon initial segments. *The Journal of Cell Biology, 166*(7), 983–990.

Laezza, F., Gerber, B. R., Lou, J.-Y., Kozel, M. A., Hartman, H., Craig, A. M., et al. (2007). The FGF14(F145S) mutation disrupts the interaction of FGF14 with voltage-gated Na$^+$ channels and impairs neuronal excitability. *The Journal of Neuroscience, 27*(44), 12033–12044.

Lai, L., & Cao, J. (2014). Spectrins in axonal cytoskeletons: dynamics revealed by extensions and fluctuations. *The Journal of Chemical Physics, 141*(1), 015101.

Lasiecka, Z. M., Yap, C. C., Vakulenko, M., & Winckler, B. (2009). Compartmentalizing the neuronal plasma membrane from axon initial segments to synapses. *International Review of Cell and Molecular Biology, 272*, 303–389.

Le Bras, B., Fréal, A., Czarnecki, A., Legendre, P., Bullier, E., Komada, M., et al. (2013). In vivo assembly of the axon initial segment in motor neurons. *Brain Structure and Function, 219*(4), 1433–1450.

Lemaillet, G., Walker, B., & Lambert, S. (2003). Identification of a conserved ankyrin-binding motif in the family of sodium channel alpha subunits. *The Journal of Biological Chemistry, 278*(30), 27333–27339.

León-Espinosa, G., DeFelipe, J., & Muñoz, A. (2012). Effects of amyloid-β plaque proximity on the axon initial segment of pyramidal cells. *Journal of Alzheimer's Disease, 29*(4), 841–852.

Leterrier, C., & Dargent, B. (2014). No Pasaran! Role of the axon initial segment in the regulation of protein transport and the maintenance of axonal identity. *Seminars in Cell and Developmental Biology, 27C*, 44–51.

Leterrier, C., Brachet, A., Dargent, B., & Vacher, H. (2011). Determinants of voltage-gated sodium channel clustering in neurons. *Seminars in Cell and Developmental Biology, 22*(2), 171–177.

Leterrier, C., Vacher, H., Fache, M.-P., Angles d'Ortoli, S., Castets, F., Autillo-Touati, A., et al. (2011). End-binding proteins EB3 and EB1 link microtubules to ankyrin G in the axon initial segment. *Proceedings of the National Academy of Sciences of the United States of America, 108*(21), 8826–8831.

Leterrier, C., Potier, J., Caillol, G., Rueda Boroni, F., Debarnot, C., & Dargent, B. (2015). Nanoscale architecture of the axon initial segment reveals an organized and robust scaffold. *Cell Reports.* http://dx.doi.org/10.1016/j.celrep.2015.11.051.

Leussis, M. P., Madison, J. M., & Petryshen, T. L. (2012). Ankyrin 3: genetic association with bipolar disorder and relevance to disease pathophysiology. *Biology of Mood and Anxiety Disorders, 2*(1), 18.

Leussis, M. P., Berry-Scott, E. M., Saito, M., Jhuang, H., de Haan, G., Alkan, O., et al. (2013). The ANK3 bipolar disorder gene regulates psychiatric-related behaviors that are modulated by lithium and stress. *Biological Psychiatry, 73*(7), 683—690.

Lewis, T. L., Mao, T., Svoboda, K., & Arnold, D. B. (2009). Myosin-dependent targeting of transmembrane proteins to neuronal dendrites. *Nature Neuroscience, 12*(5), 568—576.

Lewis, T. L., Mao, T., & Arnold, D. B. (2011). A role for Myosin VI in the localization of axonal proteins. *PLoS Biology, 9*(3), e1001021.

Lewis, T. L., Courchet, J., & Polleux, F. (2013). Cell biology in neuroscience: cellular and molecular mechanisms underlying axon formation, growth, and branching. *The Journal of Cell Biology, 202*(6), 837—848.

Lewis, D. A. (2011). The chandelier neuron in schizophrenia. *Developmental Neurobiology, 71*(1), 118—127.

Li, Y.-C., Zhai, X.-Y., Ohsato, K., Futamata, H., Shimada, O., & Atsumi, S. (2004). Mitochondrial accumulation in the distal part of the initial segment of chicken spinal motoneurons. *Brain Research, 1026*(2), 235—243.

Li, X., Kumar, Y., Zempel, H., Mandelkow, E.-M., Biernat, J., & Mandelkow, E. (2011). Novel diffusion barrier for axonal retention of Tau in neurons and its failure in neurodegeneration. *The EMBO Journal, 30*(23), 4825—4837.

Li, T., Tian, C., Scalmani, P., Frassoni, C., Mantegazza, M., Wang, Y., et al. (2014). Action potential initiation in neocortical inhibitory interneurons. *PLoS Biology, 12*(9), e1001944.

Liu, W., & Devaux, J. J. (2014). Calmodulin orchestrates the heteromeric assembly and the trafficking of KCNQ2/3 (Kv7.2/3) channels in neurons. *Molecular and Cellular Neurosciences, 58*, 40—52.

Liu, S. C., Derick, L. H., & Palek, J. (1987). Visualization of the hexagonal lattice in the erythrocyte membrane skeleton. *The Journal of Cell Biology, 104*(3), 527—536.

Liu, Y., Zhang, Y., & Wang, J.-H. (2014). Crystal structure of human Ankyrin G death domain. *Proteins, 82*(12), 3476—3482.

Lorenzo, D. N., Badea, A., Davis, J. Q., Hostettler, J., He, J., Zhong, G., et al. (2014). A PIK3C3-Ankyrin-B-Dynactin pathway promotes axonal growth and multiorganelle transport. *The Journal of Cell Biology, 207*(6), 735—752.

Lorincz, A., & Nusser, Z. (2008). Cell-type-dependent molecular composition of the axon initial segment. *The Journal of Neuroscience, 28*(53), 14329—14340.

Lorincz, A., & Nusser, Z. (2010). Molecular identity of dendritic voltage-gated sodium channels. *Science, 328*(5980), 906—909.

Lou, J.-Y., Laezza, F., Gerber, B. R., Xiao, M., Yamada, K. A., Hartmann, H., et al. (2005). Fibroblast growth factor 14 is an intracellular modulator of voltage-gated sodium channels. *The Journal of Physiology, 569*(Pt 1), 179—193.

Maglione, M., & Sigrist, S. J. (2013). Seeing the forest tree by tree: super-resolution light microscopy meets the neurosciences. *Nature Neuroscience, 16*(7), 790—797.

Mainen, Z. F., Joerges, J., Huguenard, J. R., & Sejnowski, T. J. (1995). A model of spike initiation in neocortical pyramidal neurons. *Neuron, 15*(6), 1427—1439.

Malhotra, J. D., Koopmann, M. C., Kazen-Gillespie, K. A., Fettman, N., Hortsch, M., & Isom, L. L. (2002). Structural requirements for interaction of sodium channel beta 1 subunits with ankyrin. *The Journal of Biological Chemistry, 277*(29), 26681—26688.

Maniar, T. A., Kaplan, M., Wang, G. J., Shen, K., Wei, L., Shaw, J. E., et al. (2012). UNC-33 (CRMP) and ankyrin organize microtubules and localize kinesin to polarize axon-dendrite sorting. *Nature Neuroscience, 15*(1), 48—56.

Martin, P.-M., Carnaud, M., Garcia del Caño, G., Irondelle, M., Irinopoulou, T., Girault, J.-A., et al. (2008). Schwannomin-interacting protein-1 isoform

IQCJ-SCHIP-1 is a late component of nodes of Ranvier and axon initial segments. *The Journal of Neuroscience, 28*(24), 6111—6117.

Maurer, S. P., Fourniol, F. J., Bohner, G., Moores, C. A., & Surrey, T. (2012). EBs recognize a nucleotide-dependent structural cap at growing microtubule ends. *Cell, 149*(2), 371—382.

McEwen, D. P., & Isom, L. L. (2004). Heterophilic interactions of sodium channel beta1 subunits with axonal and glial cell adhesion molecules. *The Journal of Biological Chemistry, 279*(50), 52744—52752.

Meeks, J., & Mennerick, S. (2007). Action potential initiation and propagation in CA_3 pyramidal axons. *Journal of Neurophysiology, 97*(5), 3460.

Michaely, P., & Bennett, V. (1995). Mechanism for binding site diversity on ankyrin. Comparison of binding sites on ankyrin for neurofascin and the Cl^-/HCO^{3-} anion exchanger. *The Journal of Biological Chemistry, 270*(52), 31298—31302.

Millecamps, S., & Julien, J.-P. (2013). Axonal transport deficits and neurodegenerative diseases. *Nature Reviews Neurosciences, 14*(3), 1—16.

Miyazaki, H., Oyama, F., Inoue, R., Aosaki, T., Abe, T., Kiyonari, H., et al. (2014). Singular localization of sodium channel β4 subunit in unmyelinated fibres and its role in the striatum. *Nature Communications, 5*, 5525.

Mohler, P. J., Yoon, W., & Bennett, V. (2004). Ankyrin-B targets beta2-spectrin to an intracellular compartment in neonatal cardiomyocytes. *The Journal of Biological Chemistry, 279*(38), 40185—40193.

Molnár, G., Oláh, S., Komlósi, G., Füle, M., Szabadics, J., Varga, C., et al. (2008). Complex events initiated by individual spikes in the human cerebral cortex. *PLoS Biology, 6*(9), e222.

Montersino, A., Brachet, A., Ferracci, G., Fache, M.-P., Angles d'Ortoli, S., Liu, W., et al. (2014). Tetrodotoxin-resistant voltage-gated sodium channel Nav 1.8 constitutively interacts with ankyrin G. *Journal of Neurochemistry, 131*(1), 33—41.

Morris, C. (2001). Mechanoprotection of the plasma membrane in neurons and other non-erythroid cells by the spectrin-based membrane skeleton. *Cellular and Molecular Biology Research, 6*(3), 703—720.

Muir, J., & Kittler, J. T. (2014). Plasticity of GABAA receptor diffusion dynamics at the axon initial segment. *Frontiers in Cellular Neuroscience, 8*, 151.

Nakada, C., Ritchie, K., Oba, Y., Nakamura, M., Hotta, Y., Iino, R., et al. (2003). Accumulation of anchored proteins forms membrane diffusion barriers during neuronal polarization. *Nature Cell Biology, 5*(7), 626—632.

Nakajima, Y. (1974). Fine structure of the synaptic endings on the Mauthner cell of the goldfish. *Journal of Comparative Neurology, 156*(4), 379—402.

Nakata, T., & Hirokawa, N. (2003). Microtubules provide directional cues for polarized axonal transport through interaction with kinesin motor head. *The Journal of Cell Biology, 162*(6), 1045—1055.

Nakata, T., Niwa, S., Okada, Y., Perez, F., & Hirokawa, N. (2011). Preferential binding of a kinesin-1 motor to GTP-tubulin-rich microtubules underlies polarized vesicle transport. *The Journal of Cell Biology, 194*(2), 245—255.

Namba, T., Funahashi, Y., Nakamuta, S., Xu, C., Takano, T., & Kaibuchi, K. (2015). Extracellular and intracellular signaling for neuronal polarity. *Physiological Reviews, 95*(3), 995—1024.

Nans, A., Mohandas, N., & Stokes, D. L. (2011). Native ultrastructure of the red cell cytoskeleton by cryo-electron tomography. *Biophysical Journal, 101*(10), 2341—2350.

Nishimura, K., Akiyama, H., Komada, M., & Kamiguchi, H. (2007). betaIV-spectrin forms a diffusion barrier against L1CAM at the axon initial segment. *Molecular and Cellular Neurosciences, 34*(3), 422—430.

Nusser, Z., Sieghart, W., Benke, D., Fritschy, J. M., & Somogyi, P. (1996). Differential synaptic localization of two major gamma-aminobutyric acid type A receptor alpha subunits on hippocampal pyramidal cells. *Proceedings of the National Academy of Sciences of the United States of America, 93*(21), 11939—11944.

Ogawa, Y., Schafer, D. P., Horresh, I., Bar, V., Hales, K., Yang, Y., et al. (2006). Spectrins and ankyrinB constitute a specialized paranodal cytoskeleton. *The Journal of Neuroscience, 26*(19), 5230—5239.

Ogawa, Y., Horresh, I., Trimmer, J. S., Bredt, D. S., Peles, E., & Rasband, M. N. (2008). Postsynaptic Density-93 clusters Kv1 channels at axon initial segments independently of Caspr2. *The Journal of Neuroscience, 28*(22), 5731—5739.

Ogawa, Y., Oses-Prieto, J., Kim, M. Y., Horresh, I., Peles, E., Burlingame, A. L., et al. (2010). ADAM22, a Kv1 channel-interacting protein, recruits membrane-associated guanylate kinases to juxtaparanodes of myelinated axons. *The Journal of Neuroscience, 30*(3), 1038—1048.

Ogiwara, I., Miyamoto, H., Morita, N., Atapour, N., Mazaki, E., Inoue, I., et al. (2007). Nav1.1 localizes to axons of parvalbumin-positive inhibitory interneurons: a circuit basis for epileptic seizures in mice carrying an Scn1a gene mutation. *The Journal of Neuroscience, 27*(22), 5903—5914.

O'Malley, H. A., & Isom, L. L. (2015). Sodium channel β subunits: emerging targets in channelopathies. *Annual Review of Physiology, 77*, 481—504.

Osorio, N., Cathala, L., Meisler, M. H., Crest, M., Magistretti, J., & Delmas, P. (2010). Persistent Nav1.6 current at axon initial segments tunes spike timing of cerebellar granule cells. *The Journal of Physiology, 588*(Pt 4), 651—670.

Paez-Gonzalez, P., Abdi, K. M., Luciano, D., Liu, Y., Soriano-Navarro, M., Rawlins, E., et al. (2011). Ank3-dependent SVZ niche assembly is required for the continued production of new neurons. *Neuron, 71*(1), 61—75.

Palay, S., Sotelo, C., Peters, A., & Orkand, P. (1968). The axon hillock and the initial segment. *The Journal of Cell Biology, 38*(1), 193—201.

Palmer, L., & Stuart, G. (2006). Site of action potential initiation in layer 5 pyramidal neurons. *The Journal of Neuroscience, 26*(6), 1854—1863.

Palmer, L. M., Clark, B. A., Gründemann, J., Roth, A., Stuart, G. J., & Häusser, M. (2010). Initiation of simple and complex spikes in cerebellar Purkinje cells. *The Journal of Physiology, 588*(Pt 10), 1709—1717.

Pan, Z., Kao, T., Horvath, Z., Lemos, J., Sul, J.-Y., Cranstoun, S., et al. (2006). A common Ankyrin-G-based mechanism retains KCNQ and NaV channels at electrically active domains of the axon. *The Journal of Neuroscience, 26*(10), 2599—2613.

Panzanelli, P., Gunn, B. G., Schlatter, M. C., Benke, D., Tyagarajan, S. K., Scheiffele, P., et al. (2011). Distinct mechanisms regulate GABAA receptor and gephyrin clustering at perisomatic and axo-axonic synapses on CA1 pyramidal cells. *The Journal of Physiology, 589*(Pt 20), 4959—4980.

Papandréou, M.-J., Vacher, H., Fache, M.-P., Klingler, E., Rueda-Boroni, F., Ferracci, G., et al. (2015). CK2-regulated schwannomin-interacting protein IQCJ-SCHIP-1 association with AnkG contributes to the maintenance of the axon initial segment. *Journal of Neurochemistry, 134*(3), 527—537.

Pei, Z., Lang, B., Fragoso, Y. D., Shearer, K. D., Zhao, L., Mccaffery, P. J. A., et al. (2014). The expression and roles of NDE1 and NDEL1 in the adult mammalian central nervous system. *Neuroscience, 271*(C), 119—136.

Peters, A., Proskauer, C. C., & Kaiserman-Abramof, I. R. (1968). The small pyramidal neuron of the rat cerebral cortex. The axon hillock and initial segment. *The Journal of Cell Biology, 39*(3), 604—619.

Petersen, J. D., Kaech, S., & Banker, G. A. (2014). Selective microtubule-based transport of dendritic membrane proteins arises in concert with axon specification. *The Journal of Neuroscience, 34*(12), 4135—4147.

Pielage, J., Cheng, L., Fetter, R. D., Carlton, P. M., Sedat, J., & Davis, G. (2008). A presynaptic giant Ankyrin stabilizes the NMJ through regulation of presynaptic microtubules and transsynaptic cell adhesion. *Neuron, 58*(2), 195–209.

Popovic, M. A., Foust, A. J., McCormick, D. A., & Zecevic, D. (2011). The spatio-temporal characteristics of action potential initiation in layer 5 pyramidal neurons: a voltage imaging study. *The Journal of Physiology, 589*(Pt 17), 4167–4187.

Prokop, A. (2013). The intricate relationship between microtubules and their associated motor proteins during axon growth and maintenance. *Neural Development, 8*(1), 17.

Rasband, M. N. (2010). The axon initial segment and the maintenance of neuronal polarity. *Nature Reviews Neurosciences, 11*(8), 552–562.

Rasband, M. N. (2013). Cytoskeleton: axons earn their stripes. *Current Biology, 23*(5), R197–R198.

Rasmussen, H. B., Frøkjaer-Jensen, C., Jensen, C. S., Jensen, H. S., Jørgensen, N. K., Misonou, H., et al. (2007). Requirement of subunit co-assembly and ankyrin-G for M-channel localization at the axon initial segment. *Journal of Cell Science, 120*(Pt 6), 953–963.

Reed, N. A., Cai, D., Blasius, T. L., Jih, G. T., Meyhofer, E., Gaertig, J., et al. (2006). Microtubule acetylation promotes kinesin-1 binding and transport. *Current Biology, 16*(21), 2166–2172.

Rivera, J., Chu, P.-J., & Arnold, D. B. (2005). The T1 domain of Kv1.3 mediates intracellular targeting to axons. *The European Journal of Neuroscience, 22*(8), 1853–1862.

Rolls, M. M., & Jegla, T. J. (2015). Neuronal polarity: an evolutionary perspective. *The Journal of Experimental Biology, 218*(Pt 4), 572–580.

Rowan, M. J. M., Tranquil, E., & Christie, J. M. (2014). Distinct kv channel subtypes contribute to differences in spike signaling properties in the axon initial segment and presynaptic boutons of cerebellar interneurons. *The Journal of Neuroscience, 34*(19), 6611–6623.

Sanchez-Mut, J. V., Aso, E., Panayotis, N., Lott, I., Dierssen, M., Rabano, A., et al. (2013). DNA methylation map of mouse and human brain identifies target genes in Alzheimer's disease. *Brain: A Journal of Neurology, 136*(Pt 10), 3018–3027.

Sanchez-Ponce, D., Tapia, M., Muñoz, A., & Garrido, J. J. (2008). New role of IKK alpha/beta phosphorylated I kappa B alpha in axon outgrowth and axon initial segment development. *Molecular and Cellular Neurosciences, 37*(4), 832–844.

Sanchez-Ponce, D., DeFelipe, J., Garrido, J. J., & Muñoz, A. (2011). In vitro maturation of the cisternal organelle in the hippocampal neuron's axon initial segment. *Molecular and Cellular Neurosciences, 48*(1), 104–116.

Sanchez-Ponce, D., Muñoz, A., & Garrido, J. J. (2011). Casein kinase 2 and microtubules control axon initial segment formation. *Molecular and Cellular Neuroscience, 46*(1), 222–234.

Sanchez-Ponce, D., Blázquez-Llorca, L., DeFelipe, J., Garrido, J. J., & Muñoz, A. (2012). Colocalization of α-actinin and synaptopodin in the pyramidal cell axon initial segment. *Cerebral Cortex, 22*(7), 1648–1661.

Sanchez-Ponce, D., DeFelipe, J., Garrido, J. J., & Muñoz, A. (2012). Developmental expression of kv potassium channels at the axon initial segment of cultured hippocampal neurons. *PloS One, 7*(10), e48557.

Santuccione, A. C., Merlini, M., Shetty, A., Tackenberg, C., Bali, J., Ferretti, M. T., et al. (2012). Active vaccination with ankyrin G reduces β-amyloid pathology in APP transgenic mice. *Molecular Psychiatry, 18*(3), 358–368.

Sarmiere, P. D., Weigle, C. M., & Tamkun, M. M. (2008). The Kv2.1 K$^+$ channel targets to the axon initial segment of hippocampal and cortical neurons in culture and in situ. *BMC Neuroscience, 9*, 112.

Sasaki, S., & Iwata, M. (1996). Impairment of fast axonal transport in the proximal axons of anterior horn neurons in amyotrophic lateral sclerosis. *Neurology, 47*(2), 535−540.

Sasaki, S., Shionoya, A., Ishida, M., Gambello, M. J., Yingling, J., Wynshaw-Boris, A., et al. (2000). A LIS1/NUDEL/cytoplasmic dynein heavy chain complex in the developing and adult nervous system. *Neuron, 28*(3), 681−696.

Sasaki, S., Warita, H., Abe, K., & Iwata, M. (2005). Impairment of axonal transport in the axon hillock and the initial segment of anterior horn neurons in transgenic mice with a G93A mutant SOD1 gene. *Acta Neuropathologica, 110*(1), 48−56.

Schafer, D. P., Jha, S., Liu, F., Akella, T., McCullough, L. D., & Rasband, M. N. (2009). Disruption of the axon initial segment cytoskeleton is a new mechanism for neuronal injury. *The Journal of Neuroscience, 29*(42), 13242−13254.

Schmidt-Hieber, C., Jonas, P., & Bischofberger, J. (2008). Action potential initiation and propagation in hippocampal mossy fibre axons. *The Journal of Physiology, 586*(7), 1849−1857.

Schultz, C., König, H.-G., Del Turco, D., Politi, C., Eckert, G. P., Ghebremedhin, E., et al. (2006). Coincident enrichment of phosphorylated IkappaBalpha, activated IKK, and phosphorylated p65 in the axon initial segment of neurons. *Molecular and Cellular Neuroscience, 33*(1), 68−80.

Shah, M., Migliore, M., Valencia, I., Cooper, E. C., & Brown, D. (2008). Functional significance of axonal Kv7 channels in hippocampal pyramidal neurons. *Proceedings of the National Academy of Sciences of the United States of America, 105*(22), 7869−7874.

Shams'ili, S., de Leeuw, B., Hulsenboom, E., Jaarsma, D., & Smitt, P. S. (2009). A new paraneoplastic encephalomyelitis autoantibody reactive with the axon initial segment. *Neuroscience Letters, 467*(2), 169−172.

Shavkunov, A. S., Wildburger, N. C., Nenov, M. N., James, T. F., Buzhdygan, T. P., Panova-Elektronova, N. I., et al. (2013). The fibroblast growth factor 14·voltage-gated sodium channel complex is a new target of glycogen synthase kinase 3 (GSK3). *The Journal of Biological Chemistry, 288*(27), 19370−19385.

Sheng, M., & Hoogenraad, C. C. (2007). The postsynaptic architecture of excitatory synapses: a more quantitative view. *Annual Review of Biochemistry, 76*, 823−847.

Shirahata, E., Iwasaki, H., Takagi, M., Lin, C., Bennett, V., Okamura, Y., et al. (2006). Ankyrin-G regulates inactivation gating of the neuronal sodium channel, Nav1.6. *Journal of Neurophysiology, 96*(3), 1347−1357.

Shu, Y., Duque, A., Yu, Y., Haider, B., & McCormick, D. A. (2007). Properties of action-potential initiation in neocortical pyramidal cells: evidence from whole cell axon recordings. *Journal of Neurophysiology, 97*(1), 746−760.

Shu, Y., Yu, Y., Yang, J., & McCormick, D. A. (2007). Selective control of cortical axonal spikes by a slowly inactivating K^+ current. *Proceedings of the National Academy of Sciences of the United States of America, 104*(27), 11453−11458.

Sloper, J. J., & Powell, T. P. (1979). A study of the axon initial segment and proximal axon of neurons in the primate motor and somatic sensory cortices. *Philosophical Transactions of the Royal Society of London. Series B, Biological Sciences, 285*(1006), 173−197.

Smith, K. R., Kopeikina, K. J., Fawcett-Patel, J. M., Leaderbrand, K., Gao, R., Schürmann, B., et al. (2014). Psychiatric risk factor ANK3/Ankyrin-G nanodomains regulate the structure and function of glutamatergic synapses. *Neuron, 84*(2), 399−415.

Sobotzik, J.-M., Sie, J. M., Politi, C., Del Turco, D., Bennett, V., Deller, T., et al. (2009). Ankyrin G is required to maintain axo-dendritic polarity in vivo. *Proceedings of the National Academy of Sciences of the United States of America, 106*(41), 17564−17569.

Somogyi, P., & Hámori, J. (1976). A quantitative electron microscopic study of the Purkinje cell axon initial segment. *Neuroscience, 1*(5), 361−365.

Somogyi, P., Freund, T. F., & Cowey, A. (1982). The axo-axonic interneuron in the cerebral cortex of the rat, cat and monkey. *Neuroscience, 7*(11), 2577−2607.

Song, A.-H., Wang, D., Chen, G., Li, Y., Luo, J., Duan, S., et al. (2009). A selective filter for cytoplasmic transport at the axon initial segment. *Cell, 136*(6), 1148–1160.

Song, Y., Kirkpatrick, L. L., Schilling, A. B., Helseth, D. L., Chabot, N., Keillor, J. W., et al. (2013). Transglutaminase and polyamination of tubulin: posttranslational modification for stabilizing axonal microtubules. *Neuron, 78*(1), 109–123.

Sorbara, C. D., Wagner, N. E., Ladwig, A., Nikić, I., Merkler, D., Kleele, T., et al. (2014). Pervasive axonal transport deficits in multiple sclerosis models. *Neuron, 84*(6), 1183–1190.

Sotelo, C. (2003). Viewing the brain through the master hand of Ramon y Cajal. *Nature Reviews Neurosciences, 4*(1), 71–77.

Srinivasan, Y., Elmer, L., Davis, J. Q., Bennett, V., & Angelides, K. J. (1988). Ankyrin and spectrin associate with voltage-dependent sodium channels in brain. *Nature, 333*(6169), 177–180.

Srinivasan, Y., Lewallen, M., & Angelides, K. J. (1992). Mapping the binding site on ankyrin for the voltage-dependent sodium channel from brain. *The Journal of Biological Chemistry, 267*(11), 7483–7489.

Stephan, R., Goellner, B., Moreno, E., Frank, C. A., Hugenschmidt, T., Genoud, C., et al. (2015). Hierarchical microtubule organization controls axon caliber and transport and determines synaptic structure and stability. *Developmental Cell, 33*(1), 5–21.

Sun, X., Wu, Y., Gu, M., & Zhang, Y. (2014). miR-342-5p decreases ankyrin G levels in Alzheimer's disease transgenic mouse models. *Cell Reports, 6*(2), 264–270.

Sun, X., Wu, Y., Gu, M., Liu, Z., Ma, Y., Li, J., et al. (2014). Selective filtering defect at the axon initial segment in Alzheimer's disease mouse models. *Proceedings of the National Academy of Sciences of the United States of America, 111*(39), 14271–14276.

Sweet, E. S., Previtera, M. L., Fernández, J. R., Charych, E. I., Tseng, C.-Y., Kwon, M., et al. (2011). PSD-95 alters microtubule dynamics via an association with EB3. *The Journal of Neuroscience, 31*(3), 1038–1047.

Szabadics, J., Varga, C., Molnár, G., Oláh, S., Barzó, P., & Tamas, G. (2006). Excitatory effect of GABAergic axo-axonic cells in cortical microcircuits. *Science, 311*(5758), 233–235.

Tai, Y., Janas, J. A., Wang, C.-L., & Van Aelst, L. (2014). Regulation of chandelier cell cartridge and bouton development via DOCK7-mediated ErbB4 activation. *Cell Reports, 6*(2), 254–263.

Takeda, S., Yamazaki, H., Seog, D. H., Kanai, Y., Terada, S., & Hirokawa, N. (2000). Kinesin superfamily protein 3 (KIF3) motor transports fodrin-associating vesicles important for neurite building. *The Journal of Cell Biology, 148*(6), 1255–1265.

Tapia, M., Wandosell, F., & Garrido, J. J. (2010). Impaired function of HDAC6 slows down axonal growth and interferes with axon initial segment development. *PloS One, 5*(9), e12908.

Tapia, M., del Puerto, A., Puime, A., Sanchez-Ponce, D., Fronzaroli-Molinieres, L., Pallas-Bazarra, N., et al. (2013). GSK3 and β-catenin determines functional expression of sodium channels at the axon initial segment. *Cellular and Molecular Life Sciences, 70*(1), 105–120.

Thome, C., Kelly, T., Yanez, A., Schultz, C., Engelhardt, M., Cambridge, S. B., et al. (2014). Axon-carrying dendrites convey privileged synaptic input in hippocampal neurons. *Neuron, 83*(6), 1418–1430.

Tian, C., Wang, K., Ke, W., Guo, H., & Shu, Y. (2014). Molecular identity of axonal sodium channels in human cortical pyramidal cells. *Frontiers in Cellular Neuroscience, 8*, 297.

Tran, B. T., Gavrilis, J., Cooper, E. C., & Kole, M. H. P. (2014). Heteromeric kv7.2/7.3 channels differentially regulate action potential initiation and conduction in neocortical myelinated axons. *The Journal of Neuroscience, 34*(10), 3719–3732.

Tretter, V., Jacob, T. C., Mukherjee, J., Fritschy, J.-M., Pangalos, M. N., & Moss, S. J. (2008). The clustering of GABA(A) receptor subtypes at inhibitory synapses is facilitated via the direct binding of receptor alpha 2 subunits to gephyrin. *The Journal of Neuroscience, 28*(6), 1356—1365.

Trimmer, J. S. (2015). Subcellular localization of K^+ channels in mammalian brain neurons: remarkable precision in the midst of extraordinary complexity. *Neuron, 85*(2), 238—256.

Trunova, S., Baek, B., & Giniger, E. (2011). Cdk5 regulates the size of an axon initial segment-like compartment in mushroom body neurons of the *Drosophila* central brain. *The Journal of Neuroscience, 31*(29), 10451—10462.

Tseng, W.-C., Jenkins, P. M., Tanaka, M., Mooney, R., & Bennett, V. (2015). Giant ankyrin-G stabilizes somatodendritic GABAergic synapses through opposing endocytosis of GABAA receptors. *Proceedings of the National Academy of Sciences of the United States of America, 112*(4), 1214—1219.

Uemoto, Y., Suzuki, S.-I., Terada, N., Ohno, N., Ohno, S., Yamanaka, S., et al. (2007). Specific role of the truncated betaIV-spectrin Sigma6 in sodium channel clustering at axon initial segments and nodes of ranvier. *The Journal of Biological Chemistry, 282*(9), 6548—6555.

Vacher, H., Yang, J.-W., Cerda, O., Autillo-Touati, A., Dargent, B., & Trimmer, J. S. (2011). Cdk-mediated phosphorylation of the Kvβ2 auxiliary subunit regulates Kv1 channel axonal targeting. *The Journal of Cell Biology, 192*(5), 813—824.

Van Wart, A., Trimmer, J. S., & Matthews, G. (2007). Polarized distribution of ion channels within microdomains of the axon initial segment. *Journal of Comparative Neurology, 500*(2), 339—352.

Veres, J. M., Nagy, G. A., Vereczki, V. K., Andrási, T., & Hájos, N. (2014). Strategically positioned inhibitory synapses of axo-axonic cells potently control principal neuron spiking in the basolateral amygdala. *The Journal of Neuroscience, 34*(49), 16194—16206.

Wang, L., Wang, H., Yu, L., & Chen, Y. (2011). Role of axonal sodium-channel band in neuronal excitability. *Physical Review. E, Statistical, Nonlinear, and Soft Matter Physics, 84*(5—1), 052901.

Wang, C., Yu, C., Ye, F., Wei, Z., & Zhang, M. (2012). Structure of the ZU5-ZU5-UPA-DD tandem of ankyrin-B reveals interaction surfaces necessary for ankyrin function. *Proceedings of the National Academy of Sciences of the United States of America, 109*(13), 4822—4827.

Wang, C., Wei, Z., Chen, K., Ye, F., Yu, C., Bennett, V., et al. (2014). Structural basis of diverse membrane target recognitions by ankyrins. *eLife, 3*.

Wang, X., Hooks, B. M., & Sun, Q.-Q. (2014). Thorough GABAergic innervation of the entire axon initial segment revealed by an optogenetic 'laserspritzer'. *The Journal of Physiology, 592*(Pt 19), 4257—4276.

Watanabe, K., Al-Bassam, S., Miyazaki, Y., Wandless, T. J., Webster, P., & Arnold, D. B. (2012). Networks of polarized actin filaments in the axon initial segment provide a mechanism for sorting axonal and dendritic proteins. *Cell Reports, 2*(6), 1546—1553.

Wefelmeyer, W., Cattaert, D., & Burrone, J. (2015). Activity-dependent mismatch between axo-axonic synapses and the axon initial segment controls neuronal output. *Proceedings of the National Academy of Sciences of the United States of America, 112*(31), 9757—9762.

Westrum, L. E., & Gray, E. G. (1976). Microtubules and membrane specializations. *Brain Research, 105*(3), 547—550.

Westrum, L. E. (1970). Observations on initial segments of axons in the prepyriform cortex of the rat. *Journal of Comparative Neurology, 139*(3), 337—356.

Wildburger, N. C., Ali, S. R., Hsu, W.-C. J., Shavkunov, A. S., Nenov, M. N., Lichti, C. F., et al. (2015). Quantitative proteomics reveals protein—protein interactions with fibroblast growth factor 12 as a component of the voltage-gated sodium channel 1.2

(nav1.2) macromolecular complex in mammalian brain. *Molecular and Cellular Proteomics: MCP, 14*(5), 1288–1300.

Wimmer, V. C., Reid, C. A., Mitchell, S., Richards, K. L., Scaf, B. B., Leaw, B. T., et al. (2010). Axon initial segment dysfunction in a mouse model of genetic epilepsy with febrile seizures plus. *The Journal of Clinical Investigation, 120*(8), 2661–2671.

Wimmer, V. C., Reid, C. A., So, E. Y.-W., Berkovic, S. F., & Petrou, S. (2010). Axon initial segment dysfunction in epilepsy. *The Journal of Physiology, 588*(Pt 11), 1829–1840.

Wimmer, V. C., Harty, R. C., Richards, K. L., Phillips, A. M., Miyazaki, H., Nukina, N., et al. (2015). Sodium channel β1 subunit localizes to axon initial segments of excitatory and inhibitory neurons and shows regional heterogeneity in mouse brain. *Journal of Comparative Neurology, 523*(5), 814–830.

Winckler, B., & Poo, M. M. (1996). No diffusion barrier at axon hillock. *Nature, 379*(6562), 213.

Winckler, B., Forscher, P., & Mellman, I. (1999). A diffusion barrier maintains distribution of membrane proteins in polarized neurons. *Nature, 397*(6721), 698–701.

Wirgenes, K. V., Tesli, M., Inderhaug, E., Athanasiu, L., Agartz, I., Melle, I., et al. (2014). ANK3 gene expression in bipolar disorder and schizophrenia. *The British Journal of Psychiatry: The Journal of Mental Science, 205*(3), 244–245.

Wisco, D., Anderson, E., Chang, M. C., Norden, C., Boiko, T., Fölsch, H., et al. (2003). Uncovering multiple axonal targeting pathways in hippocampal neurons. *The Journal of Cell Biology, 162*(7), 1317–1328.

Wollner, D. A., & Catterall, W. A. (1986). Localization of sodium channels in axon hillocks and initial segments of retinal ganglion cells. *Proceedings of the National Academy of Sciences of the United States of America, 83*(21), 8424–8428.

Woo, T. U., Whitehead, R. E., Melchitzky, D. S., & Lewis, D. A. (1998). A subclass of prefrontal gamma-aminobutyric acid axon terminals are selectively altered in schizophrenia. *Proceedings of the National Academy of Sciences of the United States of America, 95*(9), 5341–5346.

Woodruff, A., Xu, Q., Anderson, S. A., & Yuste, R. (2009). Depolarizing effect of neocortical chandelier neurons. *Frontiers in Neural Circuits, 3*, 15.

Xiao, M., Bosch, M. K., Nerbonne, J. M., & Ornitz, D. M. (2013). FGF14 localization and organization of the axon initial segment. *Molecular and Cellular Neurosciences, 56*, 393–403.

Xu, M., & Cooper, E. C. (2015). An Ankyrin-G N-terminal gate and protein kinase CK2 dually regulate binding of voltage-gated sodium and KCNQ2/3 potassium channels. *The Journal of Biological Chemistry, 290*(27), 16619–16632.

Xu, X., & Shrager, P. (2005). Dependence of axon initial segment formation on Na^+ channel expression. *Journal of Neuroscience Research, 79*(4), 428–441.

Xu, K., Zhong, G., & Zhuang, X. (2013). Actin, spectrin, and associated proteins form a periodic cytoskeletal structure in axons. *Science, 339*(6118), 452–456.

Yang, Y., Lacas-Gervais, S., Morest, D. K., Solimena, M., & Rasband, M. N. (2004). BetaIV spectrins are essential for membrane stability and the molecular organization of nodes of Ranvier. *The Journal of Neuroscience, 24*(33), 7230–7240.

Yang, Y., Ogawa, Y., Hedstrom, K., & Rasband, M. N. (2007). βIV spectrin is recruited to axon initial segments and nodes of Ranvier by ankyrin G. *The Journal of Cell Biology, 176*(4), 509–519.

Yap, C. C., Vakulenko, M., Kruczek, K., Motamedi, B., Digilio, L., Liu, J. S., et al. (2012). Doublecortin (DCX) mediates endocytosis of neurofascin independently of microtubule binding. *The Journal of Neuroscience, 32*(22), 7439–7453.

Yau, K. W., van Beuningen, S. F. B., Cunha-Ferreira, I., Cloin, B. M. C., van Battum, E. Y., Will, L., et al. (2014). Microtubule minus-end binding protein CAMSAP2 controls axon specification and dendrite development. *Neuron, 82*(5), 1058–1073.

Yoshimura, T., & Rasband, M. N. (2014). Axon initial segments: diverse and dynamic neuronal compartments. *Current Opinion in Neurobiology, 27C*, 96–102.

Yu, Y., Maureira, C., Liu, X., & McCormick, D. (2010). P/Q and N channels control baseline and spike-triggered calcium levels in neocortical axons and synaptic boutons. *The Journal of Neuroscience, 30*(35), 11858–11869.

Zhang, X., & Bennett, V. (1998). Restriction of 480/270-kD ankyrin G to axon proximal segments requires multiple ankyrin G-specific domains. *The Journal of Cell Biology, 142*(6), 1571–1581.

Zhang, X., Davis, J. Q., Carpenter, S., & Bennett, V. (1998). Structural requirements for association of neurofascin with ankyrin. *The Journal of Biological Chemistry, 273*(46), 30785–30794.

Zhao, H., Wang, D.-D., Xu, Y.-X., & Zhu, C.-Q. (2014). Localization and expression pattern of MDM2 in axon initial segments of neuron in rodent brain. *Sheng Li Xue Bao: Acta Physiologica Sinica, 66*(2), 107–117.

Zhong, G., He, J., Zhou, R., Lorenzo, D., Babcock, H. P., Bennett, V., et al. (2014). Developmental mechanism of the periodic membrane skeleton in axons. *eLife, 3*.

Zhou, D., Lambert, S., Malen, P. L., Carpenter, S., Boland, L. M., & Bennett, V. (1998). AnkyrinG is required for clustering of voltage-gated Na channels at axon initial segments and for normal action potential firing. *The Journal of Cell Biology, 143*(5), 1295–1304.

Zonta, B., Desmazieres, A., Rinaldi, A., Tait, S., Sherman, D. L., Nolan, M. F., et al. (2011). A critical role for neurofascin in regulating action potential initiation through maintenance of the axon initial segment. *Neuron, 69*(5), 945–956.

Index

Note: Page numbers followed by "f" indicate figures and "t" indicate tables.

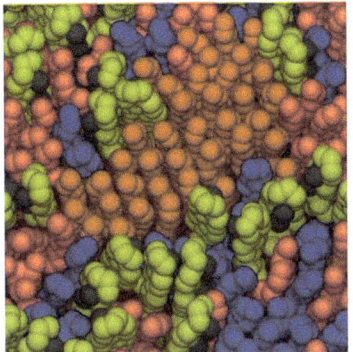

Max Berkowitz, Figure 2 The molecular details of lipid chains in Lo/Ld simulation at T = 298 K. DPPC chains are shown in red, DOPC in blue, cholesterol in yellow, and methyl groups on the β face of cholesterols are in gray. The area of locally hexagonal order is highlighted in yellow. *Reproduced with permission from Sodt et al. (2014).*

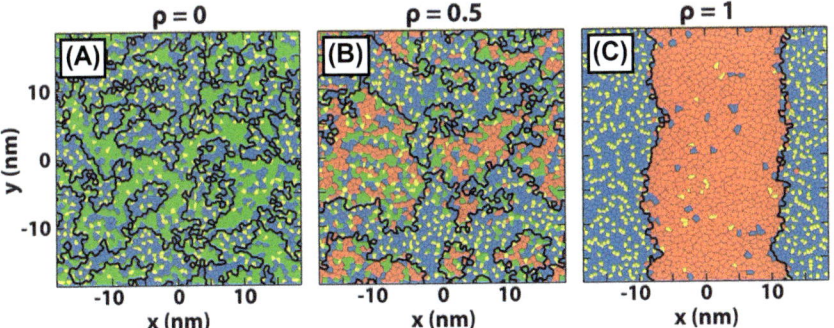

Max Berkowitz, Figure 3 Patches of Lo and Ld phases at (A) ρ = 0, (B) ρ = 0.5, and (C) ρ = 1. DPPC is shown in blue, PUPC in green, DUPC in red, and cholesterol in yellow. Black lines demarcate coexisting patches of different phases. *Reproduced with permission from Ackerman and Feigenson (2015).*

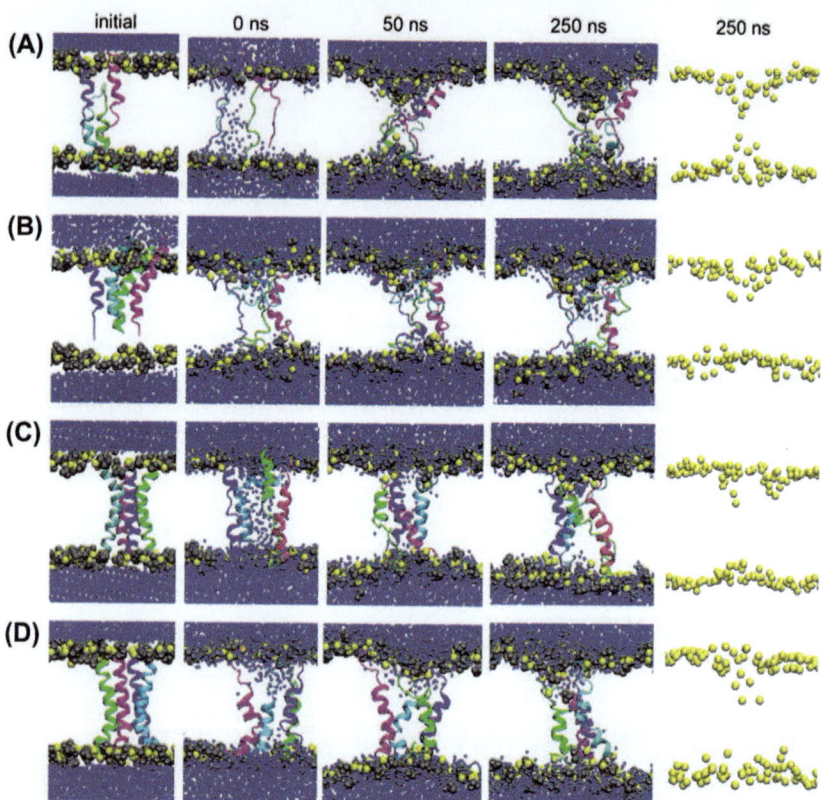

Max Berkowitz, Figure 4 Snapshots of systems A–D at various time points during the simulations. The 0 ns snapshot is the starting conformation obtained after equilibration. Peptide 1 is shown in violet, peptide 2 in magenta, peptide 3 in green, peptide 4 in cyan, phosphorus atoms in yellow, choline groups in gray, and water oxygen atoms in blue. The last column shows positions of phosphorus atoms at 250 ns, to show clearly the toroidal nature of the pore in simulation A. *Reproduced with permission from Irudayam and Berkowitz (2011).*

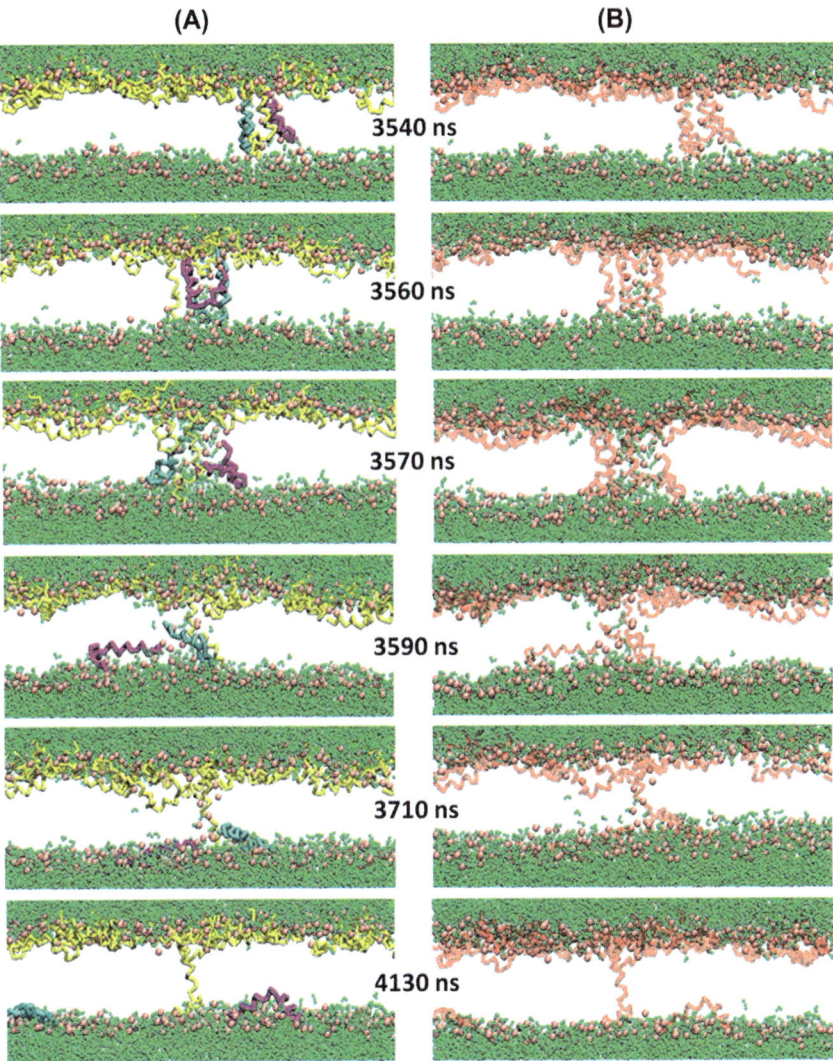

Max Berkowitz, Figure 5 Transient pore formation and translocation by melittins in a run with P/L = 1/21. In panels (A) peptides are shown in yellow, except the two peptides that translocate, which are shown in purple and cyan. In panels (B) peptides are in transparent red, so that water flux can be visible. Water particles are shown in green; pink spheres are PO$_4$ groups. *Reproduced with permission from Santo et al. (2013).*

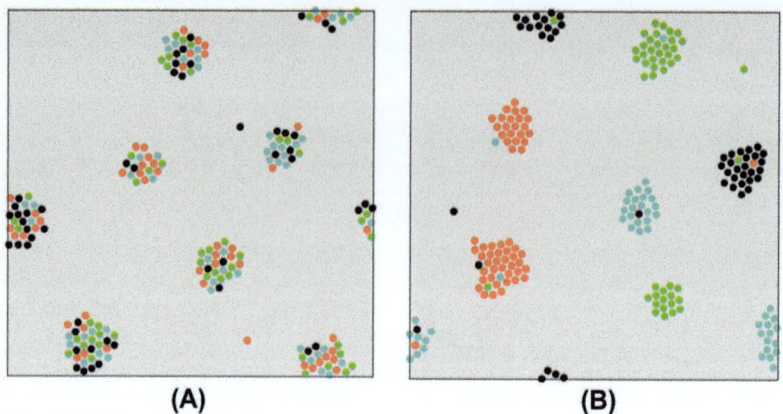

(A)   **(B)**

Nicolas Destainville *et al.,* **Figure 5** *Cluster-specificity, illustrated with four protein families (four different colors; q = 4).* (A) If proteins of different colors attract each other sufficiently with respect to proteins of the same color, they are mixed in clusters. (B) In the opposite case, clusters are essentially mono-colored, i.e., they are highly concentrated in proteins of a single color (they have "demixed"). This demixing process ensures nanodomain specialization. *Reprinted and adapted from Destainville, N. (2010) with permission from EPL.*

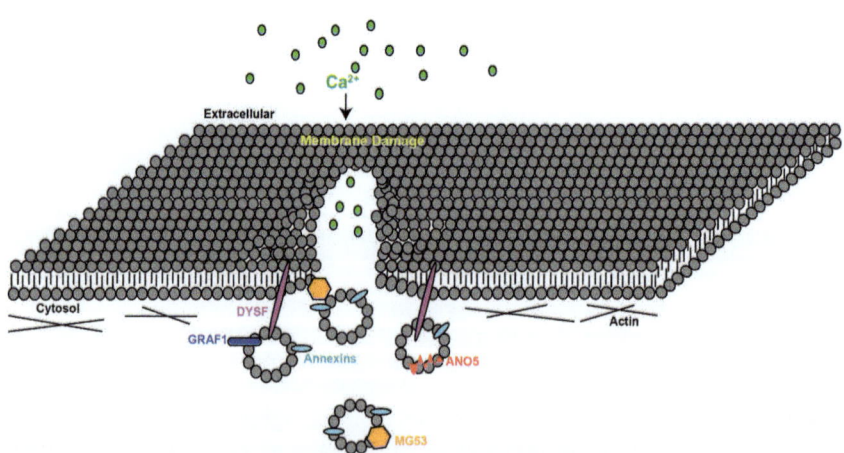

Alexis R. Demonbreun and Elizabeth M. McNally, Figure 3 *Model for plasma membrane repair.* A membrane lesion is formed within the sarcolemma creating a microhole within the membrane. The influx of calcium triggers annexin A6 (A6) localization to the repair cap. Other repair complex proteins including annexin A1, A2, A5, dysferlin (DYSF), and MG53 are recruited to the lesion aiding in the resealing of the sarcolemma.

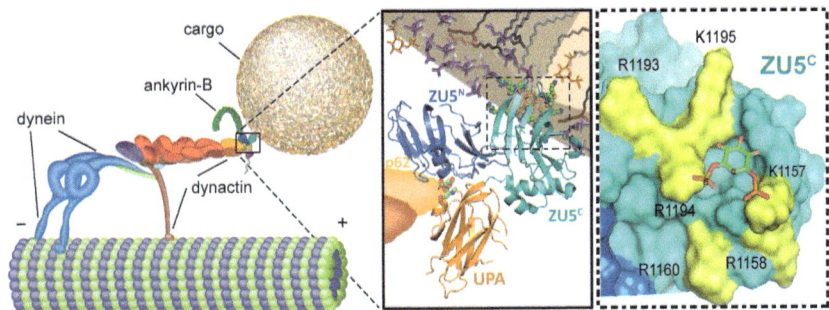

Vann Bennett and Damaris N. Lorenzo, Figure 5 *Ankyrin-B's dual interactions with the dynactin complex and membrane PtdIns(3)P lipids promote multiorganelle transport in axons.* The schematic representation illustrates a proposed model for ankyrin-B coupling of the dynein–dynactin motor complex to cargoes. The center panel shows a homology model of the ZU5^N-ZU5^C-UPA-DD region of ankyrin-B highlighting binding sites (spheres) for the dynactin subunit p62 and for PtdIns(3)P lipids in membrane cargoes. A molecular docking simulation of PtdIns(3)P binding to ZU5^C domain highlights a positively charged surface (yellow) required for the interaction with PtdIns(3)P lipids (Lorenzo et al., 2014).

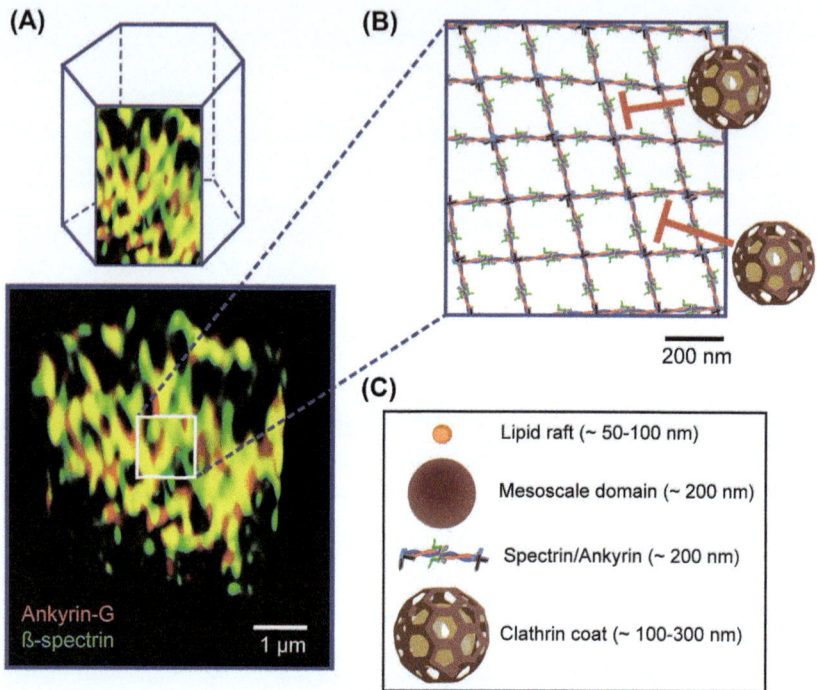

Vann Bennett and Damaris N. Lorenzo, Figure 7 *Ankyrin-G/βII-spectrin-coated microdomains promote lateral membrane height in polarized epithelial cells by opposing clathrin-mediated endocytosis.* (A) Representative deconvolved images of the lateral membrane of Madin–Darby canine kidney cell membranes stained for endogenous ankyrin-G (red) and βII-spectrin (green). (B) Schematic representation of the actin/spectrin/ankyrin lattice on membrane microdomains. (C) Representations of the relative sizes of lipid rafts, mesoscale domains, clathrin coat, and the predicted spectrin/ankyrin network coating membrane microdomains (imaged on the left), drawn at scale (Jenkins, He, & Bennett, 2015).

Protein Function	Ankyrin Partner	Ankyrin binding motif[a]	Most recent organisms	~ mybp
Cell adhesion	Dystroglycan	GVPIIFADELDDSK[1]	Cnidarians	580
	L1CAM	QFNEDGSFIGQY[2]	Bilaterians	555
	E-cadherin	KEPLLPPEDDTRDNVYYYDEE[3]	Urochordates	520
Membrane transport	Nav-α	VPIAVGESDFE[4]	Cephalochordate	500
	KCNQ2	PYIAEGESDTDSD[5]	Jawed fish	440
	Kir6.2	VPIVAEED[6]	Jawed fish	440
	CNGβ	PEPGEQILSVKMPE[7]	Jawed fish	440
	Cav2.1	NLLASREALYGDAAERWPTT[8]	Jawed fish	440
	Cav2.2	NLRASCEALYSEMDPEERLR[8]	Jawed fish	440
	Nav-β-1a	KKIAATEAAAQENASEY[9]	Jawed fish	440
	AE1	ELVMDEKNQEL...PAVLTRSGDPS[10]	Mammals	200
	RhB/G	KLPFLDSPP[11]	Mammals	200

Ankyrin-binding sites in cell adhesion molecules and membrane transporters.

[a] Minimum ankyrin-binding sequences. Mutation of critical residues marked in red results in loss of ankyrin-binding activity.

[1] Ayalon et al. (2008).

[2] Zhang et al. (1998).

[3] Kizhatil, Davis, et al. (2007) and Jenkins et al. (2013) (leucine residues marked in blue are required for endocytosis but not for ankyrin-binding).

[4] Garrido et al. (2003), Lemaillet, Walker, and Lambert (2003) and Wang et al. (2014).

[5] Pan et al. (2006).

[6] Li et al. (2010).

[7] Kizhatil, Baker, et al. (2009).

[8] Kline, Scott, Curran, Hund, and Mohler (2014).

[9] Malhotra et al. (2002).

[10] Chang and Low (2003) and Grey et al. (2012).

[11] Lopez et al. (2005).

Vann Bennett and Damaris N. Lorenzo, Table 2 Independently evolved ankyrin-binding sites

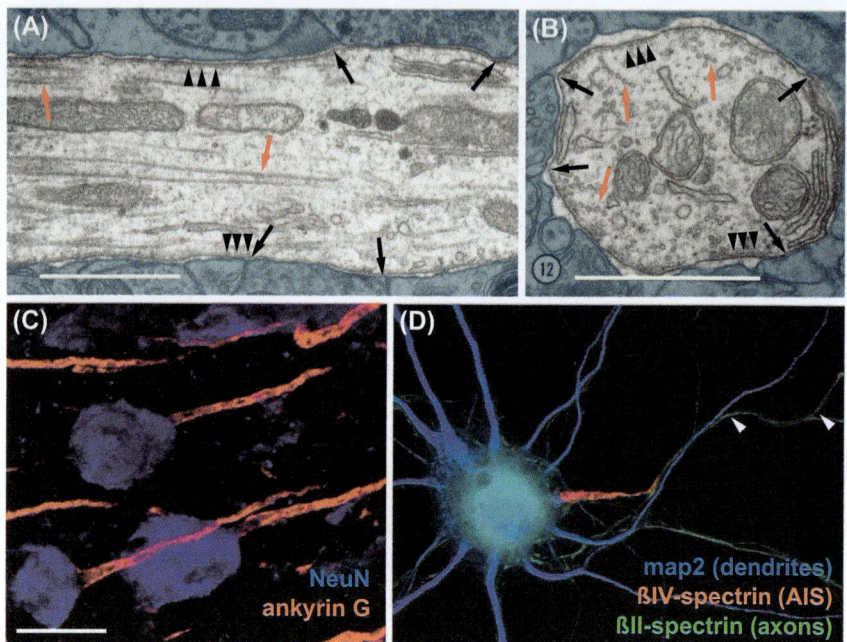

Christophe Leterrier, Figure 2 *Morphology of the axon initial segment (AIS).* (A) Electron microscopy image of a longitudinal section through the AIS of a rat hippocampal CA3 neuron. The microtubule fascicles are visible (red arrows) supporting the transport of organelles and vesicles. The dense undercoat lines the plasma membrane (arrowheads) and stops at axoaxonic synaptic contacts present on the AIS (top and bottom membrane, between arrows). Scale bar, 1 μm. (B) Electron microscopy image of a cross section through the AIS of a rat hippocampal CA3 neuron. Microtubule fascicles are seen as string of beads (red arrows). The dense undercoat can sometimes be seen as a row of small dots (arrowheads). The undercoat stops at sites of cisternal organelles apposition (arrows). Scale bar, 1 μm. *((A) and (B) are reproduced from (Kosaka, 1980) with permission from Springer.)* (C) Brain cortical section from an adult rat showing neurons (labeled with NeuN, blue) and their initial segments (ankyrin G, red). Scale bar, 10 μm. (D) Rat hippocampal neuron after 15 days in culture, labeled for the somatodendritic marker map2 (blue) the AIS scaffold βIV-spectrin (red) and the distal axon scaffold βII-spectrin (green, arrowheads). Scale bar, 20 μm.

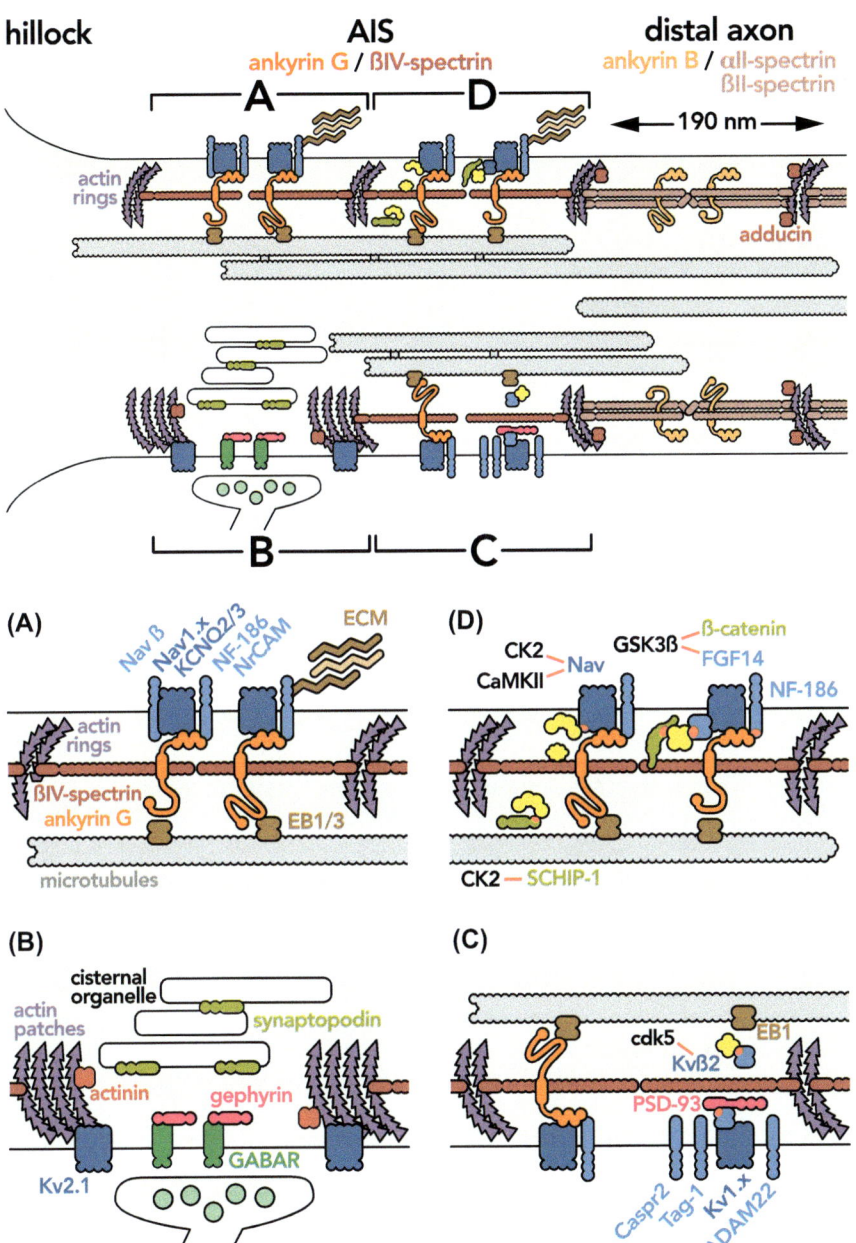

Christophe Leterrier, Figure 3 *Components of the AIS and their modulation.* The AIS, characterized by its ankyrin G/βIV-spectrin scaffold sits between the axon hillock (left) and the distal axon (right), which contains an ankyrin B/βII-spectrin scaffold. Actin rings are regularly found (190 nm period) along the AIS and distal axon, capped by adducin (left). (A) The canonical AIS complex. Actin rings are connected by dimers of βIV-spectrin, that interact with the ankyrin G scaffold. Nav α subunits and KCNQ2/3 ion channels are concentrated by their anchoring to ankyrin G. Also anchored to ankyrin G are CAMs (NF-186 and NrCAM) that bind ECM components (aggrecan, brevican, tenascin R, versican), whereas Nav β subunits are linked to their α counterpart. Ankyrin G is in turn bound to microtubules via the EB end binding proteins EB1 and EB3. (B) The AIS synapse. GABAergic synapses contain α2 and α1 GABA receptors (GABAR), anchored by gephyrin. Apposed to the synapse are cisternal organelles rich in synaptopodin. The synapses are often associated with actin patches rich in actinin, and membrane clusters of Kv2.1. (C) The distal AIS complex. In certain neuronal types, potassium channels Kv1.2 and Kv1.1 are concentrated along the distal portion of the AIS. These are associated with the CAMs Caspr2 and Tag-1, the metalloprotease ADAM22, and the PSD-93 scaffold. Kvβ2 subunits associate with Kv1 channels, and regulate their delivery via their phosphorylation by cdk2/5 that detach them from the microtubule-associated EB1. (D) Modulation of component interactions by phosphorylation: a number of kinases can phosphorylate AIS components to regulate their interactions. CK2 phosphorylates Nav channels, modulating their binding to ankyrin G. CK2 also phosphorylates SCHIP-1 to modulate its interaction with ankyrin G. CaMKII is bound to βIV-spectrin and also phosphorylates Nav channels. GSK3β phosphorylates both FGF14, a Nav channel partner and β-catenin at the AIS.